Carburizing

Microstructures and Properties

Geoffrey Parrish

The Materials Information Society

Copyright © 1999
by
ASM International®
All rights reserved

No part of this book may be reproduced, stored in a retrieval system, or transmitted, in any form or by any means, electronic, mechanical, photocopying, recording, or otherwise, without the written permission of the copyright owner.

First printing, December 1999

Great care is taken in the compilation and production of this book, but it should be made clear that NO WARRANTIES, EXPRESS OR IMPLIED, INCLUDING, WITHOUT LIMITATION, WARRANTIES OF MERCHANTABILITY OR FITNESS FOR A PARTICULAR PURPOSE, ARE GIVEN IN CONNECTION WITH THIS PUBLICATION. Although this information is believed to be accurate by ASM, ASM cannot guarantee that favorable results will be obtained from the use of this publication alone. This publication is intended for use by persons having technical skill, at their sole discretion and risk. Since the conditions of product or material use are outside of ASM's control, ASM assumes no liability or obligation in connection with any use of this information. No claim of any kind, whether as to products or information in this publication, and whether or not based on negligence, shall be greater in amount than the purchase price of this product or publication in respect of which damages are claimed. THE REMEDY HEREBY PROVIDED SHALL BE THE EXCLUSIVE AND SOLE REMEDY OF BUYER, AND IN NO EVENT SHALL EITHER PARTY BE LIABLE FOR SPECIAL, INDIRECT OR CONSEQUENTIAL DAMAGES WHETHER OR NOT CAUSED BY OR RESULTING FROM THE NEGLIGENCE OF SUCH PARTY. As with any material, evaluation of the material under enduse conditions prior to specification is essential. Therefore, specific testing under actual conditions is recommended.

Nothing contained in this book shall be construed as a grant of any right of manufacture, sale, use, or reproduction, in connection with any method, process, apparatus, product, composition, or system, whether or not covered by letters patent, copyright, or trademark, and nothing contained in this book shall be construed as a defense against any alleged infringement of letters patent, copyright, or trademark, or as a defense against liability for such infringement.

Comments, criticisms, and suggestions are invited, and should be forwarded to ASM International.

ASM International staff who worked on this project included Steve Lampman, Acquisition Editor, Bonnie Sanders, Manager of Copy Editing, Grace Davidson, Manager of Book Production, Erika Baxter, Copy Editor, Alexandra Hoskins, Copy Editor, Candace Mullet and Jill Kinson, Production Coordinators.

Library of Congress Cataloging-in-Publication Data

Parrish, Geoffrey, 1933-
Carburizing: microstructures and properties/Geoffrey Parrish.
Includes bibliographical references and index.
1. Case hardening. 2. Steel—Metallography.
3. Steel—Heat treatment. I. Title
TN752.C3 P36   1999   672.3'6—dc21   99-052675

ISBN: 0-87170-666-0
SAN: 204-7586

ASM International®
Materials Park, OH 44073-0002

Printed in the United States of America

Contents

Preface to First Edition..v
Preface to Second Edition..vii
Introduction and Perspectives..1
   Why Carburize Case-Harden?..1
   Variability..3
   Laboratory Tests..3
   Design Aspects..4
   Case-Depth Specifications...7

Chapter 1: Internal Oxidation..11
   Factors Promoting Internal Oxidation..11
   The Internal Oxidation Process...13
   Effect on Local Microstructure...18
   Influence on Material Properties...23
   Measures to Eliminate or Reduce Internal Oxidation....................30
   Summary..33
      Internal Oxidation...33
      High-Temperature Transformation Products.......................33

Chapter 2: Decarburization..37
   Decarburization Processes..37
   Testing..41
   Influence on Material Properties...43
   Control of Decarburization..47
   Summary..47

Chapter 3: Carbides..51
   Chemical Composition..51
   Massive, Network, and Dispersed Carbides................................53
   The Formation of Carbides..60
   The Effect of Network and Dispersed Carbides on Properties.........62
   Globular Carbide Dispersions and Film Carbides..........................69
      Globular Carbides and Heavy Dispersions............................69
      Film Carbides..70
      The Effect of Globular and Film Carbides on Properties..........70
   Summary..73

Chapter 4: Retained Austenite..77
   Austenite Formation..77
   Austenite in the Microstructure..81
   Effect on Material Properties...81

    Control of Retained Austenite 93
    Summary 94

Chapter 5: Influential Microstructural Features 99
    Grain Size 99
        Evaluation of Grain Size 100
        Effect of Grain Size on Properties.. 104
    Microcracking.. 107
        Factors Influencing Microcracking 108
    Microsegregation 113
        Formation of Microsegregation 113
        Effects of Microsegregation on Properties 117
    Nonmetallic Inclusions 119
        Origin of Nonmetallic Inclusions 119
        Effects of Nonmetallic Inclusions 121
        Consequences of Producing Clean Steels 128
    Summary 129
        Grain size 129
        Microcracks 129
        Microsegregation 130
        Nonmetallic Inclusions 130

Chapter 6: Core Properties and Case Depth............................ 135
    Core Factors.. 135
        Core Hardenability 135
        Core Microstructure and Hardness... 140
        Core Tensile Properties 140
        Core Toughness 143
        Effects of Core Properties 145
    Case Factors.. 148
        Case Hardenability 149
        Case Carbon Content.. 150
        Case Depth 155
    Quenching Methods 164
    Distortion 164
    Summary 165
        Core Properties 165
        Case Depth 167
        Case Carbon 168

Chapter 7: Postcarburizing Thermal Treatments 171
    Tempering 171
        Tempering Reactions... 171
        Effects of Tempering 175
        Additional Process Factors 183
    Refrigeration... 186
    Summary 194
        Tempering 194
        Refrigeration 195

Chapter 8: Postcarburizing Mechanical Treatments 199
    Grinding 199
        Grinding Action 199

- Grinding Burns and Cracks 200
- Effect of Grinding Variables 203
- Residual Stresses Caused by Grinding 207
- Effect of Grinding on Fatigue Strength 208
- Roller Burnishing 212
  - Effect on Microstructure 212
  - Effects on Material Properties 214
- Shot Peening 216
  - Process Control 216
  - Effect on Microstructures 217
  - Effects on Material Properties 218
- Summary 222
  - Grinding 222
  - Shot Peening 223

Index **227**

Preface to First Edition

This review deals with the products of carbon case-hardening—that is, the microstructural features—and their influence on the more important material properties. No intentional reference is made to the execution of the carburizing process itself. The review was compiled primarily to assess the current situation in terms of effects and perhaps show where the areas of ignorance or conflict lay. More importantly, it is to convey to the reader that because of the many variables involved in carburizing, hardening, and allied processes, the subject as a whole is complicated.

In its role as a vehicle for information, this review, as it stands, should be of value to students of engineering and of ferrous metallurgy since it covers aspects of a very important industrial process.

The works control metallurgist may more easily put into perspective the significance of some of the microstructural features observed in control test pieces, or may even look more carefully for, or at, structures hitherto ignored. Thus control procedures can be modified where necessary, and quality, perhaps, can be improved. Further, this review should be useful to those involved in service-failure analysis.

Heat treatment supervisors who, from component drawings or through company standards, work to readily measured properties such as surface hardness and case-depth requirements, might question some of the official or unofficial methods of attaining these. The result is that methods will be improved where they need to be improved.

For those engaged in carburizing research, this work should provide a useful background or platform for new work.

For the stress engineer or design engineer, an attempt has been made to provide quantitative values for some of the more important properties pertinent to carburized components; these are values that hitherto may have been assumed. Further, in view of the many complex and interrelated aspects involved and covered by this review, the engineer might exhibit more tolerance toward the metallurgist for his occasional inability to give precise answers to apparently simple questions.

The collecting of material for this book commenced in 1974. The timing for a comprehensive work seemed then to be right, not only for the practitioners but also for those engaged in industrial process computing. Now, as the 1980s begin, the stage has been reached when one might soon expect that from a steel's chemical composition and a component drawing the microstructures, hardness distributions, residual-stress gradients, and the distortion and growth behavior in the finished part will be predicted with reasonable accuracy. The main problem at the moment is in determining accurate material-properties data to feed into the programs. Such programs will link up with those for carburizing control on one hand and those for life predicitons on the other. The hope is that this review will be of value to those involved in these computer studies.

To compile a review for an intended readership of such widely varying interests has not been easy. There is, perhaps, a bias toward gears, which is not surprising in view of the fact that the carburizing process is extensively used for gears and the author is engaged in the gear industry. The bias is not really important since case hardening is case hardening, whatever the product.

This series of articles was first published in the Wolfson Heat Treatment Centre's Journal, *Heat Treatment of Metals,* during 1976–1977. Since then, the output of data regarding carburizing has continued to flow. Some of this information is relevant to the series and appears in the form of additional notes made by the author and appended at the conclusion of the chapters to which the information relates.

Geoffrey Parrish
Huddersfield
Yorkshire, England
January 1980

Preface to Second Edition

During my days as a metallurgical student, I devoted a fair amount of time to reading the many books and journals made available to students by the college library. While this activity contributed to my knowledge of metallurgy, I began to worry that I might be overburdening myself with too much disjointed information. At about that time, I purchased two books on the Production of Iron and Steel by Reginald Bashforth. These books impressed me because Bashforth had referenced hundreds of published works and provided the information in a manner that was easy to read and understand. It then dawned on me what I was doing wrong. I was collecting information in a random fashion when I should have been targeting one thing at a time. Further, I was stopping short by not assessing what the notes on a particular subject were collectively telling me. From then on, influenced by Bashforth, I would choose a topic for study, raid the library to find as much information as I could on that subject, then write myself an article with illustrations and references. That done, I would choose another topic and start all over again. It was hard work, and it was time consuming. It paid off when examinations approached. Because the articles were written earlier, one read through was all that was needed to bring that information back to the surface; pre-exam cramming was not necessary anymore.

I continued writing myself mini reviews as deemed necessary, long after I had qualified; it had become a habit.

In 1965 I joined the research and development department of David Brown Gear Industries Ltd., a leading U.K. gear manufacturing company. To a large extent, my work focused on the surface hardening of gear teeth (carbon case hardening, induction hardening, and nitride case-hardening processes; materials and properties). By about 1974, I had amassed a drawer full of journal articles relating to surface hardening and properties. Therefore, I set to out write myself a set of mini reviews concentrating on carbon case hardening; that is on the same lines as when I was a student. I did the work in my own time, and progress was not too quick. Even so, by mid-1975, I had four parts approximately complete and another four well on their way. During that summer, the department received a visit by Harry Child of the Wolfson Heat Treatment Centre, who among other things, asked if we had any information pertaining to the internal oxidation of case-hardened surfaces. I handed over a copy of my review on internal oxidation; being able to do so seemed to justify all the earlier effort and was very satisfying. Harry studied the article for a few minutes and asked if I would allow his organization to publish it in their journal, *Heat Treatment of Metals*. I pointed out that I had not really thought about publishing, and that the article was one of eight written for my own use and, therefore, hardly good enough for publishing. However, I agreed, and the eight parts were published during 1976 and 1977 with the title "The Influence of Microstructure on the Properties of Case-Carburized Components." Then during 1979, Alan Hick, the editor of *Heat Treatment of Metals* negotiated with the American Society for Metals for the work to be reissued in book form; in 1980 *The Influence of Microstructure on the Properties of Case-Carburized Components* was published.

More than twenty years after publication, I still receive an occasional compliment. Perhaps the most surprising one is that by Robert Errichello, who in *Gear Technology* (May/June 1992), listed the ASM version in "The Top Ten Books for Gear Engineers." However, the greatest compliment came in September 1996 when I was asked by ASM International to have another go at it. It is nice to see your name on a published work, but by far the greatest satisfaction, the greatest reward, comes from knowing that you have been of some use to the engineering community. To have another opportunity to do it again is especially gratifying and fortunate.

Geoffrey Parrish
March 1999

Introduction and Perspectives

Carbon case hardening, through natural evolution, commercialism, and economics, has become a process for which the possible number of variables is so large that it is hardly likely that any two companies will process exactly the same. There will always be some difference in choice of materials, equipment, or technique, and there will often be differences in the quality of the product. There may even be conflict of opinion regarding what is good practice and what is bad, and what is a valid test and what is meaningless. For each component treated, there is an optimum material and process combination, but who knows what this is for any given component? Most conflicts stem from there being too great a choice of materials or process variables and from the wide range of components that are required to be case hardened.

Despite all this, what the carburizing processes have in common is that they produce at the surface of the component a layer of carbon-rich material that after quenching, by whichever technique, should provide a surface that is hard. Regrettably, this is no indication that the case-hardening process has been successful. Additional microstructural features may exist along with, or instead of, the aimed-for martensite, and these indeed can significantly influence the properties of the component, thereby affecting its service life.

The microstructural features referred to are internal oxidation, decarburization, free carbides, retained austenite, and microcracks in the martensite.

Further modifications to the martensite in particular can be effected by tempering, and the proportions of austenite and martensite can be altered by subzero treatment after quenching. Cold working by either peening or rolling can modify the surface microstructures and have significant bearing on the life of the component, as too can surface grinding.

One must not overlook the value of the microstructure and properties of the core or of the influence of inherent features such as microsegregation, cleanliness, and grain size.

The aforementioned structural variants are the subject of this review, and where possible, examples of their effect in terms of properties are given. Those properties mainly referred to are bending-fatigue strength, contact-fatigue resistance, hardness, and wear resistance. These properties were chosen because it is to promote one or more of these properties that the carburizing treatment is employed. A gear tooth is a good example in which each of these must be considered. Some significance has been placed on the residual stresses developed during carburizing because these are additive to the applied stresses.

Why Carburize Case-Harden?

With some through-hardening steels, it is possible to develop hardnesses equal to the surface hardnesses typical of case-hardening parts; however, machine parts (for example, gears) would not be able to transmit as much load as would case-hardened parts. This is because case hardening produces significant compressive-residual stresses at the surface and within the hard case, whereas with through hardening, the residual stresses are much less predictable. Furthermore, high-hardness through-hardened steels tend to lack toughness; therefore, in general, through-hardened and tempered steels are limited to about 40 HRC to develop their best strength-to-toughness properties. To produce compressive-residual stresses to a reasonable depth in a through-hardening steel, one must resort to a local thermal hardening process, such as induction hardening, or an alternative chemicothermal treatment, such as nitriding.

When induction hardening is used for gears, for example, the preferred hardness distribution is generally to have about 55 HRC at the surface and 30 HRC in the core (Ref 1); consequently, parts so treated do not have a contact strength or wear resistance that are quite as good as in carburized and hardened parts. The induction hardening process is useful for large parts that need to be surface hardened but would distort or grow excessively if carburized and hardened. Typical gear steels surface hardened by induction are 4140 and 4340 (initially in the hardened and tempered condition), and typical case depths range from 1.0 to 3.0 mm.

Nitriding is a means of producing a hard surface with high surface compressive-residual stresses. It is a subcritical temperature process, and consequently, it is an essentially distortion- and growth-free process. The degree of hardening relates mainly to the chromium content of the steel so that a carbon steel will nitride harden only a little. Steel 4140 will harden to about 600 to 650 HV, and a 3% Cr-Mo-V steel will achieve more than 800 HV. Unfortunately, the cases that can be achieved due to nitriding are shallow (0.3 to 0.6 mm, effective), even with long processing times, for example, 80 hours. The shallowness of the case limits the range of application of nitrided steels. For gears, the limiting tooth size is about 2 mm module (12.7 dp) without downgrading. However, within its safe range of application, the case shallowness provides good bending fatigue, contact fatigue, wear, and scuffing resistance.

Carbon case hardening can be employed to achieve a wide range of effective case depths (up to greater than 4 mm) in a wide range of steels (limiting core carbon is normally 0.25%) with surface carbon contents of approximately 0.9% and hardnesses of about 60 HRC. The contact-fatigue and bending-fatigue strengths are regarded as superior to induction-hardened surfaces and to nitride-hardened surfaces (above a certain size limit). The drawbacks with carbon case hardening are distortion, growth, and costs. Distortion and growth are controlled as much as possible during heat treating (by the use of dies and plugs) and finally corrected by a limited amount of grinding. The costs are justified in the product to obtain a high power-to-weight ratio and durability.

An indication of the advantages of case hardening, compared with through hardening, is shown in the torque-speed plots of Fig. 1 (Ref 2). Here, the safe operating zone for case-hardened gear sets is much greater than it is for through-hardened steels. This means that to transmit the same power at a given speed, a set of case-hardened gears can be significantly smaller and/or lighter than a set of through-hardened gears. Alternatively, size for size, the case-hardened gear set will be much more durable.

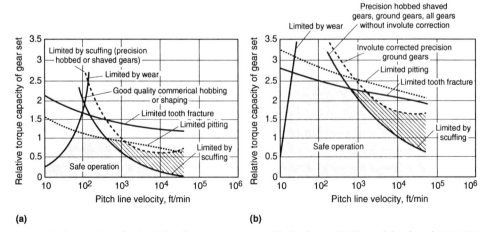

Fig. 1 Failure regions of industrial and automotive spur and helical gears. (a) Through hardened, 180–350 HB. (b) Precision gears, surface hardened

Variability

Over the past several decades, the steelmaking industry has moved from basic open-hearth steel manufacturing to processes such as VIM/VAR; consequently, the quality and consistency of steels have improved appreciably. Heat-treatment furnaces have improved, as have atmosphere and temperature control systems. Additionally, the gas-metal reactions, carbon diffusion, and other processes that take place during the carburizing and hardening of steels have become much better understood. Add to these factors the introduction of quality systems that favor process and product consistency, and, all in all, there has been considerable improvement (a far cry from the days of pack carburizing). Having said that, absolute precision is not attained because, among other reasons, exact steel compositions are impossible to achieve, and atmosphere control during carburizing is, at best, often only able to produce surface carbon contents of ±0.05% of the target value. Therefore, some metallurgical variability must be tolerated.

The grade of steel for a given machine component design, the carburized case depth, and the target values of surface carbon adopted by a manufacturer/heat treater are based on experience, design procedures, and guidelines provided in national or international standards, and perhaps on adjustments indicated by laboratory test results. It is difficult to determine the optimum metallurgical condition for a given situation; what is optimum in terms of surface carbon or case depth for a gear tooth fillet is different from what is optimum for a gear tooth flank. In fact, even if the optimum condition is known for any given situation (and this can vary from situation to situation), the heat treater probably could not provide it due to the variability described in the previous paragraph and the fact that most heat treaters are happy to get surface hardnesses within a fairly wide 58 to 62 HRC range, and effective case depths within a 0.25 mm range. Further, without considering section size, the previously mentioned composition variability could give batch-to-batch core-strength variations within a 20 ksi band. Hence, the ideal and the achievable are often different. Gear standards cater to different classes of gears, and these different classes require different degrees of dimensional precision and finish, as well as different standards of inspection. It is unlikely, however, that the heat treater will be lax for the lowest grade and fastidious for the precision gear. In most cases, the heat-treatment procedures will be to the same standard, and the heat treater will perform in the best way possible every time.

Laboratory Tests

Laboratory tests to determine the effect of metallurgical variables, for example, carbides, retained austenite, and core strength, are very useful and have contributed appreciably to the understanding of the influences of metallurgical features on material properties. However, there are problems associated with laboratory testing that must be recognized and, where possible, allowed for. One problem is that the test specimen and method of loading often bear little relationship to the machine part and service conditions they are supposed to represent. Apart from that, test pieces are often small in section so that the proportion of case to core can be high, and the microstructure can be martensitic throughout the test section. The effect of these factors on the residual stress distribution and on the contribution of metallurgical features can limit the value of the test findings. Another problem is isolating the metallurgical feature to be studied; generally, when conducting a test to determine the effect of a process variation or metallurgical feature on some property, the researcher attempts to isolate that test subject. Sometimes this is easy, for example, when determining the effects of tempering or subzero treatment. Other times, it is not so easy. For example, to determine the influence of retained austenite on bending-fatigue strength, a large batch of test pieces are prepared. Half are left as carburized and hardened with a high retained austenite content at the surface; the other half is refrigerated to transform much of the surface retained austenite. This is a common method of arriving at two retained austenite levels, but what exactly is being studied? Is it the effect of retained austenite, or is it the effect of subzero treatment? It is agreed that there are two austenite levels. Is it the difference in austenite levels that causes a difference of fatigue strength, or is it the effect of the new martensite and its associated short-range stresses induced by refrigeration that are responsible for the difference? The manufacture of batches of test pieces that are identical apart from the presence or absence of network carbides is another example. One can standardize surface carbon content and vary the heat treatment, or one can standardize heat treatment and vary the surface carbon content. Either way, there will be differences other than the carbide network. Nevertheless, laboratory testing provides trends and indicates whether a metallurgical feature will have a small or a large effect on the property under study.

Design Aspects

Laboratory test pieces are designed and loaded to fail. Machine parts, on the other hand, are designed and loaded not to fail. The basic allowable stresses used by gear designers have been conservative in order to acknowledge that design procedures are not precise enough to cater to the very wide range of gear designs, and that material variability and process variability do exist. These basic allowable stresses are derived from actual gear tests and are set at a lower value than that of failure stress. For example, in Fig. 2, the surface-hardened test gears failed due to tooth pitting at contact stresses of 1400 to 1500 MPa. These tests represent nitrided marine and industrial gears that have, in this instance, a design limit of about 1000 MPa (Ref 3). Comparable gear tests have been conducted for case-hardened automotive gears and aerospace gears. From these tests, appropriate allowable stress values (for both bending fatigue and contact fatigue) have been derived that are somewhat less than the actual failure values. The basic allowables are published in the gear standards (e.g., ANSI/AGMA 2001 or ISO 6336) (Table 1a and b). One should consider that for full-scale gear testing, the metallurgy of the tested gears might be typical of one heat treater's quality, which could rate either high or low against other heat treaters' qualities. This is another reason for setting the design allowables lower.

Designers also incorporate into a design safety factors that will account for any adverse effects of material and manufacturing variability. Therefore, there are probably numerous case-hardened parts performing satisfactorily in service with surface microstructures that contain adverse metallurgical features. For example, the high-temperature transformation products that accompany internal oxidation tend to be frowned upon, yet there are numerous case-hardened gears in service with unground roots that, therefore, contain degrees of internal oxidation. If the test gears from which the basic allowable stresses were derived had unground roots and fillets, then internal oxidation will be accounted for anyway. A metallurgical feature might indeed lower the strength of a part (according to laboratory test results), but the applied service stresses must be high enough for that feature to be significant and cause failure. If the basic allowable stress and the gear designer's safety factor together reduce the service stresses to, say, half the failure strength of the part, but

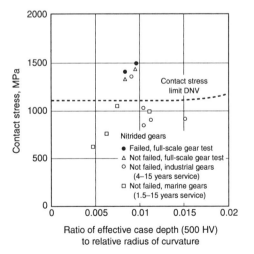

Fig. 2 The results of full-scale gear tests (failure by tooth pitting) and the typical design stresses used for industrial and marine gears. DNV, Det Norske Veritas. Source: Ref 3

Table 1(a) Basic allowable stress numbers for gears, ISO 6336-5 1996

Quality grade	Contact stress limit (σ_H), MPa	Bending stress limit (σ_F), MPa	Hardness, HV
Carburized and hardened			
ME	1650	525	670–745
MQ	1500	452–500(a)	645–745
ML	1300	315	615–800
Induction hardened			
ME	1275–1330	375–405	515–620
MQ	1160–1220	360–270	515–620
ML	960–1090	225–275	490–655
Gas nitrided, through hardened and tempered			
ME	1210	435	500–650
MQ	1000	360	500–650
ML	785	255	450–650
Gas nitrided, nitriding steels			
ME	1450	470	700–850
MQ	1250	420	700–850
ML	1125	270	650–850

Stresses are shown in MPa, and all hardness values are converted to HV. Designers should refer to the appropriate standard. (a) Varies with core hardness and/or core strength

the heat treatment has induced a serious adverse metallurgical feature with a strength reduction potential of, say, 30%, there still might not be a problem (Fig. 3). However, if something should go wrong, for example, if a bearing begins to deteriorate or the gear is slightly misaligned, increasing the tooth stress, then failure is more likely to occur.

It is not suggested here that one should ignore the metallurgical condition, or that quality control should be relaxed because design, to some extent, accommodates metallurgical variability. On the contrary. It could be that on many occasions the designer's generosity has, in effect, "saved face" for those responsible for the metallurgical quality. If the metallurgical variability could be reduced across the board, and improved quality and quality consistency could be guaranteed, then perhaps the basic allowable stresses could be increased a little. If nothing else, product reliability would be improved. Designers strive to improve their design procedures, manufacturers aim to produce levels of accuracy and finish the designer specifies, and lubrication engineers seek to improve their products. Together these efforts will lead to better power-to-weight ratios and, hopefully, reduced costs. Therefore, the metallurgists and heat treaters must continue to contribute to the cause.

Currently, it is believed that the limitations of the conventional case-hardening steels are fairly well understood. Any other gains must be made through design and process refinements (consistency and accuracy) sufficient to enable revision of the design allowables.

The future might never provide a case-hardening steel that is superior in all respects to the conventional grades. Even if it did, the cost of the steel might limit its use to very specialized applications. However, it is possible to design a steel that is superior with respect to one property. The newer grades of special-purpose aerospace gear steels for use at high operating speeds and temperatures exemplify this designing for purpose. Examples of such steel are Pyrowear Alloy 53 (Carpenter Technology Corp., Wyomissing, PA), CBS-1000M VIM-VAR (Timken Latrobe Steel Co., Latrobe, PA), CBS-600 (Timken Co., Canton, OH), Vasco X2-M, and Latrobe CFSS-42L, for which the steel compositions and heat-treatment operations depart sufficiently from the conventional. Previously, SAE 9310 steel was preferred by the aerospace industry for

Table 1(b) Basic allowable stress numbers for gears, AGMA 2001-C95

Quality grade	Contact stress limit (SAC), MPa	Bending stress limit (SAT), MPa	Hardness, HV
Carburized and hardened			
3	1900	520	650–800
2	1550	450 or 480(a)	650–800
1	1240	380	600–800
Induction hardened			
2	1310	152	515 minimum
	1345	152	580
1	1172	152	515
	1210	152	580
Gas nitrided, hardened and tempered, 4140 and 4340			
3	1210	...	460 minimum
	1240	...	485
2	1125	317–372(b)	460
	1160	317–372(b)	485
1	1030	234–276(b)	460
	1070	234–276(b)	485
Gas nitrided, 2½ Cr steel			
3	1300	420–440(b)	580 minimum
	1490	420–440(b)	690
2	1190	395–400(b)	580
	1350	395–400(b)	690
1	1070	280–310(b)	580
	1210	280–310(b)	690

Stresses are shown in MPa, and all hardness values are converted to HV. Designers should refer to the appropriate standard. This table is for spur and helical gears. (a) Depends on bainite content. (b) Varies with core hardness and/or strength

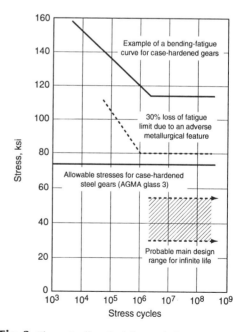

Fig. 3 Theoretically a "safe" gear design can accommodate the presence of an adverse metallurgical feature; however, there may be other adverse factors involved that also erode the difference between the fracture stress and the allowable stress.

gears, but its limitations (questionable hot strength, for example) inhibited design progress. The high-temperature limitations of lubricants for high-speed, high-temperature gearing is another factor to consider. The new grades of steel are designed to maintain their strength at operating temperatures and resist scoring and scuffing, which have a high potential to occur in high-speed, high-temperature gearing (Fig. 1). This resistance may, to some extent, make up for the limitations of the lubrication.

Metallurgy is only one factor in a bigger picture that includes machine and component design, manufacturing accuracy, machine

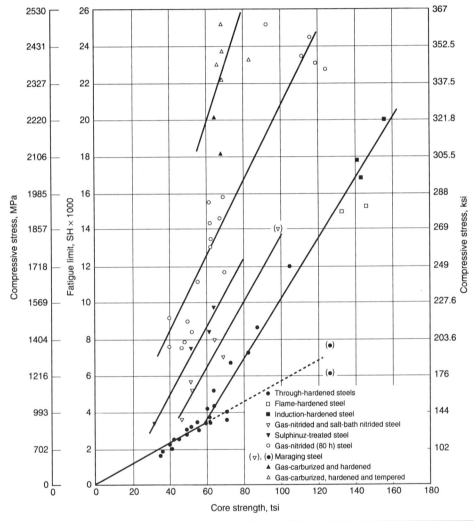

Steel	Effective case depth, mm (in.)
Through hardened (various)	...
Flame hardened (PCS)	...
Induction hardened (4340)	3.75 (0.15)
Gas nitrided and salt-bath nitrided	0.14 (0.005)
Sulphinuz treated	0.17 (0.007)
Gas nitrided (80 h) (3%Cr-Mo)	0.35 (0.015)
Maraging (x)	0.14 (0.005)
Gas carburized, hardened, and tempered (Ni-Cr)	1–1.5 (0.04–0.06)
Gas carburized and hardened (Ni-Cr)	1.5 (0.06)

Fig. 4 Effect of core strength and case depth on the rolling-contact fatigue limit of gear steels. Tests involved two 4 in. disks driven by a 2 in. roller. Test piece may have been either one of the disks or the roller. Relative radius of curvature, 2/3. SH units = lb/in. of face width divided by the relative radius of curvature.

assembly, lubrication, application, machine use or abuse, and maintenance (or lack of it). This book considers some of the current knowledge regarding the metallurgy of case-hardened steel parts and what effects or trends the various metallurgical features have on the properties of such parts. However, it focuses on conventional case-hardening steels and processing and, therefore, might not be as helpful to designers and users of new alloy grades.

Case-Depth Specifications

At the dedendum-pitch line area of a gear tooth, there is a smaller radius of curvature than at locations above the pitch line. Consequently, the contact band there tends to be narrower than at the addendum so that for a given load, the contact stresses will be higher. For that reason, the chosen case depth must be adequate to resist the stress at the dedendum-pitch line area.

The contact stress increases with transmitted load so, strictly speaking, the case depth should be determined by the load. Using the shear-fatigue strength (ultimate tensile strength × 0.34) of the material as opposed to shear stresses due to loading appears to give some conflicting results; therefore, it is not clear on which shear stresses the case depth requirement should be based. For example, if the 45° shear stresses (τ_{yz}) are considered in conjunction with the test results shown in Fig. 4, it is found that, for the 80 hour-nitrided surfaces, the predicted fatigue limit is about half of the value determined by testing. On the other hand, the fatigue limits for the carburized, hardened, and tempered surfaces (100 to 200 °C) and for induction-hardened surfaces are better predicted (Fig. 5). The orthogonal shear stresses (τ_{ortho}), however, predict fairly well the fatigue

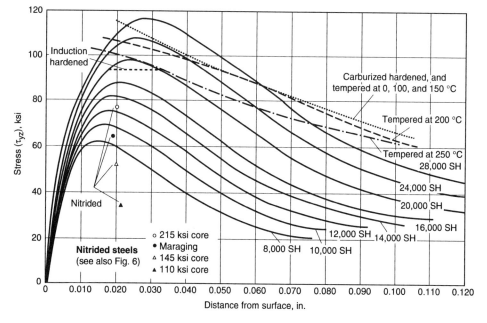

Process	Predicted fatigue limit, SH	Actual fatigue limit, SH
Carbon case hardened, untempered	24,000	18,000
Tempered at 100 °C	24,000	22,000
Tempered at 150 °C	24,000	23,000
Tempered at 200 °C	22,000	25,000
Tempered at 250 °C	20,000	26,800
Induction hardened	18,000	~18,000
Nitrided steels	Prediction equals about one half of actual	

Fig. 5 Plots of shear-fatigue strength (from hardness) against plots of shear stresses, τ_{yz}, in rolling-contact tests. Predicted and actual fatigue limit values are in close agreement for carburized steels but not for the four nitrided steels. Relative radius of curvature, 2/3. SH units = lb/in. of face width divided by the relative radius of curvature.

limits for the nitrided surfaces but overestimate the fatigue limits for the case-hardened and the induction-hardened surfaces (Fig. 6). From these apparently conflicting results, it is difficult to draw any meaningful conclusions that would help determine the appropriate hardness profile and case depth for a given application.

The relationship of residual stresses to rolling contact fatigue is also unclear. The table in Fig. 4 shows that for the case-hardened tests, the untempered roller produced the lowest fatigue limit, and the roller that had been tempered at 250 °C produced the highest value. Although residual stresses were not measured in either instance, it is nevertheless likely that the roller tempered at 250 °C had the lowest compressive-residual stress in the case, and the untempered roller had the highest (see Fig. 7.12). This implies that compressive-residual stresses might not be beneficial where rolling contact is involved—where the fatiguing actions are subsurface but still in the case. Therefore, this further complicates arriving at a theoretical solution for determining adequate hardness profiles and case depths. Fortunately, there is still the well used case depth-to-tooth diametrical pitch relationship to fall back on, even if it is not strictly correct (Fig. 7).

Interestingly, with rolling-contact fatigue tests of shallow-cased surfaces (i.e., when the depth of maximum hertzian shear stress is deeper than the effective case depth), there is no work hardening at the case-core junction up to the fatigue limit. At stresses above the fatigue limit, work hardening does occur, and the extent of the working (hardness and depth) increases with the contact stresses.

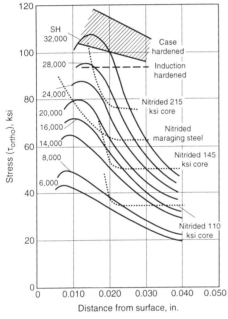

Process	Predicted fatigue limit, SH	Actual fatigue limit, SH
Carbon case hardened	>31,000	20,000–24,000
Induction hardened	27,000	18,000
(a) Nitrided for 80 h, 215 ksi core	25,000	25,000
(b) Nitrided for 80 h, maraging steel	16,500	14,000
(c) Nitrided for 80 h, 145 ksi core	14,500	15,000
(d) Nitrided for 80 h, 110 ksi core	7,000	7,000–9,000

Fig. 6 Plots of shear-fatigue strength against plots of shear stresses, τ_{ortho}, in rolling-contact tests. In contrast to Fig. 5, predicted and actual fatigue limit values are in good agreement for the four nitrided steels but not the other steels. Relative radius of curvature, 2/3. Shear fatigue strength is ultimate tensile strength × 0.34. SH units = lb/in. of face width divided by the relative radius of curvature.

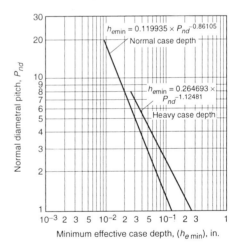

Fig. 7 Minimum effective case depth for carburized gears, $h_{e\ min}$. The values and ranges shown on the case-depth curves are to be used as guides. For gearing in which maximum performance is required, detailed studies must be made of the application, loading, and manufacturing procedures to obtain desirable gradients of both hardness and internal stress. Furthermore, the method of measuring the case, as well as the allowable tolerance in case depth, may be a matter of agreement between the customer and the manufacturer. Effective case depth is defined as depth of case with a minimum hardness of 50 HRC; total case depth to core carbon is approximately 1.5 × effective case depth. See ANSI/AGMA 2001-C 95.

Eutectoid Carbon Content

The requirements and information in any standard are, in general, readily understandable and realistic, as they should be. Unfortunately, there are exceptions. For example, the surface carbon requirement for carburized gears as set out in ISO 6336-5 1996 is "Eutectoid carbon % +0.20%, –0.1%." The standard does not justify the use of the term *eutectoid*. It does not provide a list of case-hardening steels along with a representative value of eutectoid carbon for each steel, nor does it provide an empirical formula for determining the eutectoid carbon. It is, therefore, unhelpful and unworkable as it stands. However, it is understood that the standard is to be revised to correct the problem.

The term *eutectoid carbon content* refers to the carbon content that produces only a pearlitic matrix microstructure as a result of an extremely slow cool through the Ac_3 or Ac_{cm} to Ac_1 temperature range. A steel with less than the eutectoid carbon content (hypoeutectoid) contains pearlite with some ferrite, whereas a steel with more carbon than the eutectoid carbon content (hypereutectoid) contains some carbide along with pearlite, again due to very slow cooling. Each steel grade has its own eutectoid carbon content, and considering the whole range of conventional case-hardening steels, the eutectoid carbon contents could easily vary between 0.45 and 0.8%. In case-hardening practice, the cooling rates employed, even slow cooling from carburizing, are much faster than the cooling rates researchers would use to determine the eutectoid carbon for an equilibrium diagram. Rapid cooling, typical of commercial quenching, can suppress the formation of ferrite in lean-alloy steels within about 0.2% C less than the eutectoid and suppress the carbide formation in that steel when the carbon is up to about 0.2% above the eutectoid. Suppression of ferrite or of carbide means that the carbon will be in solution in the martensite and in any retained austenite. Consider then: is a eutectoid carbon martensite the best to provide all the properties sought for a given application? Or is it the best carbon content for holding the retained austenite to a low value or for developing a better case toughness? Would a case-hardened 9310 steel gear with a surface carbon content of 0.55% be regarded as fit for service even though it might satisfy the case carbon requirements of ISO 6336-5 (1996)?

To establish where the eutectoid carbon content figures in deliberation regarding property optimization for case-hardened parts (and indeed it may have a place), there is little alternative but to establish eutectoid carbon data for each steel. For this, it may not be necessary to go through the complex procedure of determining accurate equilibrium diagrams. Instead, a set procedure could be devised in which, for example, a 30 mm bar is carburized to, say, greater than 1% surface carbon content and cooled, or heat treated to precipitate the excess carbon as carbides. The bar is then cut into two: one half is used to determine the carbon gradient and the other is used as a metallographic sample to determine the depth of carbide penetration. The two sets of data are then brought together to give a value of carbon at which, under the set conditions, carbides just appear. This could then be referred to as the "apparent eutectoid." Only with such information could the merits of the case carbon requirement of the ISO 6336 standard be assessed.

REFERENCES

1. G. Parrish, D.W. Ingham, and J.M. Chaney, The Submerged Induction Hardening of Gears, Parts 1 and 2, *Heat Treat. Met.*, Vol 25 (No. 1) 1998, p 1–8, and Vol 25 (No. 2), p 43–50
2. M. Jacobson, Gear Design: Lessons from Failures, *Automot. Des. Eng.*, Aug 1969
3. I.T. Young, T*he Load Carrying Capacity of Nitrided Gears,* BGMA, London, 1982

Chapter 1

Internal Oxidation

The presence of internal oxidation at the surfaces of parts that are case hardened by pack or gas carburizing has been known of for fifty years or more. The high-temperature transformation products (HTTP), which can form as a direct consequence of internal oxidation, have subsequently been found to have adverse influences on certain strength properties of affected parts; therefore, these products are of some concern to metallurgists and engineers.

The use of oxygen-free gas-carburizing atmospheres or vacuum-carburizing processes is known to eliminate the oxidation process, and nitrogen-base atmospheres are said to reduce it. However, conventional gas carburizing using the endothermic carrier gas is still the most popular method of case hardening, and its use will continue for many years. Thus, the problems related to internal oxidation will persist as long as the conventional process lives. Therefore, it is important to understand how internal oxidation comes about, what its likely effects are on material properties, and what should be done about it, bearing in mind that it has generally been tolerated in the past.

Factors Promoting Internal Oxidation

Endothermic Atmosphere. Gas carburizing is normally carried out at a temperature within the range of 900 to 950 °C using an endothermic carrier gas generated by the controlled combustion of another gas (such as natural gas, liquid propane gas, or towns gas) with air in the presence of a catalyst at a high temperature. Prepared from natural gas (methane), the endothermic atmosphere has a typical composition of 40% H, 20% CO, 0.46% CH_4, 0.27% CO_2, and 0.77% H_2O (vapor; dew point, 4 °C), with a balance of nitrogen. Such a mixture will have a carbon potential for iron of approximately 0.4% at 925 °C; therefore, in order to effect the carburization of steel components to the required surface carbon levels, endothermic gas must be enriched by controlled additions of a suitable hydrocarbon, such as propane or methane.

The balance of the component gases ensures that the endothermic atmosphere is reducing to iron, the parent metal of the steel, noting that the steel will be in the austenitic state at the temperature for carburizing. However, for those alloying elements in solid solution in the steel that have a greater affinity for oxygen than iron does, the atmosphere is potentially oxidizing.

Elements That Oxidize. Water vapor and carbon dioxide are the offending component gases in the endothermic atmosphere that provide the oxygen for the internal oxidation processes. The oxidation potentials of the main elements used for alloying can be derived from the ratios of partial pressures of the oxidizing and reducing constituents in the atmosphere, that is, pH_2O to pH_2 and pCO_2 to pCO. The results of such calculations, as presented by Kozlovskii et al. in Ref 1 for a temperature of 930 °C, are shown graphically in Fig. 1.1. This diagram shows that of the elements studied, titanium, silicon, manganese, and chromium are likely to oxidize, whereas iron, tungsten, molybdenum, nickel, and copper will not oxidize. This, of course, refers to elements that are not combined (i.e., those in solid solution). Two possibly important omissions from this diagram are aluminum and vanadium, both of which are common additions to steels. According to Fig. 1.2 (Ref 2), these elements will oxidize in an endothermic atmosphere; thermodynamically, aluminum appears to be slightly more ready to oxidize than does titanium, whereas vanadium will have an oxidation potential somewhere between those of silicon and manganese. It can be seen that elements favoring internal oxidation are generally necessary to the steel to impart characteristics such as hardenability, toughness, and grain refinement; but ironically, in some cases, their function is to assist in the deoxidization process during steel melting and casting operations.

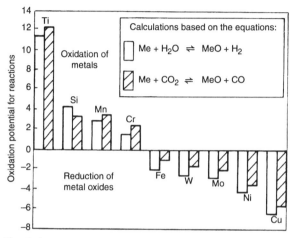

Fig. 1.1 Oxidation potential of alloying elements and iron in steel heated in endothermic gas with an average composition of 40% H_2, 20% CO, 1.5% CH_4, 0.5% CO_2, 0.28% H_2O (Dewpoint, 10 °C), and 37.72% N_2. Source: Ref 1

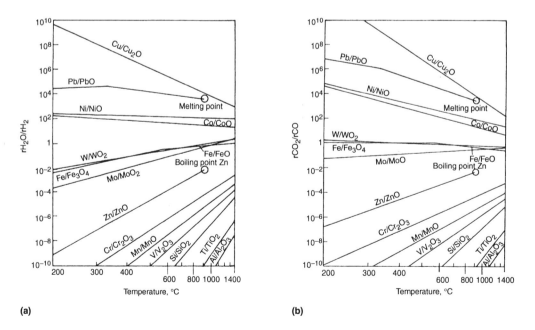

Fig. 1.2 Critical requirements for the oxidation of selected metals with indicated temperatures in atmospheres containing (a) water vapor and hydrogen and (b) carbon dioxide and carbon monoxide. Source: Ref 2

The Internal Oxidation Process

Oxygen Penetration. Oxygen is an interstitial element in iron, having an atomic size approximately 33% smaller than that of iron (noting that carbon atoms are ~34% smaller than iron atoms). However, iron has a low solubility for oxygen, and the diffusion oxygen through the ferrous matrix is relatively slow (10^{-9} cm^2/s). As Fig. 1.3 illustrates, the depth of total penetration of oxygen due to 6 hours at 930 °C in a typical endothermic atmosphere is only about 75 μm. This figure suggests that from about a 5 μm depth there is a steady fall in the oxygen content. On the other hand, a second high oxygen peak some distance from the surface has been observed (Ref 4, 5); for example, the second peak has been found at 17.5 μm (Ref 4).

At a given temperature, the oxygen content and the depth of oxygen penetration are strongly influenced by the oxygen potential of the atmosphere (the limiting oxygen potential being that at which iron begins to oxidize). However, as the carbon potential rises, the oxygen potential falls; consequently, with high-carbon potential carburizing, the oxidizing effect is reduced depending on the duration of carburizing. The relationship between time and temperature, with respect to internal oxidation, is shown in Fig. 1.4.

In the oxidation process, oxygen atoms, released by the gas-metal reactions that take place during carburizing, are adsorbed onto the metallic surface. From there, the oxygen atoms diffuse inward along grain and subgrain boundaries and into the lattice. Once there, they can chemically combine with available substitutional elements that have a high oxidation potential and form oxides. Meanwhile, carbon and hydrogen, the other interstitials released by the gas-metal reactions, similarly penetrate the surface to diffuse more quickly inward because they do not react to form compounds unless the carbon potential is unduly high enough to form carbides. When the carbon potential is high enough to form carbides at the surface, the oxidation of elements, such as chromium, titanium, and manganese, takes place within the carbide particles or at the carbide-austenite interfaces (Ref 5).

The Depth of Oxidation. In commercial case-hardening steels, the depths at which the oxides are detected by conventional optical microscopy are typically less than 25 μm (i.e., for carburized total case depths of 1 to 2 mm). Deeper cases will produce deeper penetrating oxides; for example, an 8 mm total case depth in a Cr-Ni-Mo steel would likely have an oxide penetration depth of 75 to 100 μm.

As Fig. 1.3 indicates, the depth of oxygen penetration is much greater than the depth to which the oxides form. In this instance, the depth of oxygen penetration is about three times as great; the oxides formed in the first 25 μm. (i.e., in that layer where there is a high oxygen plateau). If oxides had formed at a greater depth, they were too small to be resolved optically.

The depth of oxide penetration is influenced by a number of variables. For example, the depth of oxidation increases with the case depth, and for a given carburizing time, it increases with the temperature of carburizing. The depth of oxidation also increases with a lowering of the carbon potential (i.e., with an increase of carbon dioxide and water vapor), and also with an increase of the grain size. A most influential factor is, of course, the chemical composition of the steel.

Oxide Morphology. Metallographically, two oxide morphologies generally form. That oxide nearest to the surface (typically to a depth of 8

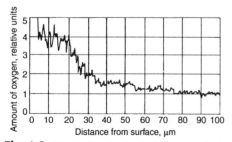

Fig. 1.3 Qualitative distribution of oxygen in the surface of a Cr-Mn-Ti steel (25KhGT) after carburizing in an endothermic atmosphere. Source: Ref 3

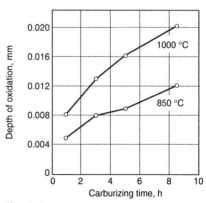

Fig. 1.4 Depth of the oxidized zones vs. carburizing time at different carburizing temperatures for SAE 1015. Source: Ref 6

μm) appears as globular particles of about 0.5 μm in diameter. This oxide resides mainly in the grain and subgrain boundaries and, to a lesser extent, within the grains themselves. Sometimes it occurs along the surface. Within this surface zone, the grains are likely to be subdivided into volumes of 0.5 to 1 μm across, although diameters of 2 to 4 μm have been quoted (Ref 3). The second type of oxide resides at typical depths of 5 to 25 μm and mainly occupies the prior austenite grain boundaries where it appears as a continuous "dark phase" (Ref 7), dark enough to resemble a void.

One can envisage that, as the oxygen gradient begins to develop during the carburizing process, the globular precipitates will start to form at the boundaries nearest to the surface and continue to grow as adequate quantities of reactants are brought together. However, Van Thyne and Krauss (Ref 8) have shown that the formation of the globular boundary oxides takes place by a discontinuous, lamellar growth process; that is, rods of the oxide form, each separated from the next by a band of alloy-depleted austenite. These rods tend to grow in the direction of the oxygen gradient. The oxides appear as rows of spheres when, in reality, the cross sections of rods are being viewed. At greater depths within the oxidized layer, the oxides appear to be continuous and at the prior austenite grain boundaries (Fig. 1.5).

The grain size at the surface of the steel is thought to influence oxide formation in that as the grain size decreases, the probability of forming oxides within the grains increases (Ref 3). However, it is suggested that the effect of temperature on penetration depth, as illustrated in Fig. 1.4, might also be affected by grain size (Fig. 1.6). One can imagine for a given oxidizing potential of the atmosphere that the more grain boundaries there are at which to distribute the available oxygen, the less the penetration will be.

Thus, steel composition and grain size are involved in the internal oxidation process. But what about carburizing conditions? What is certain is that when carburizing in an endothermic atmosphere containing 20% CO and 0.2 to 1% CO_2, the formation of internal oxidation is unavoidable. Dawes and Cooksey (Ref 10) estimate that 0.2% would be the maximum value of CO_2 that could be tolerated to prevent the internal oxidation of a 1% Cr steel, and that 0.01% CO_2 would be the limit for a 1% Mn steel. Mitchell, Cooksey, and Dawes illustrate how internal oxidation increases with manganese content and add that the severity of attack is related to the total case depth (Ref 11). Chatterjee-Fischer agrees with this, stating that samples having comparable case depths, arrived at by carburizing at two entirely different temperatures, would have the same depth of internal oxidation, even though the morphology of the oxides might differ somewhat (Ref 6). For a given temperature,

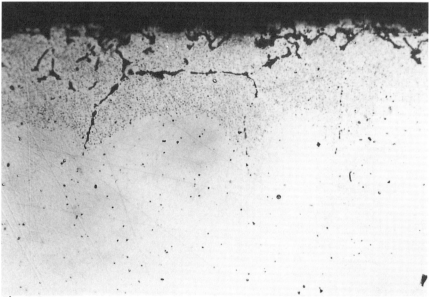

Fig. 1.5 Internal oxidation of a Ni-Cr steel carburized in a laboratory furnace, showing both grain boundary oxides and oxide precipitates within grains. 550×

the increase of depth of internal oxidation is proportional to the square root of the carburizing time. Edenhofer found that when carburizing a 16MnCr5 steel with the carbon potential and carburizing duration each held essentially constant, doubling the carbon monoxide content from 20 to 40% doubles the depth to which internal oxidation penetrates (Ref 12).

The Oxidation of Two-Component Alloys. Whereas Fig. 1.1 indicates which elements of a steel are likely to oxidize during carburizing, it gives no clues regarding how much of any one element is needed for the oxidation reaction to take place. Employing pure two-component alloys, for example Fe-Si, Fe-Mn, and Fe-Cr, Chatterjee-Fischer confirms that those alloys containing elements with a propensity to oxidize do indeed oxidize, provided that a sufficient amount of that element is present (Ref 6). Figure 1.7 summarizes her results and provides information regarding atomic number and size. It can be noted that the elements with the larger atomic sizes, or smaller atomic number, need only be present in amounts of less than 0.1 vol% to promote oxidation, whereas significantly more is needed of those elements whose atoms are of a similar size to that of iron. Alloy contents in amounts greater than those threshold values shown in Fig. 1.7 will lead to more of the oxide being produced. An increase in the silicon content, everything else being equal, will influence the depth of oxide penetration in a negative way and will increase the amount of grain-boundary oxide formed. In this respect, it was shown that increasing the silicon content of the iron-silicon alloy from 0 to 1.83% produced isolated fine precipitates to a depth of ~20 μm when the silicon was equal to 0.09%; dense globular and grain-boundary oxides were produced to a depth of only ~10 μm when silicon was equal to 1.83% (Ref 6). Between these two amounts of silicon, the quantity of oxide increased with silicon content, but the depth at which it formed decreased.

The Oxidation of Multicomponent Alloys. With multicomponent alloys (and commercial grades of carburizing steels), the situation is rather more complicated. In such alloys the silicon content to cause internal oxidation is about half that for a straight iron-silicon alloy (i.e., 0.05%), which is well below the 0.2 to 0.3% silicon content typical of case-hardening steels. This suggests that while these typical silicon contents are used, internal oxidation during conventional gas carburizing will be impossible to prevent. It does seem, however, that by limiting the manganese

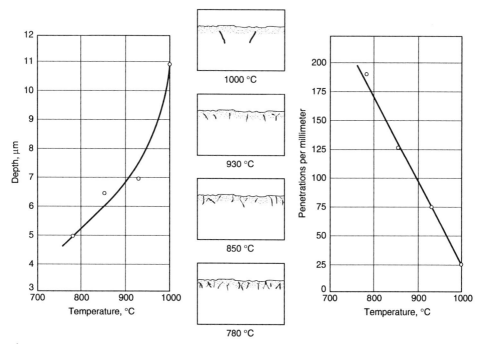

Fig. 1.6 Influence of carburizing temperature on the depth of oxide penetration and penetration frequency per millimeter of grain boundary oxidation for steel 17NiCrMo14. Adapted from Ref 9

and chromium to less than 1% in total, the depth of oxide penetration will be essentially that due to the silicon content.

The other elements with atomic numbers less than that of iron should be considered. Aluminum, for deoxidization and grain size control, is usually present in soluble form in amounts of approximately 0.01 to 0.06%. Though hardly enough to cause oxidation in an iron-aluminum pure alloy, this could well be enough to do so in a commercial grade, according to the behavior of silicon. To be effective as a grain refiner, titanium must be in excess of 0.1%. According to Okasaki, in steels with up to 0.1% titanium, the oxygen, carbon, nitrogen, and sulfur, which inhibit grain growth, combine with titanium to free the grain boundaries (0.1% titanium having the maximum effect). More titanium is needed to repin them (Ref 14). Therefore, a titanium content in excess of 0.1% would be expected to support the internal oxidation reaction. Vanadium is not normally added to case-hardening grades of steel for the purpose of, for instance, hardenability or strength, for which something in excess of 0.1% would be expected. For grain refinement, the amount could be much less than that, but whether or not it would be below the threshold for oxidation is not known.

The Oxidation of Commercial Case-Hardening Alloys. With commercial grades of steel, the observations regarding internal oxidation are at times confusing, which is not too surprising when one considers that different researchers have employed different steel compositions and carburizing conditions.

Arkhipov, employing a 18Kh2N4VA steel, found that the silicon and manganese did not oxidize, whereas the chromium did (Ref 15). The chromium content in this case was 1.65%, the silicon was 0.28%, and the manganese only 0.4%; with another nickel-chromium steel where the chromium content was 0.8%, however, the internal oxidation was less extensive. In yet another study by Arkhipov, this time using a Cr-Mn-Ti steel, the larger oxides observed in the surface (up to 6–8 µm) contained manganese, chromium, and silicon; at greater depths, however, only the oxides of silicon persisted (Ref 7). Essentially the same observations were made by Murai et al. (Ref 16) and Chatterjee-Fischer (Ref 6). Figure 1.8 shows a typical distribution of oxides.

This reflects on the fact that the composition of an oxide phase at any depth from the surface is primarily governed by its energy of formation, and the higher this energy is, the deeper the zone is in which that particular oxide will form. Again, for energy reasons, some oxides tend to form at the grain boundaries while others tend to form at sites within the grains, grain size perhaps having an influence.

With regard to the quantity of an oxidizable element, Mitchell et al. showed in Ref 11 that with C-Mn-B alloys carburized at 925 EC for 48 hours, internal oxidation was light to a depth of ~20 µm when the manganese content was 1 to 1.5%. Internal oxidation was heavy and to a

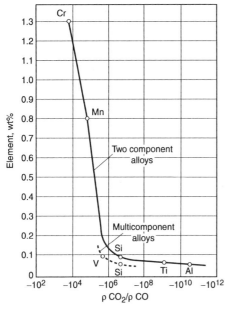

Element	Atomic number	Atomic size relative to iron, %
Interstitial elements		
Hydrogen	1	−58
Carbon	6	−34
Nitrogen	7	−36
Oxygen	8	−33
Solid-solution elements		
Molybdenum	42	10
Copper	29	1
Nickel	28	−1
Iron	26	0
Manganese	25	1
Chromium	24	1
Vanadium	23	6
Titanium	22	−36
Silicon	14	7
Aluminum	13	14

Source: Ref 13

Fig. 1.7 The limiting amount of added element (of atomic number less than that of iron) to promote internal oxidation. See also Fig. 1.2(b). Adapted from Ref 6

depth of ~40 μm when the manganese content was 1.5 to 2%

The Composition of Oxides. The actual compositions of the oxides have been variously quoted. Employing chromium-manganese steels, Kalner and Yurasov, who detected oxides to a depth of 30 μm, identified them to be of a complex compound that contained at least two metallic elements with a spinel structure: $nFeOM_2O_3$ (where M is either manganese or chromium) (Ref 3). It was found that the total M content of the oxide was, according to Table 1.1, up to 15% Cr or 11.3% Mn. It was also found that some part of the M could be replaced by small amounts of titanium and molybdenum in those steels which contained them. This latter observation is clearly of interest because titanium is present in case-hardening grades of steel in amounts of less than 0.1%, and molybdenum is one of those alloying elements regarded as being unlikely to be involved in the oxidation process. Perhaps the atomic sizes of titanium and molybdenum compared with that of iron (Ti, +36%; Mo, +10%) makes their inward or outward diffusion rates sluggish. The same authors observed that with samples of the same steel, carburized in the same heat, one sample produced the chromium oxide, whereas the other sample produced manganese oxide. The reason for this difference in behavior was not established, although it was considered that alloy segregation differences or the presence of varying amounts of carbide might have been responsible. Agreeing with Kalner and Yurasov, Sun Yitang also identified the oxide as $FeOCr_2O_3$ in a 20CrMnTi steel (Ref 4). The oxide Si_nO at grain boundaries and $(Cr_x Mn_y)_nO$ as precipitates were detected in a carburized Mn-Cr-B steel (Ref 18), and again it was found that the silicon oxide penetrated to a greater depth than did the manganese-chromium precipitates. In another study, the near-surface, grain-boundary precipitates were identified as $(Cr_2 Mn)_3O_4$, with SiO_x at a greater depth, confined to the fine grain boundaries (Ref 16). In Ref 19, Preston identified chromium, manganese, and vanadium in the internal oxidation of a 0.6% Cr, 0.7% Mn, 0.07% V steel in which silicon, titanium, and aluminum were also present. However, in Ref 20, Dowling et al. found Mn_2SiO_4 in a carburized SAE 4615 steel and something akin to $Mn_2Cr_3O_8$ in a carburized SAE 8620 steel.

Composition Gradients. The metal-oxygen reactions that lead to the precipitation of oxides must produce local composition gradients of the participating elements between the oxidized and unoxidized layers. This is because the atoms of the elements involved feed down the gradient in an attempt to compensate for those that have been utilized to form the oxides. Such an effect, involving both chromium and manganese, has been reported by a number of researchers (Fig. 1.9) (Ref 17, 18, 21, 22). However, it is not always the case that chromium and manganese composition gradients are present together. In Ref 17, Gunnerson determined the composition of the oxides at the surface of a carburized 15CrNi6 steel and came up with three different results (Table 1.1). This suggests that manganese and chromium can jointly or independently form oxides; much depends upon their respective quantities in the steel as a whole, or locally due to microsegregation. If only the oxide of one element forms, it is reasonable to assume that the matrix of the steel will be locally depleted of only that element. Arkhipov, using a 3½% Ni-Cr-W steel, determined that whereas the chromium migrated toward the surface, the nickel and tungsten migrated into the body, and it was observed that in a discrete zone, the nickel content exceeded 5% (Ref 15). Manganese appeared to diffuse only a short distance to the grain-boundaries. Robinson observed negative composition gradients involving chromium and

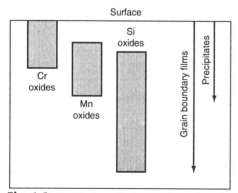

Fig. 1.8 Distribution of oxides at the surface of a Cr-Mn alloy steel carburized at 930 °C for 5 h. Adapted from Ref 6, 7

Table 1.1 Electron probe analysis of certain internal oxides

Depth below the surface, μm	Composition, %		
	Cr	Mn	Ni
2	15	11.3	Undetectable
2	7.5	...	Undetectable
2	15	...	<0.1

Source: Ref 17

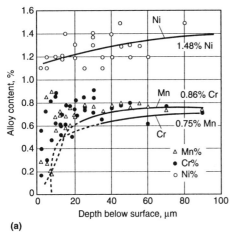
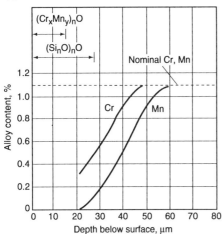
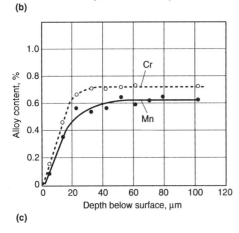

Fig. 1.9 Composition gradients associated with internal oxidation. (a) Electron probe microanalysis of manganese, chromium, and nickel within the surface zone of 15C4rNi6 steel. Source: Ref 17. (b) Chromium and manganese concentration gradients beneath the internally oxidized surface of a 20MnCrB5 steel. Source: Ref 18 and 22. (c) Chromium and manganese profiles measured by microprobe analysis of steel SIS 2515. Source: Ref 21

Table 1.2 Analysis of the surface material in SAE 8620 steel carburized in endothermic gas with natural gas additions

Position	Composition, %		
	Cr	Mo	Ni
Surface	0.79	0.18	0.55
Subsurface	0.52	0.14	0.57

Source: Ref 23

molybdenum at the surface of an SAE 8620 steel (Table 1.2) (Ref 23), whereas Colombo et al. found that there were no diffusion gradients associated with the internal oxidation in a carburized SAE 94B17 steel (Ref 24).

It is understood that much of an element migrating to the surface is utilized in forming the oxide, and that the matrix material in the vicinity of the oxide remains, if not completely depleted of that element, substantially below the average for the steel in question.

Apart from what happens to the alloying elements of a steel during the oxidation process, it may be found that the carbon content is also affected. In Ref 1, Kozlovskii et al. related that a sample of a carburized 25KhGT steel exhibited a low carbon content in the layer of internal oxidation (Fig. 1.10a). Other samples of different steel compositions used in the same investigation did not show the same decarburization effect, which makes it tempting to believe that this was a case of normal decarburization. However, using an SAE 94B17 steel, Colombo et al. determined the carbon content within the layer of internal oxidation (i.e., within the outermost 20 μm) to be 0.53% (Fig. 1.10b) (Ref 24). Shcherbedinskii and Shumakov also suggested that there was a reduced carbon content in the oxidized layer (Fig. 1.11), but only where the oxides have formed (Ref 5). These researchers considered that internal oxidation took place within the carbides, or at the carbide-austenite interfaces, which implies that a high carbon potential is a necessary requirement of the oxidation process. Other researchers have shown that a low carbon potential (high-carbon dioxide atmosphere) most favors internal oxidation. It also favors decarburization (see Chapter 2).

Effect on Local Microstructure

As yet, only the formation of the actual oxides has been discussed. However, a consequence of internal oxidation and the composition gradients

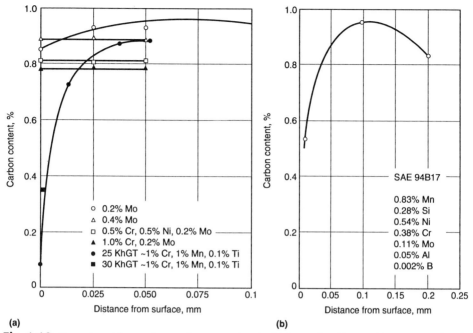

Fig. 1.10 Examples of low-carbon surfaces associated with internal oxidation. (a) Source: Ref 1. (b) Source: Ref 24

that develop during oxide formation is that the material adjacent to the oxides will have its transformation behavior modified. Thus, instead of the expected martensite, high-temperature transformation products (HTTP) can develop (Fig. 1.12). The nonmartensitic microstructures, which occupy the same area affected by internal oxidation, are variously described as pearlite or quenching pearlite, or either or both lower and upper bainites, or mixtures of all of them. It is likely, however, that the hardenability of the layer and the cooling rate are each significant. A lean-alloy steel or heavy section will tend toward a surface containing pearlite, whereas a more alloyed steel or lighter section will tend toward a bainitic microstructure being formed on quenching. The situation is to some extent confused by the presence of oxides that offer substrates on which new phases can nucleate, and by any local stresses that develop during the quench. Whichever nonmartensitic microstructure is formed, it will be comprised of ferrite and carbides, and the rate of cooling will dictate how the carbides precipitate. There is a chance that no HTTP will form when the cooling rate is high or when there are sufficient amounts of nickel and molybdenum in the matrix adjacent to the oxides. An example of this was provided by Dowling et al.

who found that the HTTP associated with the internal oxidation in a carburized SAE 8620 steel consisted of both pearlite and bainite. Neither of these, only martensite, was observed in the surface of a carburized SAE 4615 steel (Ref 20). Table 1.3 indicates the extent of alloy depletion within the matrix of the internally oxidized layers of these two steels.

The hardenability effect is illustrated for a 17CrNiMo6 steel in Fig. 1.13(a). This steel has a good case hardenability and is recommended for use in "driving pinions and high stressed cog

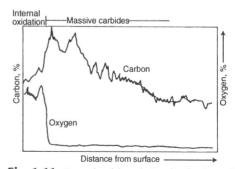

Fig. 1.11 Example of the relative distributions of carbon and oxygen at the surface of a carburized, highly alloyed steel that contains a high density of carbide phase in the outer case. Source: Ref 5

wheels" (Ref 25). The illustration indicates that even with a section equivalent to a 50 mm diameter bar, some bainite will form if the manganese and chromium are reduced to half of their original amount as a result of internal oxidation. This excludes the likelihood of the internal oxidation providing favorable sites for the nucleation of HTTP.

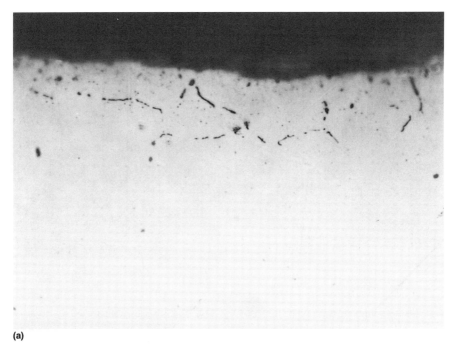

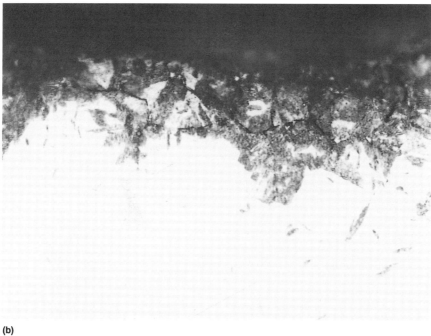

Fig. 1.12 Etching to reveal the presence of high-temperature transformation products associated with internal oxidation. (a) Unetched. 500× (b) Lightly etched in 2% nital. 500× (c) Medium etched in 2% nital. 500×

If the carbon content is reduced to 0.5%, for example, as it was in the case-hardened sample examined by Colombo (Ref 24), then for the 17CrNiMo6 steel in question, the largest section to avoid bainite formation in the low-carbon surface layer is 100 mm (4 in.) (Fig. 1.13b). If the low-surface carbon content is accompanied by a 50% reduction of both manganese and chromium, the limiting section will be approximately 37 mm (1½ in.). If, however, the manganese and chromium are completely removed to form oxides, even light sections will likely have bainite associated with the internal oxidation. In this instance, the pearlite nose will be in excess of 1000 seconds.

Figure 1.13(c) considers the situation where both manganese and chromium are removed from solid solution by oxide formation, and how this affects the ruling section when the carbon is also reduced. The indication here is that below ~0.25% C, some ferrite will be produced at the surface in all except the very lightest sections. Note that with normal transformation behavior, this steel would be unlikely to form pearlite at the cooling rates being considered.

To complete the set, Fig. 1.13(d) illustrates the effect that different levels of surface decarburization will have on the steel transformation characteristics when no manganese and chromium depletion takes place. This suggests that with carbon contents over ~0.15%, free ferrite is unlikely to be produced in sections equivalent to ~400 mm (16 in.), but low-carbon bainite will form.

Many case-hardening steels have hardenabilities less than that of the steel used for compiling Fig. 1.13, and therefore will have more of a tendency to form HTTP adjacent to any internal oxidation formed at the surface, including the formation of ferrite. Figure 1.14 depicts the condition for a carburized lean-alloy SAE 8620 steel with manganese and chromium depletion and degrees of surface decarburization. By comparing this with Fig. 1.13(c) it can be seen that the

Table 1.3 Semiquantitative analysis of elements in material adjacent to oxides

Steel	Composition, %		
	Cr	Mn	Si
8620(a)			
Case martensite	0.67	1.04	0.3
Bainite (at oxide)	0.37	0.6	0.2
Pearlite (at oxide)	0.24	0.26	0.3
4615(b)			
Case martensite	0.2	0.51	0.4
Martensite (at oxide)	0.1	0.22	0.3

(a) 0.92% Mn, 0.5% Cr, 0.11% Si, 0.38% Ni, 0.16% Mo. (b) 0.52% Mn, 0.12% Cr, 0.24% Si, 1.75% Ni, 0.54% Mo. Source: Ref 20

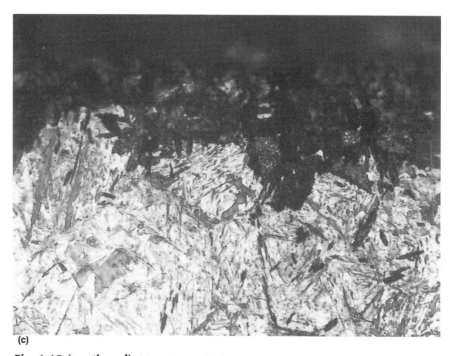

Fig. 1.12 (continued) (c) Medium etched in 2% nital. 500×

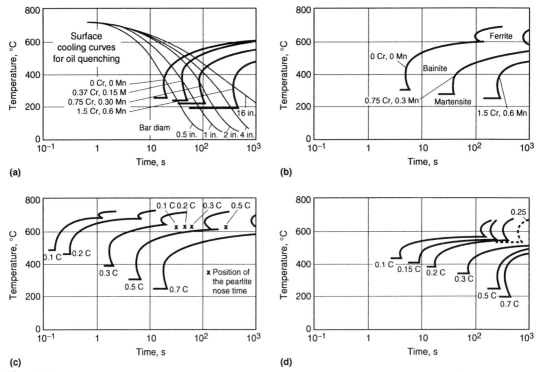

Fig. 1.13 Effect of composition variations on the transformation behavior of a case-hardening steel 17CrNiMo6. (a) Variations in Mn and Cr contents with composition 0.7 C, 1.5 Ni, 0.3 Mo. (b) Variations in Mn and Cr contents with composition 0.5 C, 1.5 Ni, 0.3 Mo. (c) Variations in C content with composition 1.5 Ni, 0.3 Mo, 0 Cr, 0 Mn. (d) Variations in C content with composition 1.5 Ni, 0.3 Mo, 1.5 Cr, 0.6 Mn

leaner grades of steel are the more likely to form pearlite at an oxidized surface. It is evident that the HTTP (ferrite, bainite, or possibly pearlite) associated with the internal oxidation depends upon:

- Steel composition and the quantities of elements remaining in solid solution following oxidation, including carbon. It was confirmed by the work of Arkhipov et al. that molybdenum reduced the amount of HTTP (Table 1.4) (Ref 26).
- Section size and the quench severity, both of which influence the cooling rate. The faster the cooling rate is, the better the chance of suppressing the formation of HTTP.
- Increase in carburized case depth, possibly. This is the case if an increase in case depth leads to a more complete depletion of alloying elements.

Alloy Depletion and the Eutectoid Carbon Content. The reduction of hardenability due to manganese and chromium depletion may not be the only consideration. The removal of manganese and chromium, or either one on its own, would be expected to raise the eutectoid carbon content in the alloy-depleted layer shown in Fig. 1.15, which means that the solubility for carbon will locally increase. Therefore, if the carbon content of the surface is at C^1, and it is sufficient to produce free carbides in the absence of any alloy depletion, then where alloy depletion has occurred and the eutectoid carbon content has been raised, there will be more carbon in solution, and, hence, less free carbide. Figure 1.16 depicts

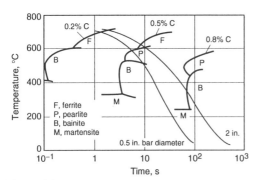

Fig. 1.14 Effect of alloy depletion and carbon content on the continuous cooling transformation behavior of an SAE 8620 steel with composition 0.5 Ni, 0.2 Mo, 0 Cr, 0 Mn

Table 1.4 Effect of depth of nonmartensitic layer on hardness, residual stress, and bending fatigue strength in case-hardened and tempered 4 mm modulus gears

Steel	Process(a)	Case depth, mm	Macrohardness, HRC Surface	Core	Depth of nonmartensitic Layer, μm	Microhardness in tooth roots, HV 0.2	Residual stress at a depth of 10 μm kg/mm²	MPa	Bending fatigue limit kg/mm²	MPa
25KhGM	CN	1.0	59/61	39/44	6	560/720	...	...	77	753
20KhNM	C	1.2	57/60	32/36	7	590/760	0–5	0–50	71	695
25MO5KhO5	CN	0.7	58/60	24/28	3	525/700	0	0	68	665
	C	1.2	60/62	22/26	6	510/575	...	...	66	646
18KhGT(1)	C	1.0	59/62	32/36	16	415/440	26–44	253–432	61	598
30KhGT	CN	1.1	60/62	40/43	15	415/510	9–12	89–119	55	540
	C	1.0	60/62	38/44	30	375/440	14–19	139–185	55	540
18KhGT(2)	C	1.2	58/60	32/36	17	380/500	...	...	54	532
18Kh2N4VA	C	1.0/1.1	58/59	40/43	17	265/575	...	...	49	482
35Kh	CN	0.7	59/62	34/37	100	320/510	24–45	234–441	42	412

(a) C, carburized; CN, carbonitrided. Source: Ref 26, 29

such a situation where the absence of carbides in a narrow band adjacent to the oxidized layer, which might be attributed to decarburization, could in fact be due to a shift of the eutectoid carbon, there being no loss of carbon.

Influence on Material Properties

The chemical and microstructural effects so far described are concerns due to their potential influence on the properties of the carburized part.

It is not clear whether internal oxidation on its own is especially deleterious, or if it simply behaves much in the same way as does surface roughness, bearing in mind that the effect of surface roughness may be offset by the presence of compressive residual stresses in the underlying steel. Having said that, it is clear that any nonmartensitic microstructures, that is, HTTP, associated with the internal oxidation can have a

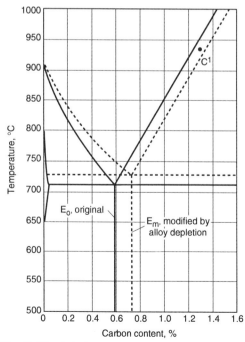

Fig. 1.15 Shift of eutectoid carbon content due to alloy depletion associated with internal oxidation

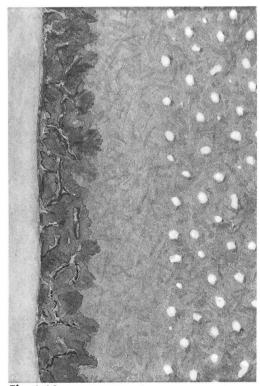

Fig. 1.16 Representation of a microstructure showing internal oxidation with associated high-temperature transformation products at the surface and spheroidized carbides some distance from the surface

deleterious effect on the strength properties of a part.

Influence on Hardness. Specified surface hardness values for carburized, hardened, and low-temperature tempered components generally fall in the range 58 to 62 HRC (the actual intermediate range used depends upon the size of the component and its application). To achieve hardness values of this order, it is necessary to produce an as-quenched microstructure of high-carbon martensite. However, a consequence of internal oxidation is that nonmartensitic microstructures are likely to be formed in a narrow zone adjacent to the oxides, resulting in a soft "skin," even though Rockwell macrohardness tests might not indicate anything other than satisfactory hardening. A file test, on the other hand, would detect its presence.

When a sample of the carburized steel is sectioned and prepared for metallographic examination, the presence of internal oxidation and of HTTP is clearly determined. Microhardness tests in the surface regions of the sample will then measure the extent of softening due to the HTTP. An example is shown in Fig. 1.17, where the structure at the surface was probably bainite, or predominantly bainite. Whereas the microstructures in the nonmartensitic layer may be mixed, or of a low or variable carbon content, the microhardness will give a clue as to the type of structure present in the layer. Figure 1.18 provides an indication of hardness against carbon content for different types of microstructures. From this the following is derived:

- *Ferrite:* due to decarburization; C ~ <0.2%; microhardness, 200 to 250 HV
- *Mixed ferrite and pearlite:* decarburization with intermediate carbon; 200 to 300 HV
- *Pearlite:* thought to exist in the HTTP of some steels; 300 HV
- *Mixed pearlite and bainite:* possibly of intermediate carbon; 300 to 400 HV
- *Bainite:* probably predominates in most HTTP layers; 400 HV

Influence on Residual Stresses. The oxides of chromium, manganese, and silicon that form in the surfaces of components during their carburization (at temperatures typically in excess of 920 °C) will likely be in compression when the temperature falls to room temperature. This is because their volume thermal contraction over the temperature range 920 to 20 °C will be less than that of the steel in which they reside.

The HTTP associated with those oxides will be in tension. This is partly due to the presence of "compressed" oxides but also due to the volume mismatch between the HTTP itself and the underlying high-carbon martensite; note that the HTTP is probably the first material to transform, while the adjacent martensite is the last to transform.

The residual stresses through the roots of case-hardened gear teeth have been determined for a number of steels (Ref 28), and the trend toward tensile residual stresses at the surface is

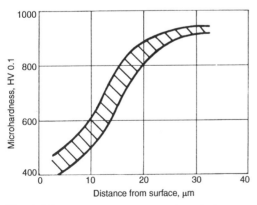

Fig. 1.17 Microhardness traverses through the internally oxidized layer of a carburized Cr-Mn-Ti steel (30KhGT). Source: Ref 7

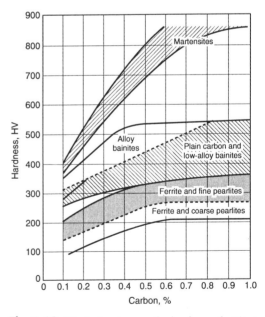

Fig. 1.18 Effect of carbon on the hardness of various microstructures observed in plain carbon and lean-alloy steels. Source: Ref 27

confirmed (Fig. 1.19). Additional work on the same gears showed that the greater the tensile residual stresses were, the lower the fatigue strength tended to be (Table 1.4) (Ref 26). Naito et al. observed a similar residual stress distribution for carburized JIS SCM415 steel with internal oxidation and HTTP at the surface (Ref 29). Dowling's results are of particular interest because they compare the residual stresses in two case-hardened surfaces: one in which pearlite and bainite formed due to internal oxidation and one where no HTTP formed (Fig. 1.20) (Ref 20).

Influence on Bending Fatigue. Internal oxidation on its own does not appear to have a great influence upon the bending fatigue strength of case-hardened parts (Ref 20, 30). If anything, it may have an effect similar to a small degree of surface roughness. If, however, the internal oxidation is accompanied by HTTP (bainite or pearlite), the bending fatigue strength will be significantly reduced.

It was earlier believed that there was a threshold value of 13 μm for the depth of HTTP below which it had no adverse effect, and that with less than 13 μm of HTTP, surface carbon content was the dominant variable (Fig. 1.21) (Ref 1, 31). The idea that there is a threshold might not be incorrect, though other work has indicated that the presence of less than 13 μm of HTTP has led to a reduction of fatigue strength (Ref 32); 11 μm of HTTP obtained a 15% reduction of fatigue strength. It could be that smaller amounts of HTTP do cause reductions of fatigue life, but the amount is sufficiently small to be regarded as insignificant. Also, there may be a size aspect where, for instance, 15 μm of HTTP affects a small test piece but has a smaller effect on a larger test piece.

The existing test data indicate that in the low-cycle region of the *S-N* plots, there is a general trend toward a small adverse effect (0 to 12% reduction of strength at 10^4 cycles) due to the presence of HTTP (Ref 7, 17, 20, 23, 29, 32). In the high-cycle region, the fatigue life can be reduced by as much as 45% due to the presence of HTTP, though 20 to 35% is more common; the amount depends upon the depth, the microstructural content, and hence, the hardness of the HTTP layer.

The influence of case hardness on the rotating-bending fatigue strength of carburized steels was studied by Weigand and Tolasch (Ref 33), and they showed that there was a progressive fall of bending fatigue strength as the case hardness fell below ~680 HV5 (Fig. 1.22). Their work dealt with test pieces where the measured hardness values represented the maximum hardness at, or close to, the surface. With internally oxidized parts, only the "skin" is soft; the underlying material is martensitic and hard. Nevertheless, the trend is the same, as Fig. 1.23 suggests. From this, a guide to the potential loss of bending fatigue strength can be estimated (Table 1.5). This might suggest that two carburized steels with dif-

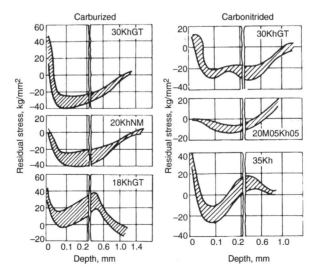

Fig. 1.19 Residual stresses at the base of the teeth in carburized and carbonitrided gears. Source: Ref 28

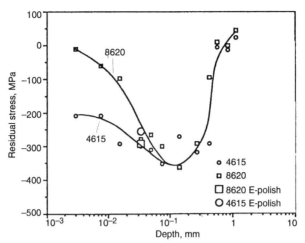

Fig. 1.20 Residual stress profiles for both 4615 and 8620 materials. Source: Ref 20

ferent alloy contents can have very different fatigue lives in the absence of HTTP, but very similar fatigue lives in the presence of HTTP. The results of Brugger's study, for example, showed that the fatigue limits for case-hardened 20MnCr5 and 15CrNi6 were 680 and 780 MPa, respectively, when there was no HTTP present at the surface (Ref 34, 35). When HTTP was present in similar amounts, the fatigue limits of the two steels fell in the range 520 to 540 MPa.

Gunnerson found no difference in high-cycle fatigue limit between samples that had HTTP reaching to a depth of 15 to 17 µm, and samples internally oxidized to ~15 µm without HTTP (achieved by inhibiting the formation of HTTP with ammonia additions late in the carburizing cycle) (Ref 17). This finding is contrary to other reported results. Unfortunately, many of Gunnerson's test pieces failed with subsurface fracture initiation points and, therefore, were not really relevant to the study of a surface condition; these test pieces said more about the subsurface properties. After replotting the data using only those points that represented surface initiated failures, a distinct difference was noted in the upper finite life part of the S-N curve, shown in Fig. 1.24(a); it is suggested that the fatigue curve representing the specimens with HTTP at the surface might have a double knee. Naito et al. also observed a possible double knee effect for the samples having HTTP at their surfaces (Fig. 1.24b) (Ref 29). These are interesting observations, the significance of which is not clear, apart from the fact that both projects employed rotating beam test pieces. However, what

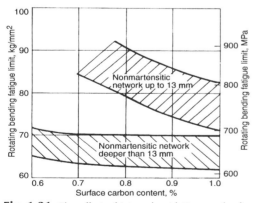

Fig. 1.21 The effect of internal oxidation on the fatigue strength of carburized 25KhGT steel. With this steel, internal oxidation in accompanied by a decrease in surface carbon. Source: Ref 1

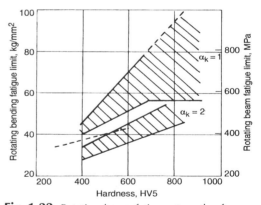

Fig. 1.22 Rotating beam fatigue strength of case-hardened 12 mm diam specimens, notched and unnotched. The line for carburized gears shown in Fig. 1.23 is superimposed (converted to rotating bending fatigue). Source: Ref 33

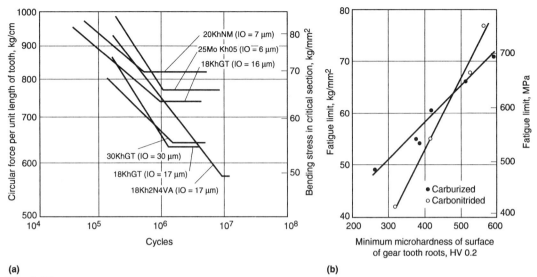

Fig. 1.23 Effect of internal oxidation and surface microhardness on the fatigue properties of 4 mm modulus gears. See also Table 1.4. IO, internal oxidation. (a) Fatigue strength plots for 4 mm modulus gears. Information on case-hardened gears given in Table 1.4. Source: Ref 26. (b) Effect of tooth root surface hardness on the bending fatigue limit of 4 mm modulus gears

the two sets of results do seem to suggest is that up to, or at, about 10^4 load cycles there would be essentially no difference between the samples with and without HTTP.

With respect to grain size, it is interesting to note the findings of Pacheco and Krauss in Ref 30: "Fine grained gas carburized specimens tolerate the presence of inter-granular oxidation and have better fatigue resistance than a coarse grained specimen without surface oxidation." In their work, which compared plasma- and gas-carburized test pieces, the gas-carburized specimens contained internal oxidation to a depth of ~13 μm without the HTTP associated with internal oxidation; this supports the idea that, on its own, internal oxidation is not particularly damaging.

Influence on Contact Fatigue. Most case-hardened gears are precision ground before going into service; therefore, the effect of internal oxidation on contact fatigue durability need not be considered for them. There are, however, a few types of gears and a number of other components that enter service in the unground, or perhaps in the lightly lapped, condition for which the influence of internal oxidation is pertinent. Unfortunately, there are little data available on the subject, and these tend to conflict. Adverse effects have been reported in Ref 3, whereas in Ref 36 Sheehan and Howes, working with case-hardened SAE 8620, regarded internal oxidation as not being detrimental, and even beneficial, to contact durability under slide-roll test conditions. Figure 1.25 presents the contact fatigue life as a function of the amount of material removed from the surface of the test disks prior to testing. It shows that there was generally no loss of contact fatigue resistance until the internal oxidation and nonmartensitic microstructure had been removed, the depth of HTTP penetration being 25 to 37 μm. Where the soft layer had been completely removed, lower values of fatigue were recorded. As a result of supporting tests, these researchers considered that plastic deformations within the HTTP layer, which would bring about a more favorable distribution of contact load, could only account for part of the difference shown. In tests where the slip was about 30% and the contact load was 2390 MPa (347 ksi), they found that the unground surface had a lower coefficient of friction than the ground surface did

Table 1.5 Loss of fatigue strength

Skin hardness, HV0.2	Fatigue limit, kg/mm²	Loss of fatigue strength, %
700(a)	80(b)	0
600	72.5	9
500	65	19
400	~58	28
300	51	36
250	48	40

(a) 700 HV is an extrapolation; see Fig. 1.23. (b) 80 kg/mm² is taken to represent 100% fatigue strength.

(0.068 and 0.080 respectively); they reasoned, however, that this difference was not large enough to account for the differences in fatigue strength that were observed. It was concluded, although the test did not really confirm it, that it was likely that an "as-heat-treated" surface would develop a more stable oil film than either a ground or electropolished surface.

Influence on Bending and Impact Fracture Strength. The fracture testing of case-hardened samples by slow bending or by impact (plane, notched, or precracked specimens) does not seem to yield results that are significantly influenced by the presence of internal oxidation and any accompanying HTTP.

Kozlovskii et al. stated that there was no negative effect of on the impact strength of case-hardened steels due to HTTP and indicated that there was no clear separation of bend test results whether the HTTP was more or less than 13 μm (Ref 1). Diesburg et al. concluded that fracture was independent of the structure state in the outer 0.25 mm of the case and was more dependent on the residual stresses and microstructure away from the surface (Ref 37). Tests showed that fracture toughness increased with distance from the surface (Ref 38), and that crack initiation was influenced by case hardenability (the ability to produce martensite throughout the case); crack propagation, however, was more influenced by the core properties and composition (Ref 39). Fett agreed that composition was important with respect to bending and impact, the resistance to these being enhanced by nickel

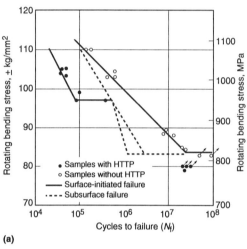

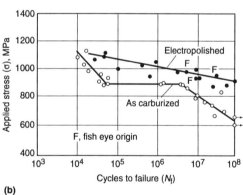

Fig. 1.24 Effect of internal oxidation and high-temperature transformation products on the high- and low-cycle bending fatigue strength. (a) Fatigue data on rotating beam tests, 6 mm outside diameter test section, quench 860 °C into oil at 200 °C. Steel composition: 0.75 Mn, 0.86 Cr, 1.48 Ni, 0.04 Mo. Samples with HTTP: 7 failed at surface, 4 ran out, 16 failed subsurface. Samples without HTTP: 10 failed at surface, 2 ran out, 23 failed subsurface. Replotted from Ref 17. (b) Fatigue data for samples with and without internal oxidation, quenched from 850 °C into oil at 70 °C, temper at 180 °C. Steel composition: 0.74 Mn, 1.01 Cr, 0.18 Mo. Source: Ref 29

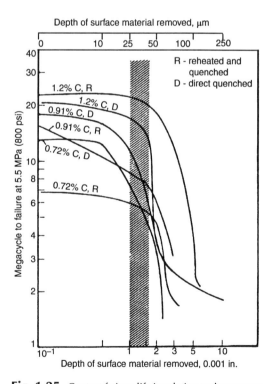

Fig. 1.25 Contact fatigue life in relation to the amount of material removed from carburized SAE 8620 samples prior to testing. The hatched band represents the depth of internal oxidation beneath the original surface of specimens. Source: Ref 36

and chromium (Ref 40). Others, however, considered that static bending and impact strength were related to the case depth (Ref 41). Ono et al., using unnotched impact and bend tests, showed that 930 °C direct-quenched, gas-carburized test pieces with 15 μm of HTTP performed better than did 1040 EC vacuum-carburized test pieces with no internal oxidation or HTTP (Ref 32). A 930 EC reheat-quenched, vacuum-carburized set of test pieces produced even better results (Fig. 1.26). Unfortunately, no data were provided for any reheat-quenched, gas-carburized samples, so a real comparison could not be made. Nevertheless, the results implied that factors other than internal oxidation and the presence of HTTP were more significant with respect to bending and impact fracture tests.

Influence on Wear Resistance. The deterioration of a surface by wear can be due to either or both of two processes: adhesive wear and abrasive wear. Adhesive wear, as the name suggests, is adhesion that occurs when the pressure and heat generated during sliding cause small areas of one of the surfaces to chemically bond to the other surface. The relative motion breaks the bond, but not necessarily (perhaps rarely) at the original junction; therefore, metal is transferred from one surface to the other. What happens to the transferred particle then depends upon a number of factors. It may remain adhered to the other surface, or it may separate to become loose debris.

Terms for adhesive wear include scoring, scuffing, galling, and seizure. Fretting is a form of adhesive wear for which the relative movement is minute, as in the vibration between a key and keyway, for example. Abrasive wear, or normal wear, involves the removal of particles from the mating surfaces by asperity shearing due to asperity collisions or collisions with loose debris passing between the sliding surfaces. Wear processes are appreciably influenced by any lubricant present, its quality, and its condition.

When internal oxidation is present on its own at a case-hardened surface, one would expect it to inhibit the adhesive wear due to its intermetallic nature. Also, as Sheehan and Howes suggested in Ref 36, the oxide could assist in the lubrication process; as it is, the oxide will lower the coefficient of friction. The oxide, however, might favor abrasive wear and eventually crumble under the action of sliding, thereby producing loose particles (debris). In such an instance, the depth of wear will likely be only the depth of internal oxidation: 10 to 20 μm.

When the internal oxidation is accompanied by the HTTP, the situation in terms of wear-resistance changes. Both adhesive and abrasive wear resistances are related to the surface hardness (Fig. 1.27, 1.28). As the hardness falls, the wear rates will increase, considering that HTTP will work harden, and martensite will soften a little as pressure and temperature conditions approach those that cause adhesive

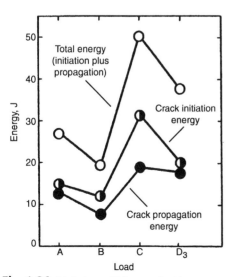
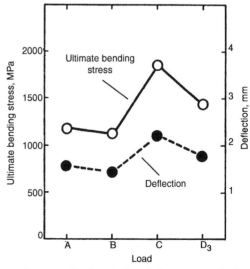

Fig. 1.26 Variations of (a) unnotched impact energy and (b) bending strength (15 × 60 × 2 mm). A: vacuum-carburized, 1040 °C, reheat quenched. B: vacuum-carburized, 1040 °C, direct quenched. C: vacuum carburized, 930 °C, reheat quenched. D3: gas-carburized, 920 °C, direct quenched. Source: Ref 32

wear. In case-hardened parts, high-carbon martensite at over 700 HV offers the best resistance to wear. Other phases, such as proeutectoid ferrite or austenite, can be tolerated in the martensite without adverse effects, provided those phases are indeed in small quantities, are fine and well distributed, and don't significantly reduce the hardness (Ref 42). If the HTTP is bainite at ~400 HV or ferrite at, for instance, 250 HV, then the wear resistance of the case-hardened layer will be seriously affected. Whether or not the wear processes stop or slow down once the HTTP has been removed by wear depends on the actual system in which the damage is taking place. Wear processes can virtually stop.

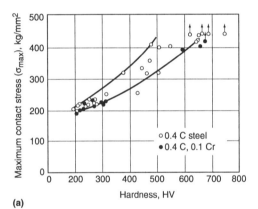

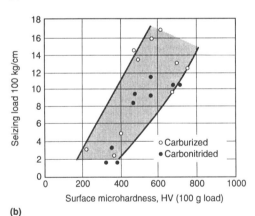

Fig. 1.27 Relationship between surface hardness and seizure. (a) Relation of hardness, HV, with maximum contact stress, σ_{max}, when destructive seizure occurs for through hardened or induction hardened steels. Source: Ref 42. (b) Variation of seizing load with microhardness of the outer layer of carburized and carbonitrided samples. Radius of curvature, 2.5 cm; slip speed, 1.05 m/s; rolling speed 2.28 m/s; specific slip, 1.7; gage width of roller, 0.5 cm. Source: Ref 43

Measures to Eliminate or Reduce Internal Oxidation

It is more or less impossible to control the endothermic carburizing atmosphere to eliminate the formation of internal oxidation at the surfaces of case-hardened parts. The use of alternative oxygen-free atmospheres or of vacuum-carburizing or plasma-carburizing processes might not be economically viable or as flexible and convenient as the conventional process. Consequently, the endothermic generator will continue to provide the carrier gas for the carburizing process for many years to come. The problem of internal oxidation will, therefore, persist. What can be done about it? The approaches can involve steel design, process control, mechanical or chemical removal, or design of components that acknowledge its presence.

Steel Design. The formation of internal oxidation is related to the presence of certain alloying elements and their quantities in solid solution within the steel. The problem is that silicon is used to kill case-hardening grades of steel, and the nominal 0.25% silicon content by far exceeds the maximum required to produce internal oxidation. It also lowers the maxima for the other elements, such as manganese and chromium. The replacement, or partial replacement, of oxide-forming elements with elements that have atomic numbers greater than that of iron could contribute to reducing the effect. Such elements are nickel and molybdenum, and maybe tungsten.

The basic alloy, therefore, would contain:

- *Carbon:* to adjust the core strength, though not enough to adversely affect the develop-

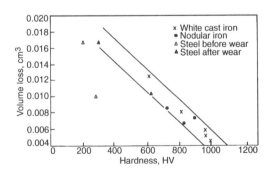

Fig. 1.28 Effect of hardness on wear. Note that wear resistance of the steel sample only fits the general pattern if the hardness of the work-hardened surface is considered. Ref 44

ment of favorable residual stresses or to encourage unwanted distortion or growth
- *Nickel:* to contribute primarily to the toughness of both the case and the core
- *Molybdenum:* to provide case and core hardenabilities, and also to suppress any HTTP should internal oxidation occur

Added to these, optimum amounts of deoxidizing and grain-refining elements must be included, bearing in mind that these may be elements with atomic numbers lower than that of iron. Hardenability can be assisted further with moderate additions of manganese and chromium, but less than 0.5% of each should be added when silicon is present in amounts typically 0.25%. The aim is to accept that internal oxidation will occur to some extent but avoid the formation of significant composition gradients.

According to Kozlovskii et al. in Ref 1, nickel at 1% appears to be incapable of providing the hardenability necessary to inhibit bainite formation at the boundary regions when internal oxidation and diffusion gradient have occurred. The effect of nickel on the hardenability of a steel picks up somewhat when its content exceeds ~1.5%, especially in the presence of molybdenum. When nickel content is higher and direct quenching techniques are used, unacceptable quantities of austenite can be retained in the outer part of the case. With reheat quenching, on the other hand, higher nickel content values can be tolerated, although a downward adjustment of the surface carbon content might be prudent.

Internal oxidation has been reported in chromium-molybdenum steels, but in such cases the molybdenum content has generally been low (~0.2%). The introduction of 0.5% (or more) molybdenum is claimed to be very beneficial, especially in those steels where molybdenum has been sensibly balanced with elements having a positive oxygen affinity that also impart necessary hardenability. For example, internal oxidation to a depth of 14 to 20 μm was observed in steels with molybdenum to chromium ratios of up to 0.4, but not in steel with a molybdenum to chromium ratio of 1 (Ref 1). With a nickel-molybdenum steel where the molybdenum to chromium ratio was 0.4 and the nickel content was over 1%, element impoverishment (indicated by dark etching constituents in the metallographic sample) was only observed with a reduced carbon concentration in the case. Therefore, it would seem that high-carbon surfaces are less prone to oxidation, bearing in mind

that as the carbon potential rises, the oxygen content of the atmosphere decreases. Contrary to this, Kalner and Yurasov found that with the lean-alloy steels they studied, 0.5% molybdenum did not prevent or reduce the internal oxidation of chromium or manganese steels (Ref 3). Chatterjee-Fischer tends to believe that as long as the accepted norm for silicon content is present in the steel, internal oxidation will occur irrespective of how the other elements are adjusted (Ref 6). Nevertheless, the presence of any alloying element must be beneficial if, for the section sizes being considered, it can inhibit the formation of HTTP.

Process Control. Prior to the heat-treatment of parts in an endothermic atmosphere, precautions should be taken to ensure that the surfaces are free from metal oxides (scale or corrosion products) and certain lubricants. Such surface impurities can contaminate the furnace atmosphere; they appear to influence the amount of oxide subsequently formed and the depth to which it penetrates (Ref 3). Furthermore, the quality of the machined surface can have a bearing on how well a surface responds to carburizing. Machining with sharp tools, along with the use of good quality cutting lubricants, makes for uniform machining and uniform carburizing (Ref 45).

Although it is clear that internal oxidation takes place during the carburizing operation, it can nevertheless be intensified during any high-temperature tempering cycle (carried out in air to facilitate intermediate machining operations) (Ref 15); it can also be intensified during reheating operations, normally under atmosphere, prior to quenching. Heating in air, especially to the quenching temperature, will cause scaling to occur, which might be found to scale off the internally oxidized layer. Subsequent shot blasting would then remove both types of oxide. This approach to internal oxide removal, while possible, is not really recommended due to the difficulty of precise process control and the risk of inducing decarburization if the reheating period is lengthy.

As stated previously, internal oxidation on its own might not be particularly damaging to fatigue resistance, whereas the presence of the HTTP can be harmful. The aim, therefore, should be to suppress the formation of HTTP. By increasing the cooling rate during the quenching operation, it may be possible to achieve this aim and thereby have only martensite associated with the oxides. Increasing the cooling rate during quenching could lead to distortion or growth problems, however. The critical cooling rates have been established for a number of steels and the optimum quenching parameters for those steels were determined (Ref 46). Unfortunately, as the size of the component increases, it becomes more difficult to achieve the cooling rates necessary to ensure the suppression of HTTP. For example, Fig. 1.29 indicates the cooling time to 400 °C for the fillets of gear teeth of different sizes (Ref 27). When these are used in conjunction with the continuous cooling transformation (CCT) diagrams of Fig. 1.13 and 1.14, one can loosely assess how component size might restrict the option to increase cooling rate to suppress the formation of HTTP.

Surface hardenability and strength can be restored by the introduction of ammonia into the carburizing chamber for a short period at the end of the carburizing cycle. Gunnerson suggested a 5 to 10% ammonia addition for ten minutes before the end of carburizing (Ref 17). Using the same guidelines, Kozlovskii et al. determined that a nitrogen content of 0.1 to 0.2% was achieved in the outer 0.05 mm of surface (Ref 1). That was effective in removing the tendency to form HTTP in all but the leanest steels, and for the leanest steels only when they were fine grained. Gu et al. determined that a nitrogen content of 0.1% in the outer 1 mm was the optimum amount (Ref 47); when nitrogen exceeded 0.1%, the crack-growth rate property increased. An alternative action was used by Zinchenko et al., who reduced or eliminated near-surface bainite by increasing the carbon potential to 0.9 to 1.2% during the last 20 minutes of carburizing (Ref 48). The method, it seems, does not form new carbides but raises the carbon content of the solid solution and, therefore, the hardenability of the outer surface.

Oxide Removal. The knowledge that internal oxidation will occur, and its confirmation by means of suitable test pieces that have been carburized and hardened along with the components they represent, enables a manufacturer/heat-treater to develop an acceptable method for removing the oxide and HTTP that have formed. Such methods as electropolishing, electrochemical machining, honing, grinding, grit blasting, shot blasting, or peening might be considered. Which method is used will depend upon what is available and what the negative effects might be; for example, grinding to remove oxides can induce tensile residual stresses of a magnitude in excess of the tensile stresses associated with internal oxidation.

If it is essential to have an oxide-free surface, but its removal is unacceptable for some reason, then it will be necessary to consider the use of alternative carburizing processes, for example, vacuum or plasma carburizing. Nitrogen-base atmospheres and exothermic-based atmospheres would be expected to reduce the amount of oxidation but would not eliminate it.

Part Design. It is difficult to say how the results of tests involving samples with test sections of 6 to 8 mm and case depths of 0.5 to 1.5 mm translate to real-life components. However, these current assumptions are not unreasonable: conventional gas carburizing produces internal oxidation, internal oxidation will likely be accompanied by HTTP, and fatigue strength will be approximately 25% lower than for an oxide-free counterpart. Designers may cope with these outcomes, as gear designers have done for many years, by having conservative basic allowable design stresses, or by accounting for the potential of this surface condition in their calculations. (Internal oxidation is acknowledged in gear standards like AGMA 2001, C95, or ISO 6336-5.2 [1995], where acceptable amounts in relation to tooth size are given. These amounts probably refer to the depth of the HTTP.) One should consider that processes, such as vacuum carburizing and plasma carburizing, while capable of producing carburized surfaces that are free from internal oxidation, can produce surfaces in which the austenitic grain size is coarse and the quan-

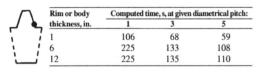

Rim or body thickness, in.	Computed time, s, at given diametrical pitch:		
	1	3	5
1	106	68	59
6	225	133	108
12	225	135	110

Rim or body thickness, in.	Computed time, s, at given diametrical pitch:		
	1	3	5
1	155	79	65
6	419	161	120
12	440	163	120

Rim or body thickness in.	Computed time, s, at given diametrical pitch:		
	1	3	5
1	155	93	81
6	880	758	733
12	2400	2290	2180

Fig. 1.29 An indication of cooling times for gear shapes. (a) Time for gear tooth fillet surface to cool from 800 to 400 °C during oil quenching. (b) Time for gear tooth center on the root circle diameter to cool from 800 to 400 °C during oil quenching. (c) Time for rim or body center to cool from 800 to 400 °C during oil quenching. Source: Ref 27

tity of retained austenite is excessive (Ref 30). The designer may have to allow for these also.

Summary

Internal Oxidation

Internal oxidation is a surface effect due to the oxidation of certain elements in the steel (mainly Mn, Cr, and Si) during carburizing. It is unavoidable in conventional gas carburizing, but can be avoided in vacuum carburizing processes.

- *Preprocess considerations:* Choice of steel grade, a small uniform grain size (normalized), and essentially clean parts are factors.
- *In-process considerations:* The depth of oxidation relates to case-depth, and the composition of the carburizing atmosphere.
- *Postprocess considerations:* Internal oxidation can be removed by grit blasting and/or shot peening.
- *Effect on properties:* Internal oxidation possibly has a minor negative effect on bending fatigue, and a positive effect on contact fatigue. See also "High-Temperature Transformation Products."

The following are ANSI/AGMA standards:

Grade 2	Grade 3	Case depth
0.0007 in.	0.0005 in.	<0.030 in.
0.0010 in.	0.0008 in.	0.030–0.059 in.
0.0015 in.	0.0008 in.	0.059–0.089 in.
0.0020 in.	0.0010 in.	0.089–0.118 in.
0.0024 in.	0.0012 in.	>0.118 in.

Grade 1 has no specification. Recovery by shot peening is acceptable with agreement of the customer. Note that ISO 6336-5.2 is similar to this (ANSI/AGMA 2001-C95). The equivalent grades are ML, MQ, and ME.

High-Temperature Transformation Products

When associated with internal oxidation, high-temperature transformation products (HTTP) are a surface effect. The formation of internal oxidation products locally denudes the matrix of certain elements. When HTTP are present, they often extend deeper than the internal oxidation, especially with leaner steel grades. HTTP can be avoided or controlled.

- *Preprocess considerations:* Consider a steel grade with sufficient matrix alloy in the depleted layer to give martensite on quenching.
- *In-process considerations:* A late increase of carbon potential, or a late addition of ammonia to the furnace chamber can offset the effect. Employ a faster quench to suppress HTTP formation, but watch for distortion.
- *Postprocess corrections:* Grinding, grit blasting and/or shot peening can be used to remove HTTP.
- *Effect on properties:* HTTP is soft and has an adverse effect on surface residual stresses and bending fatigue resistance—up to about 35% reduction in extreme eases. If shallow, HTTP might have a positive effect on the contact fatigue of unground surfaces.
- *Standards:* No specification or guideline. Assume that HTTP should not exceed the maximum depth of internal oxidation as judged metallographically. See "Internal Oxidation."

REFERENCES

1. I.S. Kozlovskii, A.T. Kalinin, A.Y. Novikova, E.A. Lebedeva, and A.I. Feofanova, Internal Oxidation during Case-Hardening of Steels in Endothermic Atmospheres, *Met. Sci. Heat Treat.(USSR)* (No. 3), March 1967, p 157–161
2. L.H. Fairbank and L.G.W. Palethorpe, "Controlled Atmospheres for the Heat Treatment of Metals," Special Report 95: Heat Treatment of Metals, Iron and Steel Institute, London, 1966, p 57–69
3. V.D. Kalner and S.A. Yurasov, Internal Oxidation during Carburising, *Met. Sci. Heat Treat. (USSR)* (No. 6), June 1970, p 451–454
4. Y. Sun, D. Xu, and J. Li, The Behavior of Internal Oxidation in Gas-Carburised 20CrMnTi Steel, *Heat Treatment and Surface Engineering, Proc. Sixth International Congress,* Chicago, 1988
5. G.V. Shcherbedinskii and A.I. Shumakov, Internal Oxidation of Excess Carbides in Alloy Steels during Carburising, *Izv. Akad. Nauk SSSR. Met.,* May/June 1979, (No. 3), p 193–199
6. R. Chatterjee-Fischer, Internal Oxidation during Carburising and Heat Treatment, *Metall. Trans. A,* Vol 9, Nov 1978, p 1553–1560
7. I.Y. Arkhipov, V.A. Batyreva, and M.S. Polotskii, Internal Oxidation of the Case on Carburised Alloy Steels, *Met. Sci. Heat Treat. (USSR),* Vol 14 (No. 6), June 1972, p 508–512

8. C. Van Thyne and G. Krauss, A Comparison of Single Tooth Bending Fatigue in Boron and Alloy Carburising Steels, *Carburising: Processing, and Performance,* G. Krauss, Ed., ASM International, 1989, p 333–340
9. R. Chatterjee-Fischer, Zur Frage de Randoxidation bei Einsatzgehärteten Teilen, *Kritische Literaturaus Wertung,* Härterei Technische Mitteilungen, (No. 28) Nov 1973, p 259–266
10. C. Dawes and R.J. Cooksey, "Surface Treatment of Engineering Components," Special Report 95: Heat Treatment of Metals, Iron and Steel Institute, London, 1966, p 77–92
11. E. Mitchell, R.J. Cooksey, and C. Dawes, "Lean-Alloy Carburising Steels," Publication 114: Low-Alloy Steels, Iron and Steel Institute, London, 1968, 31–36
12. B. Edenhofer, Progress in the Technology and Application of In-Situ Atmosphere Production in Hardening and Case-Hardening Furnaces, *Proc. of the Second International Conf. in Carburizing and Nitriding with Atmospheres,* ASM International, 1995, p 37–42
13. *ASM Metals Reference Book,* American Society for Metals, 1981
14. K. Okasaki et al., Effect of Titanium on Recrystallisation and Grain Growth of Iron, *J. Jpn. Inst. Met.,* Vol. 1 (No. 39), 1975, p 7–13
15. I.Y. Arkhipov, Internal Oxidation of Steel 18Kh2N4Va, *Met. Sci. Heat Treat. (USSR),* Vol 15 (No. 7), July 1973, Plenum Publishing, 622–624
16. N. Murai, T. Tsumura, and M. Hasebe, "Effect of Alloying Elements and Oxygen Potential at the Equilibrium Carbon Content in Gas Carburising," presented at the Tenth Congress of the International Federation for Heat Treatment of Surface Engineering, Brighton, 1996
17. S. Gunnerson, Structure Anomalies in the Surface Zone of Gas-Carburised Case Hardened Steels, *Metal Treatment and Drop Forging,* Vol 30 (No. 213), June 1963, p 219, 229
18. Report No. I.S.J.-888, Climax Molybdenum
19. S. Preston, Influence of Vanadium on the Hardenability of a Carburizing Steel, *Carburizing: Process and Performance,* G. Krauss, Ed., Conf. Proc. (Lakewood, CO), ASM International, 1989, p 191–197
20. W.E. Dowling, Jr., W.T. Donlon, W.B. Copple, and C.V. Darragh, Fatigue Behavior of Two Carburized Low-Alloy Steels, *Proc. Second Int. Conf. on Carburizing and Nitriding with Atmospheres,* ASM International, 1995, p 55–60
21. B. Hilldenwall and T. Ericsson, Residual Stresses in the Soft Pearlite Layer of Carburised Steels, *J. Heat Treat.,* Vol 1 (No. 3), June 1980, p 3–13
22. Y.E. Smith and G.T. Eldis, "New Developments in Carburised Steels," Climax Molybdenum Co., Michigan, 1975
23. G.H. Robinson, The Effect of Surface Condition on the Fatigue Resistance of Hardened Steel, *Fatigue Durability of Carburized Steel,* American Society for Metals, 1957, p 11–46
24. R.L. Colombo, F. Fusani, and M. Lamberto, On the Soft Layer in Carburised Gears, *J. Heat Treat.,* Vol 3 (No. 2), Dec 1983
25. C.W. Wegst, *Stahlschlüssel,* Verlag Stahlschlüssel Wegst KG, Marbach, Germany, 1977, p 14
26. I.Y. Arkhipov, M.S. Polotskii, A.Y. Novikova, S.A. Yurasov, and V.F. Nikanov, The Increase in the Strength of Teeth of Carburised and Carbonitrided Gears, *Met. Sci. Heat Treat. (USSR)* (No. 10), Oct 1970, Plenum Publishing, p 867–871
27. G. Parrish and G.S. Harper, *Production Gas Carburising,* Pergamon Press, 1985
28. I.Y. Arkhipov and V.A. Kanunnikova, Residual Stresses in the Teeth of Quenched Gears, *Met. Sci. Heat Treat. (USSR),* (No. 11), Nov 1970, p 909–913
29. T. Naito, H. Ueda, and M. Kikuchi, Fatigue Behaviour of Carburised Steel with Internal Oxides and Non-Martensitic Microstructure Near Surface, *Metall. Trans. A,* Vol 15, July 1984, p 1431–1436
30. J.L. Pacheco and G. Krauss, Microstructure and High Bending Fatigue Strength of Carburised Steel, *J. Heat Treat.,* Vol 7 (No. 2), 1989, p 77–86
31. W. Beumelburg, Der Einfluss von Randoxydation auf die Umlaufbiegefestigkeit und Statische Biefestigkeit einsatzgehärteter Proben, Härterei Technische Mitteilungen, Vol 25 (No. 3), Oct 1970, p 191–194
32. H. Ono, K. Okamoto, and Y. Nishiyama, "Some Mechanical Properties of Vacuum Carburised Steel," Paper 32, presented at Heat Treatment 1981, Sept 1981
33. H. Weigand and G. Tolasch, Dauerfestigkeitsverhalten Einsatzgehärter Proben, Härterei Technische Mitteilungen, Vol 22 (No.

4), Dec 1967, p 330–338 (in German; also available in English)
34. H. Brugger, Effect of Material and Heat Treatment on the Load Bearing Capacity of the Root of Gear Teeth, VDI-Berichte, Vol 195, 1973, p 135–144
35. H. Brugger, "Impact-Bending Test for the Assessment of Case-Hardening Steels," conf. paper presented to the Swiss Materials Testing Association
36. J.P. Sheehan and M.A.H. Howes, "The Effect of Case Carbon Content and Heat Treatment on the Pitting Fatigue of 8620 Steel," Paper 720268, presented to the Society of Automotive Engineers, 1972, p 16
37. D.E. Diesburg, C. Kim, and W. Fairhurst, "Microstructural and Residual Stress Effect on the Fracture of Case-Hardened Steels," Paper 23, presented at Heat Treatment 1981, Sept 1981
38. D.E. Diesburg and G.T. Eldis, "Fracture Toughness and Fatigue Behaviour of Carburised Steels," Laboratory Report No. L-193-87/88, Climax Molybdenum, 1975
39. D.E. Diesburg, High Cycle and Impact Fatigue of Carburised Steels, SAE Paper 780771, 1978
40. G. Fett, Bending Properties of Carburised Steels, *Adv. Mater. Process. inc. Metal Prog.*, April 1988, p 43–45
41. T. Aida, H. Fujio, M. Nishikawa, and R. Higashi, Influence of Impact Load on Fatigue Bending Strength of Case-Hardened Gears, *Bull. Jpn. Soc. Mech. Eng.*, Vol 15 (No. 85), 1972, p 877–883
42. Y. Terauchi and J. Takehara, Paper 152-21, *Bull. Jpn. Soc. Mech. Eng.*, Vol 21 (No. 152), Feb 1978, p 324–332
43. S.E. Manevskii and I.I. Sokolov, Resistance to Seizing of Carburised and Carbonitrided Steels, *Met. Sci. Heat Treat. (USSR)* (No. 4), April 1977
44. T.S. Eyre, The Mechanisms of Wear, *Tribol. Int.*, April 1978, p 91–96
45. V.I. Astraschenko and L.A. Karginova, Spotty Carburising of Steels *Metallovedenie I Termicheskaya Obrabotka Metallov*, June 82, p 13–15
46. I.S. Kozlovskii, S.E. Manevskii, and Kazachenko, Effect of Quenching Conditions on the Layer Structure and Antiseizing Stability of Carburised Steels, *Metallovedenie I Termicheskaya Obrabotka Metallov*, June 1980, p 7–10
47. C. Gu, B. Lou, X. Jung, and F. Shen, Mechanical Properties of Carburised Co-Ni-Mo Steels with added Case Nitrogen, *J. Heat Treat.*, Vol 7 (No. 2), 1989, p 87–94
48. V.M. Zinchenko, B.V. Georgievskaya, and V.V. Kugnetsov, Improvements of the Technological Process of Automotive Parts Carburisation, *Metallovedenie I Termi- cheskaya Obrabotka Metallov*, August, 1979, p 47–50

SELECTED REFERENCES

- R. Chatterjee-Fischer, Cause and Effect of Internal Oxidation, *Antriebstechnik*, Vol 25 (No. 6), June 1986, p 41–43
- R. Chatterjee-Fischer and H. Kunst, Some Considerations on the Property Behaviour of Salt-Bath and Gas Carburized Specimens Made of Steel 16MnCr5, *Hart.-Tech. Mitt.*, Vol 43 (No. 1), Jan/Feb 1988, p 41–44
- J.A. Colwell and R.A. Rapp, Reactions of Fe-Cr and Ni-Cr Alloys in CO/CO_2 Gases at 850 and 950 °C, *Metall. Trans. A*, Vol 17 (No. 6), June 1986, p 1065–1074
- L. Faminski, M. Golebiowski, and J. Szawlowski, Influence of High Temperature Carburizing on the Structure and Properties of Steel Alloyed With Titanium, *Inz. Materialowa*, Vol 7 (No. 6), Nov/Dec 1986, p 146–152
- K. Isokawa and K. Namiki, Effects of Alloying Elements on the Rotating Bending Fatigue Properties of Carburized Steels, *Denki Seiko (Electr. Furn. Steel)*, Vol 57 (No. 1), Jan 1986, p 13–22
- H. Kanizawa and H. Satoh, Effect of Alloying Elements on the Formation of Internal Oxidation Layer in Carburized Steels, *Netsu Shori (J. Jpn. Soc. Heat Treat.)*, Vol 37 (No. 4), 1997, p 225–230
- M. Kikuchi, A. Komine, and Y. Kibayashi, Influence of Internal Oxides on the Fatigue Strength of Carburized Steel, *J. Soc. Mater. Sci., Jpn.*, Vol 38 (No. 425), Feb 1989, p 111–116
- Y.H. Kim and S.W. Lee, Internal Oxidation Control of the Steels During Carburizing in Cracked Methanol Gas Atmospheres, *J. Korean Inst. Met. Mater.*, Vol 29 (No. 11), 1991, p 1131–1138
- T. Kimura and K. Namiki, Carburizing and Fatigue Properties of Plasma Carburized SCM 420, *Denki Seiko (Electr. Furn. Steel)*,

Vol 61 (No. 1), Jan 1990, p 32–40
- J.Y. Liu and S.C. Chang, The Oxidation and Carburization of Fe-Mn-Al Alloys in a Carbon-Containing Atmoshere, *Corros. Sci.,* Vol 39 (No. 6), June 1997, p 1021–1035
- G. Prunel, Internal Oxidation During the Carburizing of Gears, *Trait. Therm.,* Vol 197, Dec 1985, p 31–41

Chapter 2

Decarburization

Decarburization, as the term implies, is a loss of carbon atoms from the surface of the workpiece, thereby producing a surface with a lower carbon content than at some short distance beneath the surface. If carburization promotes a positive carbon gradient, then decarburization promotes a negative carbon gradient.

The useful properties developed by carburizing and hardening will not be realized if the working surface of the component becomes decarburized. Therefore, decarburization is an unwanted metallurgical feature. The optimal amount of decarburization is regarded as being zero, but in reality, a small amount will likely occur. A small amount of decarburization may be tolerable if it is partial decarburization and penetrates no deeper than any HTTP associated with the surface internal oxidation and it is within any specification for decarburization.

In well-run carburizing and hardening facilities, the decarburization of case-hardened surfaces does not seem to be a problem. This success reflects the use of good, well-maintained equipment and good process operating procedures. Nevertheless, with case hardening, the gas mixtures employed necessarily contain decarburizing agents. As long as these agents are present, there is always the possibility that a loss of furnace atmosphere control, for whatever reason, will lead to decarburization of the parts being treated.

Decarburization Processes

Decarburizing reactions can occur at temperatures above about 700 °C and when, in the furnace atmosphere, decarburizing agents are available to react with the carbon in the metal surface. The decarburizing agents used in furnace atmospheres for carburizing and reheating are carbon dioxide (CO_2), water vapor (H_2O), hydrogen (H_2), and oxygen (O_2). Under certain conditions, these gaseous molecules can react with the carbon atoms at the gas-metal interface and thereby extract them from the surface of the metal. This extraction is an attempt to establish some measure of equilibrium between gas and metal. The chemical reactions involved are:

$$C_{Fe} + CO_2 \leftrightarrow 2CO$$

$$C_{Fe} + H_2O \leftrightarrow CO + H_2$$

$$C_{Fe} + 2H_2 \leftarrow CH_4$$

When these reactions proceed from left to right they are decarburizing, and when they go right to left they are carburizing. There seems to be some disagreement regarding the third reaction (involving methane) and whether it is able to proceed in both directions as reported in Ref 1. It has been reported as being able to proceed in one direction only (Ref 2). If the latter is the case, then any hydrogen released by this reaction must combine with other component gases before it can be an effective decarburizing agent. By combining with any oxygen in the furnace atmosphere, it will produce water vapor at the metal surface and thereby enter into the second of the reactions shown, in a left to right direction.

Decarburizing Conditions. During carburizing, there are numerous carburizing and decarburizing reactions occurring simultaneously, with more carburizing than decarburizing reactions taking place (i.e., the reactions proceed from right to left). Changes of temperature, flow rate, or composition of the atmosphere can shift the balance so that the number of decarburizing reactions outstrips the number of carburizing reactions (i.e., the reactions proceed from left to right), and the overall result is decarburization.

In an endothermic gas generator, for example, if the catalyst is in good order, the methane of the output gas will be <2%. The carbon dioxide and water vapor will be present in only small amounts, and these amounts are dependent on the actual air/fuel ratio employed. If, on the other hand, the efficiency of the catalyst is impaired by an accumulation of soot, for example, then the gas produced will contain greater amounts of methane, carbon dioxide, and water vapor (Ref 3). This increase in the proportion of decarburizing agents will lead to a reduction in the number of carburizing reactions taking place at any one time and, concurrently, an increase in the number of decarburizing reactions taking place. In other words, the carbon potential will be reduced (Fig. 2.1).

Typical CO_2 contents during a carburizing cycle are shown in Fig. 2.2, where an endothermic carrier is used throughout and is enriched for the carburizing part of the operation. Normally, the atmosphere at the end of the process, just prior to the quench, will still be carburizing or at least of a high carbon potential. If, however, the catalyst is inefficient and, therefore, the generator gas contains more CO_2 than it should, then some decarburization is possible.

When the reheat quench process is being used, matching the atmosphere to the as-carburized surface carbon content might not be very precise, and a minor lowering or raising of the surface carbon can occur. This practice is normally accepted.

Following are some reported examples regarding the occurrence of decarburization during high-temperature heating.

Gutnov et al. (Ref 5) observed that samples placed in a furnace at the beginning of an 800 °C heating cycle decarburized appreciably, whereas those introduced into the furnace halfway through the heating period did not decarburize (Fig. 2.3). Gutnov et al. concluded that moisture in the furnace had been responsible for the dif-

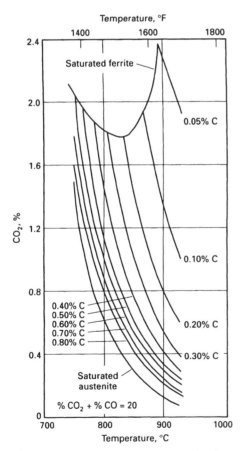

Fig. 2.1 Temperature and percentage of carbon-dioxide for equilibrium conditions with carbon steels of various carbon contents. Source: Ref 1

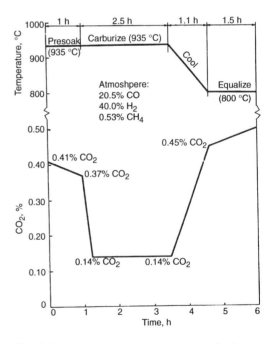

Fig. 2.2 Complex industrial carburizing cycle. Source: Ref 4

ference. With an excess of moisture in the furnace atmosphere (i.e., an increased dew point), the carbon potential would fall in accordance with Fig. 2.4.

Virta (Ref 6) presented the time-temperature relationship for decarburization of a 0.9% tool steel during reheating in air in a fluidized bed furnace (Fig. 2.5). Thus, decarburization to a depth of 200 µm (0.008 in.) was induced by heating in air at 900 °C for about 2 h. Kileeva et al. (Ref 7), studying methods of coating to protect against the decarburization of high-carbon surfaces during normalizing at 890 °C, observed that the surface carbon would fall to under 0.4% in 1 h in air at 900 °C (Fig. 2.6).

Stratton (Ref 2), whose work considered atmosphere contamination in general, showed the influence of air leakage on the decarburization and scaling of an SAE 8620 steel after heating for 2 h in a nitrogen-4% natural gas atmosphere (Fig. 2.7). Such work underlines the importance of equipment maintenance, in this instance with respect to leaks and air ingress.

Ambrus and Pellman (Ref 8), investigating nitrogen-base atmospheres for the heat-treatment of 4340 and 300M steels, found that the three atmospheres (pure nitrogen, methanol CAP-air, and methanol CAP-CO_2) produced some partial decarburization as well as some internal oxidation (Table 2.1). The interesting result of this work is that, for a given steel, the depth of decarburization was similar for all three atmospheres (300M steel, ~69 µm; SAE 4330, ~52 µm), whereas the internal oxidation was significantly less when pure nitrogen was used.

Sagaradze and Malygina (Ref 9) showed that heating in a salt bath (50% NaCl + 50% KCl) caused decarburization. The effect became worse as the salt bath aged, and deoxidation of the bath failed to overcome the problem.

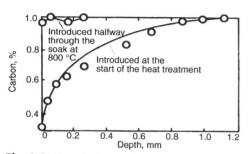

Fig. 2.3 Decarburization due to moisture in the furnace fireclay lining. Source: Ref 5

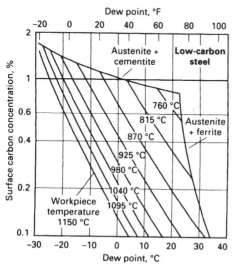

Fig. 2.4 Variation of carbon potential with dew point for an endothermic-based atmosphere containing 20% CO and 40% H_2 in contact with plain-carbon steel at various workpiece temperatures. Source: Ref 1

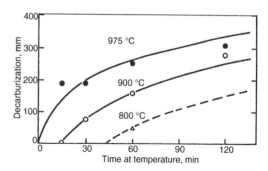

Fig. 2.5 Depth of decarburization of a cold-work steel in a fluidized bed in air. Source: Ref 6

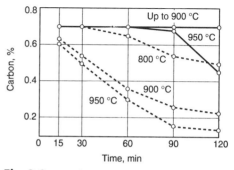

Fig. 2.6 Decarburization as a function of holding time at different temperatures. Solid lines, samples with protective coating; dashed lines, samples without protective coating. Source: Ref 7

The choice of atmosphere carbon potential for reheating prior to quenching is most important. Naisong et al. (Ref 10), with a 52100 steel (~1.02% C) and a reheating temperature of 810 °C, employed endothermic atmospheres with carbon potentials between 0.15 and 0.9% to determine which carbon potential was best suited to that steel. Theoretically, an endothermic carrier gas, perhaps with a slight enrichment (e.g., 0.5% C potential), should have been adequate. The tests showed that to obtain the best results the atmosphere had to be carburizing. Anything less resulted in a reduced hardness and residual tensile stresses at the surface (Table 2.2).

The Physical Metallurgy of Decarburization. If a carburized steel is in the austenitic condition at a temperature above the Ac_3 (e.g., 900 °C) and if the furnace atmosphere is decarburizing (i.e., the reactions proceed from left to right), then carbon atoms will leave the surface in an endeavor to restore equilibrium with the surrounding furnace atmosphere. A negative carbon gradient is produced with carbon atoms feeding down the gradient. The surface carbon content is determined by the carbon potential of the atmosphere. If the potential is low, the surface carbon content will be correspondingly low. The depth of the decarburized layer is determined by the length of time that the part resides in the outlined furnace conditions. At temperatures below the Ac_3 and above the Ac_1, including those used for reheating and quenching (800 to 840 °C), the decarburizing reaction is somewhat different (Ref 11). At these temperatures, the carbon content of the immediate surface (originally A in Fig. 2.8 and 2.9a) rapidly falls to the value shown by B. Any further lowering of the carbon content must result in a material of carbon con-

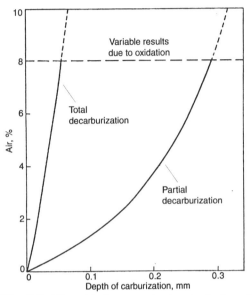

Fig. 2.7 Effect of leak rate on the decarburization of SAE 8620 after 2 h at 850 °C in a nitrogen/4% natural gas atmosphere. Source: Ref 2

Table 2.1 Laboratory results: effect of atmospheres on high-strength steels

Atmosphere	Depth of decarburization, μm (in.)		Depth of intergranular attack, μm (μin.)		Average tensile strength, MPa (ksi)
	Average	Standard deviation	Average	Standard deviation	
4330					
Nitrogen	48 (0.0019)	20.3 (0.00080)	0.5 (20)	0.5 (20)	1627 (236)
Methanol CAP-air	58 (0.0023)	7.9 (0.00031)	None visible	None visible	1658 (240.5)
Methanol CAP-CO_2	51 (0.0020)	7.9 (0.00031)	11.2 (440)	1.0 (40)	1612 (233.9)
300M					
Nitrogen	66 (0.0026)	19.8 (0.00078)	1.0 (40)	2.5 (100)	2013 (292)
Methanol CAP-air	69 (0.0027)	0	4.1 (160)	1.0 (40)	2015 (292.3)
Methanol CAP-CO_2	69 (0.0027)	0	5.8 (230)	0.8 (30)	2025 (293.7)

Source: Ref 8

Table 2.2 Effect of carbon potential on microstructure, hardness, and residual stresses of a 52100 steel (1.02% C) during reheating prior to quenching

Atmosphere carbon potential, %	Retained austenite, %	Carbides, %	Hardness(a), HK	Surface residual stresses, MPa
0.9	17.4	16	820	−40
0.7	9.8	12	811	+10
0.5	6.2	8.5	751	+140
0.3	2.5	7	695	+270
0.15	2.0	7	626	>270

Reheated to 810 °C for 1 h, oil quenched at 55 °C, tempered at 150 °C for 2 h. HK, Knoop hardness number. (a) 1 kg load. Source: Ref 10

tent C being formed in equilibrium with material of carbon content B. Therefore, because the atmosphere is decidedly decarburizing in nature, a further lowering of the average surface carbon content must result in the development of ferrite containing carbon to the equilibrium value C (curve 2, Fig. 2.9a). This layer, once it becomes continuous, reduces the rate of decarburization because, in the low-carbon ferrite layer, the carbon activity is reduced. Further, because the ferrite can have only a shallow gradient across it, the rate of flow of carbon atoms through the layer is also reduced. As the ferrite layer increases in thickness, the decarburizing rate is further reduced due to the decreasing effectiveness of the driving force responsible for moving carbon atoms through the layer. Behind the ferrite layer, the carbon gradient in the austenite becomes more gradual with time as carbon feeds down to the interface with the ferrite layer (curve 3, Fig. 2.9a).

When a controlled atmosphere is employed with a carbon potential of some value between A and B (Fig. 2.8), such as D, then a ferrite layer cannot form. Instead, a gradient is produced between carbon contents A and D (curve 1, Fig. 2.9b). With time the gradient becomes flatter (curves 2 and 3, Fig. 2.9b) until eventually no gradient to the surface exists (curve 4, Fig. 2.9b).

The two-stage carburizing method utilizes this gradient depletion process. In the first stage, a high carbon potential is employed to produce a high-carbon surface (e.g., 1.2% C). For the second stage, the carbon potential is reduced to produce the specified surface carbon content (e.g., 0.85% C). Carbon diffuses both inward and outward, and the end result is a carbon gradient with a surface carbon plateau at 0.85%. Too short a duration or too high a CO_2 content for the second stage results in a carbon gradient with a negative slope at the surface (Fig. 2.10). A negative carbon gradient formed in this way does not lead to a rejection, provided the surface carbon content and the surface hardness are each within their specified range. Nevertheless, detection of this problem should lead to a reassessment of appropriate process parameters. A negative carbon gradient of this type could produce an acceptable martensitic microstructure at the surface with a subsurface microstructure that contains excessive amounts of carbides or retained austenite. Whether or not such a combination of microstructures is accepted depends on the application.

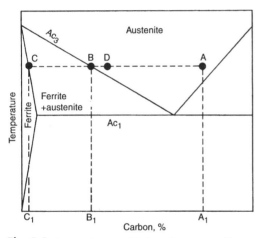

Fig. 2.8 Iron-carbon equilibrium diagram used to explain decarburization

Testing

Testing for Decarburization. Heat-treatment operators go to great pains to avoid classical decarburization because of its adverse effects. Its

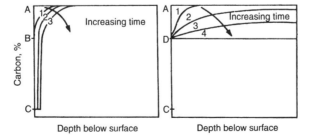

Fig. 2.9 Effect of carbon potential on the surface carbon of a decarburized surface. (a) Atmosphere decarburizing: carbon potential less than B. (b) Atmosphere decarburizing: carbon potential equals D

detection, through macrohardness and/or microhardness testing, metallography, carbon analysis, or all three, will lead to the need to rework or more probably the rejection of the affected components. An occurrence of decarburization might be suspected if something untoward happened during the heat-treatment sequence. Such a suspicion precipitates action to determine if the heat-treated parts suffered as a result of the operating problem.

Soft surfaces, determined during routine quality testing of the components (by file or hardness test) or of test pieces that accompany the furnace load, will only indicate that something is wrong. Soft surfaces can occur due to high-temperature transformation products (HTTP) that accompany internal oxidation, excessive retained austenite, nonmartensitic structures resulting from either a "slack" quench or a steel with an uncharacteristically low hardenability (extremely fine-grained), or decarburization. The first action taken must determine exactly which of the possibilities caused the soft skin effect.

The Metallography of Decarburization. Decarburization is classified as either total or partial. Total decarburization refers to the removal of essentially all the carbon from the immediate surface; the polished and etched surface microstructure is ferritic (Fig. 2.11a). It will likely have a layer of partial decarburization beneath it. Partial decarburization is any surface carbon loss that is not sufficient to produce total decarburization (a condition somewhere between curves 1 and 2 in Fig. 2.9a). Many consider partial decarburization to be that which occurs when metallography detects HTTP and grain-boundary ferrite at the surface of the part (Fig. 2.11b). For such a microstructure, the surface carbon content would be quite low. With a little more surface carbon, the microstructure might consist of HTTP only. With still more surface carbon (say, 0.4–0.5%), the microstructure of the decarburized layer might be martensitic. However, this martensite, having a low carbon content, will nital etch to be more gray in color than would a high-carbon martensite, and, therefore, this level of decarburization should be detectable metallographically. For a given loss of surface carbon, the final microstructure will depend on the alloy content of the steel and the cooling rate. (See Fig. 1.13a and b and Fig. 1.14 for examples.)

(a)

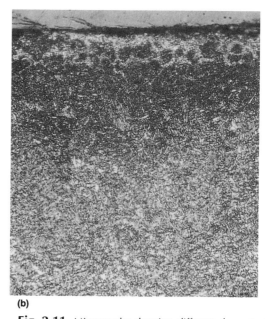

(b)

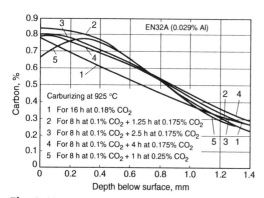

Fig. 2.10 Types of carbon profiles. Source: Ref 12

Fig. 2.11 Micrographs showing different degrees of decarburization. (a) Total decarburization caused by severe furnace leak during gas carburization of 1018 steel (1% nital etch, 500×). (b) Partially decarburized specimen. 190×

The surfaces of quenched parts that have suffered partial decarburization will likely be bainitic or even martensitic depending on the alloy content. A bainitic structure at the surface is distinct compared to the underlying martensite. Although in a well tempered structure (dark etching), it might not be too apparent at first sight. Low- to medium-carbon martensites in a quenched and tempered surface (through the microscope) appear to be more gray than the higher-carbon material beneath the affected layer.

Any indication that a surface is decarburized justifies a microhardness traverse being carried out. Small amounts of decarburization might not be too obvious, metallographically or in terms of a microhardness traverse. However, a good dense martensitic surface layer with free carbides or fair amounts of retained austenite beneath it is an indicator of a negative carbon gradient (upper curve, Fig. 2.12b). It is then necessary to conduct chemical analysis to confirm any suspicions of this condition.

Influence on Material Properties

Influence on Hardness. Shallow decarburization, or a minor reduction of the surface carbon content, does not greatly influence the surface macrohardness of a case-hardened part. In such an instance, the hardness could be either increased or decreased by some small, and possibly insignificant, amount. Hardness increases are obtained when decarburization is responsible for lowering the retained austenite content consequently increasing in the martensite produced; the carbon gradient for this is similar to the upper curve of Fig. 2.12(b). Hardness reductions are obtained when low-carbon martensites or bainites are produced from a carbon gradient such as Fig. 2.12(a). If the reduced hardness resembles a "skin" effect, it might not be detected by macrohardness testing. Microhardness testing, however, is expected to determine the presence of such microstructures. For the relationship between hardness and carbon content for the different microstructures, see Fig. 1.18.

With severe decarburization, even if proeutectoid ferrite is not formed, appreciable amounts of other low-carbon transformation products occur at the surface. This presence of decarburization can be detected by macrohardness testing. However, it should be remembered that surface softness can also be caused by high levels of retained austenite or excessive amounts of HTTP associated with internal oxidation, in which case it is necessary to utilize metallography to determine the cause. Having said that, for a given depth of HTTP associated with internal oxida-

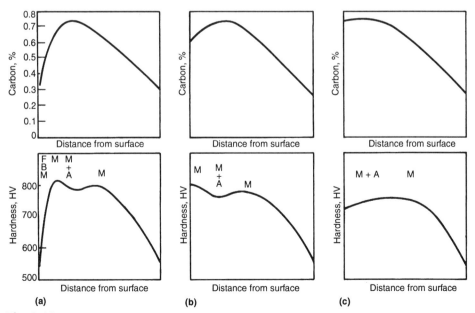

Fig. 2.12 Influence of decarburization on hardness profiles. F, ferrite; B, bainite; M, martensite; A, austenite. (a) Severe decarburization. (b) Slight decarburization. (c) No decarburization

tion and the same depth of HTTP due to decarburization, the effect on other properties might not be all that different.

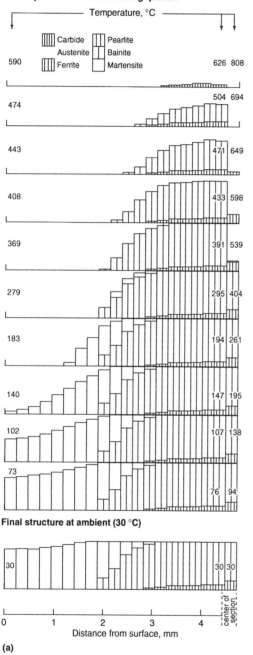

Influence on Residual Stresses. When a carburized surface without decarburization is quenched, transformation progresses inward then outward from the case/core interface (Fig. 2.13). When a decarburized component is quenched and some high-temperature transformation product is formed at the surface, transformation (besides progressing inward then outward from the case/core junction) will take place prematurely at the surface (Fig. 2.14). The decarburized layer, in this example, contains almost 50% bainite, and the volume expansion is less than that of the underlying martensite. Consequently, when the whole of the case has eventually transformed, the surface will be in less compression, or even in tension, compared with a fully martensitic microstructure.

An example of the influence of decarburization on residual stresses within a carburized surface is shown in Fig. 2.15. With no decarburization present and a surface carbon content of about 1%, the residual compressive stresses at the surface are in excess of 392 MPa (40 kg/mm^2). When sufficient decarburization occurs to lower the surface carbon content to 0.64% and to penetrate to an estimated depth of 0.3 mm, the residual stresses at the surface are virtually zero. When decarburization causes the surface carbon content to fall to 0.35% and the decarburized layer is about 0.5 mm deep, the residual stresses at the surface are tensile to 226 MPa (23 kg/mm^2). Tests on the decarburization of a 52100 bearing steel (Ref 10) suggest that

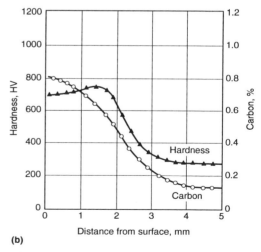

Fig. 2.13 Progression of transformation in carburized EN36A (655M13) steel. (a) Transformation occurs first in the core region before any martensitic transformation in the case region. The final structure in the case is predominantly martensitic with some retained austenite. Ruling section: 125 mm; heat treatment: oil quenched from 820 °C; case carbon: 0.8% at surface. (b) Carbon and hardness profiles. Source: Ref 13

only small reductions of surface carbon are sufficient to produce tensile residual stresses at the surface (Table 2.2). These examples imply that through hardened high-carbon surfaces are perhaps more sensitive to small amounts of decarburization than are carburized surfaces.

Influence on Bending Fatigue Strength. A consequence of decarburization and of reducing surface hardness and developing unfavorable residual stresses is to impair the bending fatigue strength of the components. Gunnerson (Ref 14), using rotating bending fatigue specimens, found that decarburization and heavy internal oxidation reduced the fatigue limit from ~81 to 53 kg/mm². Reductions of fatigue limit exceeding 50% were obtained by decarburizing the surface of a case-hardened Cr-Mn-Ti steel to 41 to 42 HRC (Fig. 2.16) (Ref 15). Figure 2.16 also shows that decarburization can nullify the benefits of carburizing by reducing the fatigue limit to approximately that of a core containing approximately 0.19% C with a hardness of 30 HRC. Sagaradze and Malygina (Ref 9), with a 20Kh2N4A steel, obtained ~50% reduction in bending fatigue by reducing the surface carbon from about 0.8 to less than 0.35%; the respective fatigue values were 784 MPa (80 kg/mm²) and 353 MPa (36 kg/mm²). The corresponding residual stress distributions are presented in Fig. 2.15. Fatigue

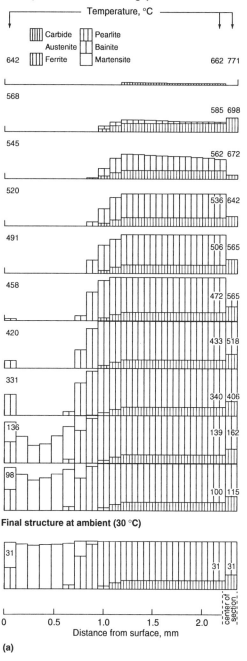

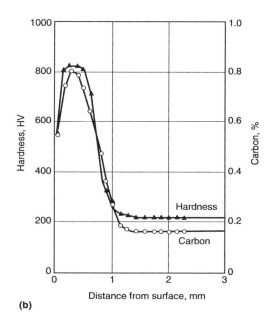

Fig. 2.14 Transformation sequence in decarburized EN34 (665M17) steel. (a) Surface decarburization causes transformation to initiate at the surface in addition to at the case/core interface. This effect reduces not only surface hardness by the presence of bainite, but it also reduces the level of surface residual compressive stress (possibly even into the tensile range) by the change in transformation sequence. Predictions are based on the following input data: ruling section, 100 mm; heat treatment, austenitized at 925 °C and direct oil quenched from 820 °C; case carbon, 0.8% falling to 0.5% at the surface. (b) Carbon and hardness profiles. Source: Ref 13

tests, using notched test bars that simulated a 13 mm modulus gear (~2 diametrical pitch), showed that the fatigue limit was reduced by almost 40% on those samples decarburized to a depth of 0.22 mm (Fig. 2.17). Kileeva et al. reported reductions of fatigue strengths of 19% in one case and 35% in another case from the decarburization that occurred during post-carburizing normalizing or hardening heat treatments (Ref 7). The microstructure of the decarburized layer has the most influence on the fatigue strength; the depth of decarburization (within reason) and the microstructure of the underlying material are probably not very significant.

Influence on Contact Fatigue. Contact fatigue resistance is related to the shear strength of the material; therefore, it would be expected that a decarburized surface would be detrimental to the contact durability of a carburized and hardened part. In the absence of test data, it is impossible to determine what would happen if a decarburized part were to enter service. Naturally, it greatly depends on how much rolling and sliding are involved, what the contact pressure is, and how severe the decarburization is.

Consider a SAE 8620 part manufactured to function as a case-hardened part, which has a decarburized surface carbon content of only 0.45% and a penetration depth corresponding to maximum shear. Early contact cycles deform the softer surface layers and thereby quickly contribute to a good load distribution. At the same time, the soft material work hardens to improve its resistance to wear. However, it can only be deformed so much before it becomes overworked and the amount of spreading gives rise to extreme tensile residual stresses. These stresses occur in a narrow zone somewhere within the decarburized layer and possibly close to the interface with harder underlying carburized material. The outcome of that circumstance is surface spalling, which involves flakes of work hardened surface being detached from the surface. The time to failure would not be long.

If the amount of decarburization was confined to the outermost 75 μm, for example, then the adverse effects of decarburization might not be great. See the corresponding section in Chapter 1, "Internal Oxidation," with attention to the work of Sheehan and Howes.

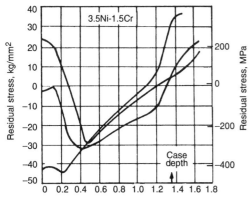

Fig. 2.15 Effect of decarburization on the residual stresses developed in carburized and hardened plates. The carbon content at 0.002 mm was estimated to be 1% (curve 1), 0.64% (curve 2), and 0.35% (curve 3). Source: Ref 9

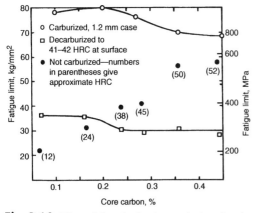

Fig. 2.16 Effect of decarburization on the bending fatigue strength of a Cr-Mn-Ti steel of varying core carbon. Source: Ref 15

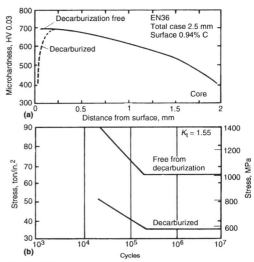

Fig. 2.17 Effect of decarburization in carburized EN36 (3.16% Ni, 0.9% Cr) steel on (a) microhardness and (b) bending fatigue strength

Influence on Bending and Impact. Decarburization does not necessarily reduce the static bending strength of case-hardened surfaces (Ref 9). As Table 2.3 shows, bend strength peaks when the surface carbon content is 0.6%, despite the almost zero residual stresses at the surface. Presumably, this corresponds to the highest carbon content that produced only martensite.

Influence on Wear. The adhesive and abrasive wear resistance of a case-hardened surface is adversely affected by decarburization to a degree relating to the surface hardness (see Fig. 1.27).

Control of Decarburization

Severe decarburization of case-hardened parts occurs only rarely, especially with modern atmosphere monitoring systems and when furnaces can be nitrogen purged at the first signs of trouble. Even improvements in health and safety monitoring mean safer handling of parts, so there is less risk of accidents occurring during transfer of parts from furnace to quench. The risk of encountering decarburization is further reduced by having good plant maintenance, good supervision, and sound process control.

The less obvious type of decarburization (i.e., caused by too short a diffusion stage or an incorrect atmosphere adjustment in the boost/diffuse method) is fairly easy to correct. It is important to be able to detect that there is a problem.

Rectification of components once they are found to be decarburized depends on a number of factors. Questions must be asked and answered. For example, does the amount of decarburization measured render the part useless? Is the item to be stressed to its undecarburized limit in service, or is there a fair safety margin built into the design? If the decarburization is shallow, can it be grit blasted in the critical areas to remove the affected layer and shot peened to ensure surface compression? If the decarburization is deeper than what can be reasonably grit blasted and shot peened, can the part be recovered by reclamation heat treatment (Ref 16)? It must be kept in mind that any reheat treatment could induce additional shape and size changes (distortions) that on their own could warrant a rejection of the part.

Each incident of decarburization must be assessed on its own; what might work for one part design might not be applicable to another. The economics of encountering decarburization is obviously an important consideration. Whatever action is taken to recover the situation, the primary consideration should be with respect to the component and its eventual fitness for service.

Summary

Decarburization can occur when surface carbon is slightly lower than at depth due to an incorrect diffuse stage of a boost-diffuse carburization program. Surface carbon is adequate if acceptable hardness is achieved. This condition is not serious. When surface carbon is moderate to very low, decarburization is detectable by hardness test and by metallographic methods. This condition is unacceptable, as it may result in a failure.

- *Preprocess considerations:* Good maintenance of equipment is essential. Decarburization might signify an equipment failure. Long reheating times in air are not recommended.
- *In-process considerations:* Atmosphere monitoring system might indicate a problem. Reheating in air is not recommended.
- *Postprocess considerations:* Post-process corrections depend on the degree of decarburization. For shallow partial decarburization, consider shot-peening. For deeper total or partial decarburization, consider restoration

Table 2.3 Effect of decarburization on the fatigue and bending strength of case-hardened 20Kh2N4A steel

Surface content, % (atmosphere)	Fatigue strength ($\sigma-1$), MPa	Bend strength (σ bend), MPa
Normalized(a)		
1.04 (not normalized)	80	319
0.8 (gas furnace)	81	355
<0.8 (nitrate salt)	77	318
0.6 (chloride bath, 20 min)	60	407
0.35% (chloride bath, 160 min)	39	318
Not normalized(b)		
0.16	45	276
0.51	50	343
0.63	49	412
0.84	57	350
1.05	51	307

(a) Carburized, normalized at 890 °C, high-temperature tempered and quenched from 800 °C, then tempered to 140 °C. (b) Carburized, high-temperature tempered, heated in chloride bath for various times to decarburize, oil quenched at 800 °C, then tempered to 140 °C. Source: Ref 9

carburizing if added distortion can be tolerated.
- *Effect on properties:* Significant decarburization leads to incorrect surface microstructures and low hardness values. If surface carbon is greater than 0.6%, the surface hardness should be acceptable. If surface carbon is approximately 0.6% or less, all the main properties will be adversely affected, for example, bending fatigue could be reduced by 50%.
- *Standards:* ANSI/AGMA 2001-C95: no specification for grade 1. For grade 2 and grade 3, no partial decarburization is apparent in outer 0.13 mm (0.005 in.) at 400× except in unground roots. ISO6336-5.2: for MQ and ME grades, the reduction of surface hardness due to decarburization in the outer 0.1 mm (0.004 in.) should not exceed 2 HRC on the test bar.

REFERENCES

1. *Heat Treating,* Vol 4, *Metals Handbook,* 9th ed., American Society for Metals, 1981
2. P.F. Stratton, Living with Furnace Atmosphere Contamination, *Heat Treat. Met.,* 1984, Vol 2, p 41–48
3. L.H. Fairbank and L.G.W. Palethorpe, "Controlled Atmospheres for the Heat Treatment of Metals," *Heat Treatment of Metals,* Special report 95, Iron and Steel Institute, 1966, p 57–69
4. F.A. Still and H.C. Child, Predicting Carburising Data, *Heat Treat. Met.,* 1978, Vol 3, p 67–72
5. R.B. Gutnov, L.P. Emel'yanenko, E.N. Samsonova, L.A. Shvartsman, and D.D. Shishlov, Sources of Decarburisation during the Restoration Case-Hardening of Steel in Bell Furnaces, *Stahl,* (No. 11), Nov 1971, p 1039–1040
6. Virta, *Heat Treatment '84* (London), The Metals Society, 1984
7. A.I. Kileeva, D.I. Potoskveva, and V.S. Sagaradze, Prevention of Decarburisation with Enamel Coat, *Met. Sci. Heat Treat. (USSR),* (No. 10), Oct 1975, p 68–70
8. Z. Ambrus and M.A. Pellman, Hardening of Aerospace Alloys in Nitrogen Based Atmospheres, *Met. Prog.,* May 1983, p 47–51
9. V.S. Sagaradze and L.V. Malygina, Influence of Decarburisation on the Properties of Case-Hardened Steel, *Metal Sci. Heat Treat. (USSR),* (No. 7), July 1966, p 560–563
10. X. Naisong, C.A. Stickels, and C.R. Peters, The Effect of Furnace Atmosphere Carbon Potential on the Development of Residual Stresses in 52100 Bearing Steel, *Metall. Trans. A,* Vol 15, Nov 1984, p 2101–2102
11. W.A. Pennington, A Mechanism of the Surface Decarburisation of Steel, *Trans. ASM,* 1946, Vol 37, p 48–91
12. R. Collin, Mathematical Model for Predicting Carbon Concentration Profiles, *J. Iron Steel Inst.,* Oct 1972
13. D.W. Ingham and P.C. Clarke, Carburise Case Hardening—Computer Prediction of Structure and Hardness Distribution, *Heat Treat. Met.,* 1983, Vol 4, p 91–98
14. S. Gunnerson, Structure Anomalies in the Surface Zone of Gas-Carburised Case-Hardening Steels, *Met. Treat. Drop Forging,* Vol 30 (No. 213), June 1963, p 219–229
15. V.S. Sagaradze, Effect of Carbon Content on the Strength of Carburised Steel, *Met. Sci. Heat Treat. (USSR),* (No. 3), March 1970, p 198–200
16. V.G. Koroshailov, E.L. Gyulikhandanov, and V.V. Kislenkov, Restoration Carburising of Steel ShKh15 in Endothermic Atmosphere, *Met. Sci. Heat Treat. (USSR),* (No. 12), Dec 1975, p 33–35

SELECTED REFERENCES

- P. Baldo and E. Duchateau, "Process for Heat Treatment Under a Gaseous Atmosphere Containing Nitrogen and Hydrocarbon," U.S. Patent US4992113, 22 Aug 1990
- M.J. Gildersleeve, Relationship Between Decarburisation and Fatigue Strength of Through Hardened and Carburising Steels, *Mater. Sci. Technol.,* Vol 7 (No. 4), April 1991, p 307–310
- H.J. Grabke, E.M. Muller, H.V. Speck, and G. Konczos, Kinetics of the Carburization of Iron Alloys in Methane-Hydrogen Mixtures and of the Decarburization in Hydrogen, *Steel Res.,* Vol 56 (No. 5), 1985, p 275–282
- B. Korousic and B. Stupnisek, Predicting of Reactions during Carburization and Decarburization of Steels in Controlled Atmospheres, *Kovine Zlitine Tehnol.,* Vol 30 (No. 6), Nov/Dec 1996, p 521–526
- G. Matamala and P. Canete, Carburization and Decarburization Kinetics of Iron in CH_4-H_2 Mixtures between 1000–1100 °C, *Mater. Chem. Phys.,* Vol 12 (No. 4), April 1985, p 313–319
- M. Renowden, O. Borodulin, and P. Baldo, Unique Atmosphere Control System for Carbon

Control, *Conf. Proc. 61st Annual Convention; 1991 Regional Meeting* (Atlanta, Georgia), Wire Association International, 1991, p 191–195
- G. Sobe and V. Polei, On the Reaction Mechanism of the Carburization and Decarburization of Alpha-Iron in CH_4/H_2- Mixtures, *Steel Res.*, Vol 57 (No. 12), Dec 1986, p 664–670
- B.S. Soroka, B.I. Bondarenko, A.E. Erinov, V.E. Nikolskii, and V. K. Bezuglyi, Thermodynamic Analysis of the Effects of the Heating Characteristics on Steel Decarburizing, *Russ. Metall.*, Vol 4, p 192–199
- A.F. Zhornyak, and V.E. Oliker, Kinetics and Thermodynamic Characteristics of the Reduction-Decarburization-Carburization of an Atomized Cast Iron Powder in Converted Gas, *Sov. Powder Metall. Met. Ceram.*, Vol 23 (No. 2), Feb 1984, p 102–107

Chapter 3

Carbides

Free carbides, or pro-eutectoid carbides, are those carbides that form in steel at temperatures above the Ac_1 transformation temperature or are rejected by the austenite during cooling and prior to the austenite-to-pearlite, -bainite, or -martensite transformations. These carbides are in excess of the transformation requirements and, therefore, can only appear when the carbon content is above the eutectoid carbon content. Carbides may be seen in the form of networks, as a dispersion of particles or spheroids, as large chunky particles, or as surface films. The forms are determined by the chemical composition of the steel, the heat treating schedule, and, obviously, the carbon content.

For most applications involving case-hardened parts, an effort is made by heat treaters to avoid the presence of free carbides in the case, or to produce cases which have no more than a dispersion of small-diameter spheroidal carbides at or near the surface. Spheroidized carbides and fine precipitated carbides tend to be more acceptable than filaments or residues of network carbides, although very few heat treaters would reject parts that contained only small quantities of such carbides. Parts with surface or near-surface carbides in fully or almost fully continuous networks at critical stress locations could be rejected. When the quantity of carbides is somewhere between a few slender filaments and a fully continuous network, rejection depends on aspects of the design (safety factors) and the application. When wear resistance approaching that of a white cast iron is the primary requirement, then the case-hardening process can be modified to intentionally produce surfaces in which the carbide content is high.

Chemical Composition

Alloying elements act either as ferrite (α) or austenite (γ) formers, and have either an affinity or repulsion for carbon in steels. Behavior trends for the elements typically contained in carburizing steels are illustrated in Fig. 3.1 (Ref 1). Those elements attractive to carbon are often referred to as the carbide formers, whereas those repulsive to carbon are sometimes referred to as graphitizers. The term "balanced composition," often used with respect to case-hardening steels, suggests a balancing of ferrite formers with austenite formers, and of carbide formers with

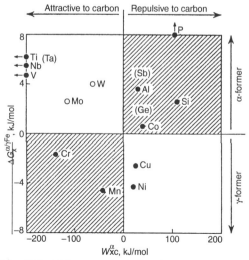

Fig. 3.1 Gibbs energy (ΔG) plotted against the interaction energy (W_X) for alloying elements added to steel.

graphitizers. Note that carbon and nitrogen are austenite formers.

In the standard range of carburizing steels, the carbides formed will have the Fe_3C (cementite) structure, in which atoms of carbide-forming elements can replace iron atoms (to a point). Table 3.1 shows examples of the chromium content in equilibrium carbides. One might expect the chromium content of a carbide formed during a commercial cooling operation to be less than the equilibrium amount. On the other hand, any carbides formed slowly during heating and holding in a carburizing atmosphere can have alloy contents approaching the equilibrium alloy content. In a SAE 94B17 steel (0.5% Cr), Sponzilli et al. (Ref 3) produced massive carbides with a chromium content of 2 to 3%, which is approaching the equilibrium chromium content of 3.1% shown in Table 3.1. The table indicates a corresponding matrix chromium content of 0.12%, which gives an indication of the amount of matrix alloy depletion that can accompany the formation of equilibrium or near-equilibrium carbide particles. The hardening potential of the matrix will thereby be impaired.

Silicon reduces the lattice parameter of iron, making the solubility of interstitials more difficult and, hence, retarding carburizing. Silicon and carbon repel one another so that during the tempering of martensite, silicon is rejected from the carbide phase. Silicon has a tendency to inhibit the formation of grain boundary carbides and to suppress the formation of massive carbides (Ref 4).

Phosphorus repels carbon from the iron lattice, yet it appears to stimulate the formation of grain boundary carbides as it segregates to the grain boundaries (Ref 5, 6). However, the segregation tendencies of phosphorus depend on what other elements are in the alloy.

Nickel does not form carbides and, like silicon, retards carburization (Ref 7). It increases the carbide nucleation rate during processing.

Chromium favors the formation of spheroidal carbides and makes cementite coarsening more difficult during cooling (Ref 8, 9). In a chromium steel of ~1% Cr, carbide formation and coarsening will occur when the carbon content exceeds the Ac_{cm} value at the carburizing temperature. However, with less than the Ac_{cm} carbon content, any carbides deposited during cooling will tend to be relatively small in size and will appear in smaller quantities as the carbon content diminishes. In many steels, carbides disappear at about 0.6 to 0.7% C. The application of a carburizing treatment that would produce massive carbides in a conventional carburizing steel with, for example, 0.5 to 1.5% Cr would produce only a fine close dispersion (0.5 μm diameter) of Cr_7C_3 carbides in a 12% Cr steel (Ref 4).

Manganese is not as potent a carbide former as is chromium; it partitions between the carbides and the matrix. It does not stimulate homogeneous nucleation of massive iron carbide (Fe_3C) in austenite and, therefore, does not favor the formation of massive carbides (Ref 8). Manganese does, however, promote cementite networks.

Molybdenum in excess of 0.5% is said to increase the solubility of carbon in austenite, thereby reducing the amount of carbide phase (Ref 10). Molybdenum favors the formation of spheroidal carbides, and, in the presence of chromium, it can be difficult to produce coarse carbides. Molybdenum partitions between carbides and the matrix, which is important because it preserves some measure of hardenability in the matrix material to counter alloy depletion (of chromium, for example) due to carbide formation.

Table 3.1 Equilibrium states of the system Fe-Cr-C at 700 °C

| Steel No. | Steel composition, % | | Composition of extracted carbides (Fe_3C), % | | In ferrite, | Partition ratio of Cr |
	C	Cr	C	Cr	% Cr	between Fe_3C and α-Fe
1	1.02	0.48	7.00	3.10	0.12	25.6:1
2	0.98	0.99	6.67	6.32	...	...
3	0.48	0.53	6.74	6.54	0.23	28.4:1
4	0.49	1.05	6.46	10.9	0.41	26.6:1
5	0.51	1.55	6.98	16.2	0.59	27.5:1
6	0.97	2.91	6.77	18.1	0.59	30.7:1

Quenched from 1300 °C and tempered at 700 °C for 120 h. Source: Ref 2

Vanadium and titanium are thought not to influence the formation of carbide; they form their own special carbides.

Massive, Network, and Dispersed Carbides

Equilibrium Conditions. In pure iron-carbon alloys, and when the process of heating to and cooling from the fully austenitic state are slow enough to encourage all reactions to continue until completion, the eutectoid carbon content is about 0.78%. In such circumstances, the amount of carbide formed depends on how much the carbon content exceeds the eutectoid value of the alloy. For example, consider an iron-carbon alloy with a carbon content of C_1 in relation to the equilibrium diagram shown in Fig. 3.2. Once cooled from the austenitizing temperature, t_0, to temperature t_1, the austenite becomes conditioned and ready to reject any excess carbon, and the carbon content of the austenite is still C_1 on average. On further cooling to t_2, some of the carbon precipitates as Fe_3C at the austenite grain boundaries, and the carbon content of the austenite becomes C_2. With further cooling to the eutectoid temperature, t_3, the rejection of carbon to form carbide at the grain boundaries ceases, and the remaining austenite, now of carbon content C_3, transforms to the eutectoid microstructure pearlite. The final structure, therefore, consists of pearlite grains in a network of iron carbide. The proportions of precipitated carbide and austenite at any temperature between the Ac_{cm} temperature and the Ac_1 temperature are, according to the lever rule, in the ratio of the lengths C_xC_1 to C_1B (see the inset in Fig. 3.2 for temperature t_x).

This discussion considers a situation where the carbon present in the alloy is less than the limiting solubility of carbon in austenite at the austenitizing temperature, t_0. When the carbon content exceeds that solubility limit (indicated by the Ac_{cm} boundary) and reaches, for example, C_4, carbides will precipitate in the austenite at the austenitizing temperature, leaving the austenite with a carbon content of C_5. On cooling to t_3, the carbon content of the austenite will progressively decrease through C_1 and C_2 to C_3, and meanwhile the amount of carbide deposited will increase, the proportion of carbide being C_xC_4 to C_4B.

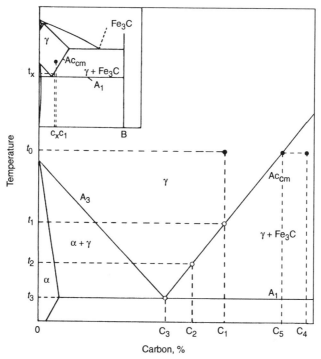

Fig. 3.2 The iron-carbon equilibrium diagram used to explain network carbide formation

Nonequilibrium Cooling. Pure iron-iron carbide alloys are not dealt with in commercial heat treating, nor does any process heat or cool steel parts at rates approaching those necessary to achieve equilibrium conditions. The steels used commercially contain varying amounts of alloying elements, and these alter the transformation characteristics of iron. In terms of the equilibrium condition, these alloying elements reduce the eutectoid carbon content and alter the Ac_1 temperature (Fig. 3.3). In turn, the Ac_3 and Ac_{cm} boundaries are moved (Fig. 3.4). Furthermore, equilibrium conditions give ferrite, plus special carbides, with more alloyed steels, rather than ferrite, plus cementite, as is the case with the $Fe-Fe_3C$ system.

Cooling rates employed for carburized parts, ranging from a slow pit cooling to a standard quench, are fast compared with cooling rates for the equilibrium condition; consequently, the equilibrium diagram ceases to be valuable for cooling. More appropriate for commercial practice is the continuous-cooling transformation (CCT) diagram. Note that each steel will have its own diagram, just as each steel has its own Jominy hardenability curve. Together the equilibrium diagram and the CCT diagram are useful for explaining the different phases found in carburized surfaces.

Unfortunately, in terms of free carbides, these diagrams cannot indicate how the carbide phase will appear to the metallographer. Generally, if the carbides form during cooling they will be grain-boundary carbides, and if reheating is involved there may also be fine dispersions. Globular carbides, on the other hand, form during heating and during holding in a carburizing atmosphere at temperatures between the Ac_1 and the Ac_3 (globular carbides are discussed in the section "Globular Carbide Dispersions and Film Carbides" in this chapter). The following sections consider the carburizing and cooling conditions that may be applied to steels and how these conditions might influence carbides.

Surface Carbon Content in Excess of the Ac_{cm} Carbon. Carburizing with a high carbon potential might cause the carbon content of the austenite (γ) at the surface to exceed the Ac_{cm} carbon value. Therefore, the carbon in excess of the Ac_{cm} value is deposited as carbide during carburizing, and the austenite has a carbon content equal to the Ac_{cm} value (see Fig. 3.5). If the

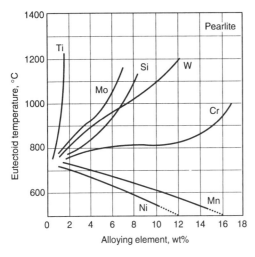

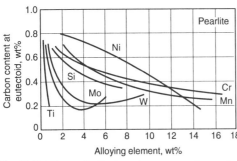

Fig. 3.3 Effects of alloying elements on the carbon content and temperature of the eutectoid point. Depression (-) or elevation (+) of eutectoid temperature (Ac_1) by 1% of alloying element: Ni, -30 EC; Mo, -25 EC; Co, -15 EC; Si, +20 to 30 EC; Mo, +25 EC; Al, +30 EC; V, + 50 EC. Source: Ref 11

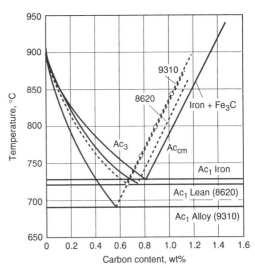

Fig. 3.4 Examples of how steel composition can affect the Ac_1, Ac_3, and Ac_{cm} phase boundaries. Data for 3%NiCr steel are experimental; data for 8620 steel are estimated. Source: Ref 12, 13

surface carbon content is controlled to be less than the Ac_{cm} value, these carbides should not form during carburizing.

Even so, the carbon content has been known to reach high values at surface features, such as corners and edges, even when the carburizing atmosphere has been controlled to a lower potential.

Such carbides tend to be massive and at grain boundaries, though only to a shallow depth, and appear to be encouraged by chromium. As the chromium content of the steel increases, there is a tendency for carbides to precipitate within the austenite grains during cooling. An example of corner buildup is presented in Ref 14 by Jones and Krauss. Following a high-temperature carburize-and-diffuse treatment producing a final surface carbon content of 0.85%, they observed that the carbon content had built up to ~1.1% at corners, resulting in the formation of heavy carbides.

Slow Cooling from the Carburizing Temperature. Slow cooling to room temperature to permit, for example, some intermediate machining causes any carbon in excess of the apparent equilibrium carbon to precipitate at the austenite grain boundaries (see Fig. 3.6). The material within the grains will have a carbon content close to that of the eutectoid carbon, and the microstructure there will likely contain pearlite and/or bainite, though martensite can form even

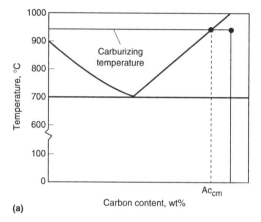

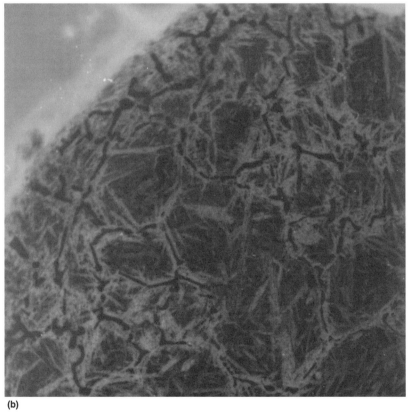

Fig. 3.5 Carbides produced during carburizing with surface carbon above Ac_{cm} carbon. (a) Iron-carbon phase diagram. (b) Micrograph of 4%Ni-C-Cr carburized steel. Courtesy of Peter Smith. 500×

with slow cooling in the higher-hardenability carburizing grades of steel.

The term *apparent eutectoid* relates to the carbon content within a slowly cooled, double-quenched carburized case at which free carbides are just detected. The apparent eutectoid carbon is therefore higher in value than the true eutectoid carbon for which the cooling conditions are necessarily much slower. Nevertheless, the apparent eutectoid is applicable to case hardening and for estimating the apparent eutectoid carbon. In Ref 15, Haynes suggests the following for each 1% of added alloying element:

Element	Apparent eutectoid carbon, %
Manganese	–0.08
Nickel	–0.04
Chromium	–0.13
Molybdenum	–0.40

The reduction of carbon content for each element (shown above) is deducted from the established iron-iron carbide eutectoid of 0.78%. Thus, for a carburized and slow-cooled sample of a 0.57Mn-3.06Ni-0.87Cr-0.07Mo alloy steel, carbides may be detected to the depth at which the carbon content is 0.47%.

Direct Quenching from the Carburizing Temperature. Quenching directly from the carburizing temperature tends to suppress the precipitation of carbides, though much depends on the cooling rate (see Fig. 3.7). For a given quench, the larger the part is (and therefore the slower the cooling rate is), the more opportunity there is for the austenite to reject carbon as grain boundary carbides (Fig. 3.7c). For a given high carbon content, the less carbide deposited due to suppression, the more carbon there will be in solution in the austenite. This means that the martensite start and finish temperatures will be lowered, and, hence, the retained austenite of the as-quenched surface will be increased.

Slow Cooling to Just above the Ac_{cm}. Slow cooling from the carburizing temperature to a temperature just above the Ac_{cm} (typically 830 to 850 °C) prior to quenching is common (see Fig. 3.8). It aims to produce a smaller martensite plate size and a lower retained austenite content, and to reduce distortion levels (compared with those in direct-quenched products). Also, at such temperatures the core will be nearly or fully austenitized prior to quenching; therefore, in steels of adequate hardenability, there should be no ferrite precipitation within the core.

This procedure can produce more carbides than direct quenching does, presumably because at temperatures close to the Ac_{cm}, carbon atoms in excess of the eutectoid carbon begin to segregate to preferred sites, such as grain boundaries. When quenching takes place, some carbide deposition could occur.

If lowering the temperature causes the carbon in the carburized surface to exceed its Ac_{cm} carbon content at that temperature, then carbide precipitation will occur ahead of and during the dwell at the lower temperature (see B in Fig. 3.8).

Reheat Quenching from above the Ac_{cm} Temperature. Lean-alloy steel parts that are slowly

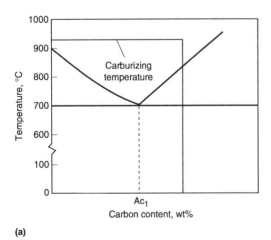
(a)

(b)

Fig. 3.6 Carbides formed after slow cool from carburizing temperature. (a) Pase diagram. (b) Micrograph of 4% Ni steel carburized and slow cooled. 196×

cooled from the carburizing temperature will likely have a surface microstructure comprised of a mixture of pearlite, bainite, and possibly some martensite enclosed in a carbide network (see Fig. 3.9). Reheating to the Ac_{cm} (for example, to 830 °C) will more or less reaustenitize the pearlite and bainite but will not necessarily dissolve all the grain boundary carbides and any large carbide particles. Though carbides do begin to dissolve above the Ac_1 temperature, carbide solution is also time dependent; therefore, adequate time must be allowed. Reporting on a plain carbon steel with a high-carbon carburized surface, Jenkins showed that whereas carbide solution increased with reheat temperature, traces of carbide network could nevertheless survive reheating at 900 °C (Ref 17). Alloyed steels might require higher temperatures or longer times to dissolve the carbides if the surface carbon content is high.

In reheated and quenched lean-alloy steel parts, grain boundary filaments do not usually exist at carbon contents below about 0.85%,

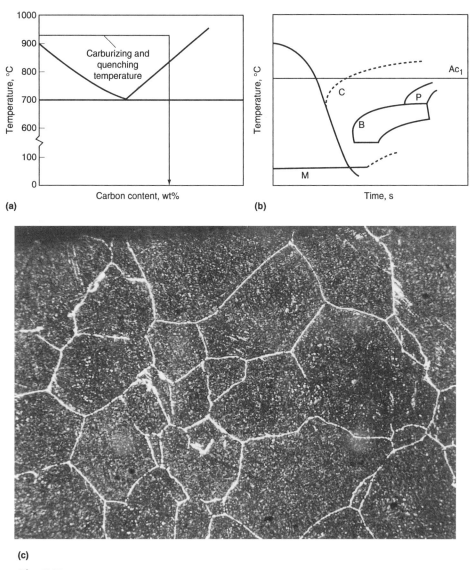

Fig. 3.7 Direct quenching from carburizing temperature. (a) Phase diagram relations. (b) Continuous-cooling transformation curve for a high-carbon surface. (c) Micrograph of direct quenched 3%Ni-Cr carburized steel. 500×. Source: Ref 16

whereas in more highly alloyed case-hardening grades, that limit is about 0.75% C. Given that the soak time at temperature is adequate, most network carbides in conventional case-hardening steels should be dissolved at temperatures just above the Ac_{cm}; naturally, the higher the temperature is above the Ac_{cm}, the more effective the carbide solution will be. That said, the best matrix microstructures for high-stress applications are attained by reheat quenching from about the Ac_{cm} temperature. Such reheat temperatures should also be adjusted so they are sufficient to austenitize the core material.

Reheat Quenching from Temperatures between the Ac_1 and the Ac_{cm}. For the most part, network carbides will survive reheating to within the A_1 to Ac_{cm} temperature range (see Fig. 3.10). However, the nearer the reheat temperature is to the Ac_{cm}, the greater the chance becomes of carbide solution.

When reheating carburized lean-alloy grades of steel to any quenching temperature, the amount of carbide dissolved is controlled by the temperature and duration of reheating (assuming that there are carbides in the microstructure). The manipulation of these

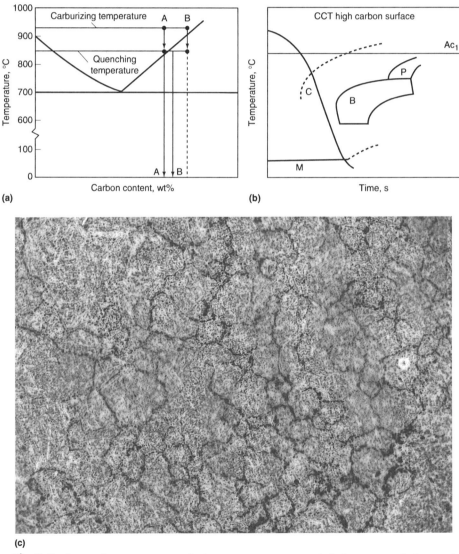

Fig. 3.8 Slow cooling to approximately the Ac_{cm} temperature. Carbide formation with slow cool prior to quench. (a) Phase diagram relations. (b) Continuous cooling diagram. (c) Micrograph of 3%Ni-Cr steel cooled from carburizing temperature to below Ac_{cm} prior to quenching

variables provides a potential for controlling the amount of carbon in the final as-quenched martensite. For example, while working with bearing steels, Averbach varied the temperature within the Ac_1 to Ac_{cm} range to control the amount of undissolved carbide, thereby controlling the amount of carbon remaining in the austenite (Ref 18). Subsequent quenching of the lower-carbon austenite led to a lower-carbon martensite and a reduction in the number of microcracks in the final martensitic structure.

In determining the optimum microstructure for contact fatigue, Vinokur demonstrated the importance of developing the correct surface carbon content and of reheat quenching from the Ac_{cm} temperature (Ref 13). The results of that work are presented in Table 3.2, which relates how the microstructure of a carburized surface is influenced by carbon content and reheating temperatures ranging from 760 to 860 °C. An alternative for selecting a reheat quench temperature is to choose conditions that will just austenitize the core material.

Double Reheat Quenching. Double reheat quenching was used extensively before grain-refined steels were available (see Fig. 3.11). Though not usually practiced now, double reheat quenching is nevertheless a means of producing high-quality components. The first reheat is carried out at a high temperature (approximately 870 °C) so that both the case and the core materials are austenitized and the carbides are dissolved. Quenching then fully hardens and refines the core and inhibits network carbide precipitation in the case, but leaves the case relatively coarse and containing retained austenite. The second reheating operation to about 780 °C, followed by the quench, leaves the core refined but softened and somewhat toughened. This reheat operation refines the case and causes the precipitation of a fine dispersion of carbides along with less retained austenite. This procedure likely produces more dispersed carbides to a greater depth than single reheat quenching does. The drawbacks of double reheat quenching are the time required for processing, the costs, and the distortion problems that can arise due to austenitizing twice.

Fig. 3.9 Reheat quenching from temperatures between Ac_1 and Ac_{cm}. (a) Fe-C diagram of reheat condition. (b) Micrograph of reheat near 890 °C. (c) Micrograph of reheat near 840 °C

Subcritical Annealing. Some carburized parts are slow cooled from carburizing (see "Slow Cooling from the Carburizing Temperature") to permit additional machining for removal of unwanted case. The parts are subsequently reheated for hardening off by one of the methods referred to in the previous three sections.

To make the intermediate machining easier, particularly with more alloyed case-hardening grades of steel, carburized and cooled parts are usually subcritically annealed. Slowly cooled cases may contain austenite, martensite, and bainite (and maybe some pearlite in the leaner grades of steel), along with carbide precipitates (some of them quite coarse) and grain boundary films or networks. Subcritical annealing isothermally transforms austenite to bainite, converts any martensite to ferrite, and precipitates out as carbides any carbon that has been held in solution in the martensite. These carbides, and those fine carbides that decorate the bainite, tend to eventually coalesce, grow, and spheroidize, whereas the matrix ferrite recrystallizes. Also, the cementite plates of any pearlite present will disintegrate to form spheroids. Any "special" alloy carbides, and those coarser carbides and grain boundary carbides rejected by the austenite during the cool from carburizing, might resist becoming spheroidized; this resistance increases with increasing amounts of carbide-forming alloying elements in the steel. Naturally, the duration and temperature of the annealing treatment affect the degrees of carbide spheroidization and softening, but, even so, some of the original carbides can survive the treatment. It is important to consider whether the surface is sufficiently softened to permit the intermediate machining operation to proceed satisfactorily.

The Formation of Carbides

The development of the carbide phase in a hypereutectoid 4.4% Cr steel was studied by Ruxanda and Florian (Ref 19). Three carbides were identified: massive, film, and intragranular carbides. This study did not consider the an-

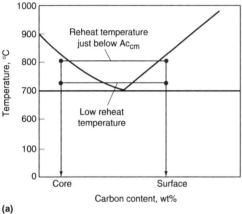

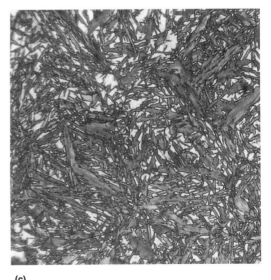

Fig. 3.10 Reheating between Ac_1 and Ac_{cm}. (a) Phase diagram plot. (b) Reheat just below Ac_{cm}. (c) Low reheat temperature

nealed carbides, only those formed isothermally or during cooling.

Massive carbides form isothermally in the carbon-supersaturated austenite at the junctions of a number of austenite grains (Fig. 3.5). Such carbides form readily in three directions, involving step and ledge growth where the heights of the ledges are many times greater than the heights of the steps (Fig. 3.12a).

Film carbides form during cooling below the Ac_{cm} (see Fig. 3.7). The excess carbon migrates to the austenite grain boundaries where it is deposited as small thin platelets that make contact with one another, forming a film. These carbides are similar to massive carbides—the ledges and steps are of similar height—yet they differ due to the thermal conditions (Fig. 3.12b). Such carbides can be seen as isolated grain boundary particles, a row of particles (necklace effect), a film of a length equivalent to that of a prior austenite grain boundary facet, through what is considered to be a continuous carbide network. Figure 3.6 shows that at first, film carbides can have a jagged appearance but will become smoother with thermal processes, such as intermediate annealing and reheating for quenching.

Intragranular carbides precipitate on planes within the grains, and differ greatly from massive or film carbides. They have an orientation tendency and usually have a cylindrical form and a smooth gently spiraling surface (Fig. 3.12c).

Table 3.2 Effect of the austenitization temperature of heating before quenching in the ratio of structural components and the composition of the matrix of the case-hardened layer on the contact fatigue

					Content, %			Element content in solid solution(a), %				N, million
C_{la}, %	A_{ks}, %	Ac_{cm}(a), °C	t_q, °C	K_d(a), %	Martensite	Residual austenite	Carbides	C	Mn	Cr	Mo	cycles
0.69	13.9	760	725	5	87	0	13	0.04	0.84	0.50	0.08	0
			750	92	87	10	3	0.57	1.18	1.17	0.19	7
			775	100	86	14	0	0.69	1.30	1.32	0.25	22
			800	100	84	16	0	0.69	1.30	1.32	0.25	40
			825	100	81	19	0	0.69	1.30	1.32	0.25	35
			850	100	78	22	0	0.69	1.30	1.32	0.25	35
0.81	15.0	790	725	5	84	2	14	0.04	0.83	0.48	0.07	3
			750	56	84	10	6	0.45	1.08	0.63	0.18	9
			775	96	83	16	1	0.77	1.28	1.27	0.24	25
			800	100	79	21	0	0.81	1.30	1.32	0.25	68
			825	100	75	25	0	0.81	1.30	1.32	0.25	63
			850	100	72	28	0	0.81	1.30	1.32	0.25	55
0.92	16.4	820	725	5	80	5	15	0.05	0.81	0.44	0.07	7
			750	21	77	11	13	0.19	0.89	0.59	0.10	15
			775	60	75	18	7	0.45	1.04	0.85	0.15	30
			800	87	71	26	3	0.78	1.23	1.18	0.23	65
			825	100	70	30	0	0.92	1.30	1.32	0.25	90
			850	100	67	33	0	0.92	1.30	1.32	0.25	82
0.98	17.8	840	725	5	76	7	17	0.05	0.79	0.43	0.07	3
			750	12	70	14	16	0.12	0.82	0.49	0.08	10
			775	25	66	21	13	0.25	0.90	0.61	0.11	20
			800	68	65	29	6	0.68	1.13	1.01	0.19	50
			825	90	64	34	2	0.90	1.25	1.21	0.23	80
			850	100	63	37	0	0.98	1.30	1.32	0.25	71
1.12	19.2	860	725	5	73	9	18	0.06	0.75	0.39	0.06	0
			750	8	76	16	18	0.09	0.77	0.42	0.07	7
			775	11	61	23	16	0.19	0.83	0.51	0.08	11
			800	32	57	30	13	0.38	0.95	0.72	0.12	29
			825	67	57	35	8	0.74	1.11	0.99	0.18	33
			850	90	57	39	4	0.99	1.24	1.21	0.23	35

C_{la}, carbon content in carburized layer; A_{ks}, amount of carbide annealed structures; t_q, quenching temperature; K_d, amount of dissolved carbides; N, number of cycles to appearance of contact fatigue dimples. The nickel content in the solid solution was 0.82% for all carbon content in the layer and all heating temperatures for quenching. The temperature of the point Ac_1 was 720 °C for all carbon concentrations. The tempering temperature was 170 °C in all cases. (a) The studies were conducted on simulation steels. Source: Ref 13

The Effect of Network and Dispersed Carbides on Properties

For most applications, the aim of the heat treater is to produce carburized and hardened cases with a surface microstructure consisting of high-carbon martensite with an acceptable amount of retained austenite. Incidental carbides, for example, occasional filaments of grain boundary carbides or a loose distribution of small spheroidal carbides, are usually permitted. The properties of such structures are dealt with here. For wear applications the carburizing process can be adjusted to intentionally produce surfaces containing large quantities of globular or spheroidal carbides. The properties of such structures are discussed in the section dealing with globular carbides.

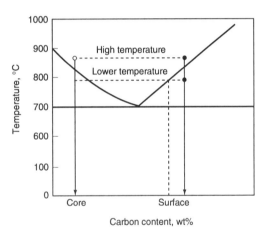

Fig. 3.11 Double reheat quench. A high temperature first reheat quench refines the core, and a lower temperature second reheat quench refines the case.

The effects of network carbides and dispersed spheroidal carbides in a carburized and hardened case are not clear cut. With carbide networks, much depends on the continuity of the network, the depth of penetration, and the thickness of the carbide. Whether or not network formation has denuded the adjacent matrix material of the carbide-forming elements, thereby encouraging the formation of nonmartensitic transformation products, is also a factor. With spheroidal carbides, the mean diameter, interparticle spacing, and penetration depth are expected to influence certain properties as well. In forming carbides, the carbon content of the matrix will be reduced, modifying the proportions of martensite and austenite in the matrix, and potentially affecting the martensite morphology.

Influence on Hardness. The macrohardness of low-temperature tempered, case-hardened surfaces is primarily determined by the martensite-austenite aggregate. The hardness of any carbide phase is greater than that of the martensite-austenite in which it resides, but because the amount of carbide (as network and/or spheroidal carbides) is normally less than about 10%, it is unlikely that the hardness will be noticeably increased. In other words, a routine hardness test on a case-hardened part might not readily indicate the presence of carbides. The surface macrohardness could be significantly reduced if the amount of carbide is greater, and if there are high-temperature transformation products (HTTP) associated with the carbides (due to local alloy depletion during carbide formation) and, hence, local softening (400 to 500 HV). This was demonstrated by Jones and Krauss when an excessive buildup of carbon at a corner reduced the corner hardness of the quenched and tempered test piece to 50 to 54 HRC (Ref 14);

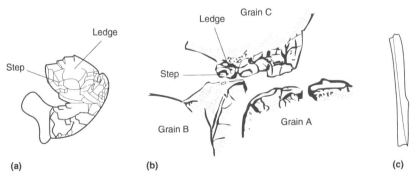

Fig. 3.12 Geometric models of carbides formed during case-hardening. A,B, and C are 3 grains (parted). (a) Massive carbide, 4000×. (b) Film carbide, 2000×. (c) Intragranular carbide, 4000×. Source: Ref 19

Table 3.3 Effect of carbide state on fracture strength, toughness, residual stresses, and fatigue life

Tempering temperature, °C	Amount of carbides	Fracture strength, kg/mm²	Absorbed energy, kg·mm/mm²	Impact toughness kg·m/cm²	Residual stress at Surface k g/mm²	Fatigue life, 10³ cycles
160	Few carbides	148.5	8.50	1.3	−25.6	...
	Large quantity network	144.5	8.81	0.9	+4.4	...
	Large quantity spheroidal	155.7	9.62	1.3	−8.9	...
190	Few carbides	141.3	7.09	1.89	−13.3	113
	Large quantity network	160.2	9.11	1.23	+3.0	54
	Large quantity spheroidal	167.4	10.35	1.95	−9.9	82

1 kg/mm² = 9.80665 N/mm² (MPa). Source: Ref 20

this reduction was attributed to nonmartensitic microstructures (retained austenite and HTTP).

Influence on Residual Stresses. In test pieces with surfaces containing network carbides, Wang et al. found that the surface macroresidual stresses were tensile, whereas for those test pieces with few carbides or with a large quantity of spheroidal carbides, the surface residual macrostresses were compressive (Table 3.3) (Ref 20).

One might imagine that in parts containing carbides at the immediate surface, the short-range stress distribution (microstresses) will be quite complicated, more so as the carbide shape departs from being spherical. The carbides form at relatively high temperatures. During cooling to the martensite start temperature, they contract at a different rate than the adjacent austenite, encouraging tensile stresses to develop in the austenite alongside the carbides. These stresses are sufficient to locally yield the austenite and possibly stimulate transformation at the carbide-austenite interfaces. Once the M_s is exceeded, the austenite-to-martensite reaction takes place that involves a transformation expansion while, at the same time, the carbide continues to thermally contract. Consequently, in microstructures that contain spheroidized carbides the carbides are in compression and in microstructures where the carbide is of the grain boundary (network) type, the carbide is in compression through its thickness and in tension along its length. The more austenite transformation there is, the greater the microstresses will be. The stress magnitude also depends on the composition of the carbide, which in turn depends on the composition of the steel. In many case-hardening steels, the carbide is of the Fe_3C type where the Fe component can be substituted by carbide-forming elements. With the more alloyed steels, alloy carbides may form, and these have different properties to one another and to Fe_3C. Stuart and Ridley determined the average coefficient of expansion/contraction (α) of a number of carbides for the temperature range 20 to 700 °C (Table 3.4) (Ref 21). The greater the difference between the carbide and the matrix, the larger the accompanying residual microstresses in the quenched alloy will be (Table 3.5).

Influence on Bending Fatigue. Continuous carbide networks are thought to reduce the bending fatigue strength of case-hardened gears as Fig. 3.13 implies (Ref 22), whereas partial carbide networks, according to Robinson, do not have a detrimental effect (Fig. 3.14) (Ref 23). The dimensions of carbides in the immediate surface seem to have a bearing on the resistance to bending fatigue, as the results of Evanson et al. suggested (Ref 24). Comparing two procedures for vacuum carburizing, these researchers found that a surface containing grain-boundary massive carbides of about 10 μm maximum dimension had a lower fatigue limit and developed a greater scatter of failure points in the finite life regime

Table 3.4 Mean linear expansion coefficients (20 to 700 °C), Young's modulus, and Poisson's ratio for carbides and matrices

Nominal composition	α mean (20–700 °C), ×10⁻⁶	Young's modulus, ×10³ tons/in²	Poisson's ratio
WC	4.50	39.3	0.185
ZrC	6.03	21.5	0.257
Mo_2C	6.08	21.2	...
TaC	6.15	26.5	0.206
NbC	6.95	25.3	0.229
VC	7.26	17.2	...
TiC	7.39	22.7	0.186
M_6C	9.24	...	...
$M_{23}C_6$	10.36	...	...
Fe_3C	12.63	12.9_6	0.361
α-Fe	14.79	13.4_7	0.283
γ 20Cr/25Ni	17.72	13.2	0.30
γ 18Cr/10Ni	18.08	13.2	0.30

Source: Ref 21

Table 3.5 Calculated tessellated stresses for WC and Fe$_3$C in ferrite, cooled rapidly from 700 to 20 °C

Type of stress	Carbide	Calculated stress, tons/in.^{2} Spherical case	Cylindrical case
Ferrostatic pressure in carbide	WC	−109	−72
	Fe$_3$C	−21	−16
Radial stresses in matrix	WC	−109	−72
	Fe$_3$C	−21	−16
Circumferential stresses in matrix	WC	+54	+72
	Fe$_3$C	+10	+16
Axial stress in carbide	WC	...	−280(a)
	Fe$_3$C	...	−30(a)
Axial stress in matrix	WC	...	+15(a)
	Fe$_3$C	...	+1.6(a)

Tons/in.^{2} × 2240 = ksi; tons/in.^{2} × 15.45 = MPa (N/mm^2). (a) 5% volume fraction of carbide. Source: Ref 21

than did a surface containing carbides of 2 μm maximum dimension (Fig. 3.15).

When carburizing conditions are controlled to avoid the formation of carbides in most surfaces of a part, sufficient carbon buildup can occur at certain features, such as external edges and corners, to produce carbides in those features (Ref 14). The presence of corner carbides can adversely affect the bending fatigue strength of a part according to the quantity and type of carbide, network carbides being more detrimental than others (Fig. 3.16). The effect can be eliminated by adequately rounding the corners and edges prior to carburizing; corner "breaking" and chamfering might not be sufficient.

Fatigue resistance is divided into stages that include initial crack formation, growth to critical crack size, and crack propagation. If a carbide particle is likened to a hard oxide inclusion, in the absence of residual macrostresses, the critical particle diameter for fatigue crack initiation close

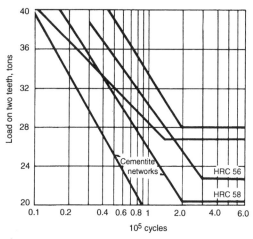

Fig. 3.13 A comparison of the bending fatigue strengths of carburized 12KhN3 gears illustrating the adverse influence of cementite networks. Note that these tests were carried out over a period of about 10 years; therefore, there may have also been other factors involved in producing the differences observed. Source: Ref 22

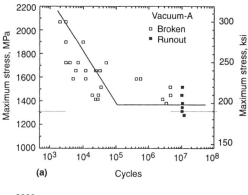

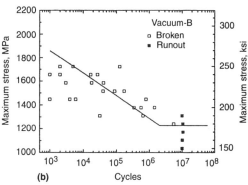

Fig. 3.15 Bending fatigue S-N curves at a stress ratio of R = 0.1 for (a) Vacuum-A showing an endurance limit of 1370 MPa (200 ksi) and for (b) Vacuum-B showing an endurance limit of 1235 MPa (180 ksi). Considerable scatter in the data exists for both Vacuum-A and Vacuum-B conditions. Carbides in outer 10 μm: upper, 2 μm maximim dimension; lower, 10 μm maximim dimension. Source: Ref 24

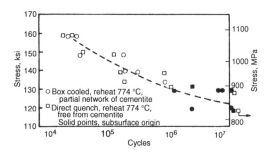

Fig. 3.14 The bending fatigue strength of carburized samples of SAE 6120, comparing those containing a partial network of cementite with those free from cementite. Source: Ref 23

to the surface will be approximately 10 μm. When macroresidual stresses are present and compressive, the critical defect size will be larger than 10 μm, whereas if the macroresidual stresses are tensile, the critical defect size will be smaller than 10 μm. Most spheroidal carbides have diameters of less than 1 μm and, therefore, might not contribute much to the crack initiation process. Massive carbides come in a range of sizes up to a maximum dimension of about 10 μm, and consequently the larger particles will provide potential sites for crack initiation. Network carbides have thicknesses of less than 5 μm, but in well-formed, heavy, continuous carbide networks there are large areas of carbide surface that could provide crack initiation sites when so orientated.

The coherency between the carbide and the matrix is important; however, the magnitude of tensile microstresses associated with each type of carbide is perhaps the most important factor. In the previous section, it was discussed that spheroidal carbides would be in compression and the surrounding matrix material would be in tension. However, these carbides are so small that the tensile zone around them might not be too harmful; that is, the size of the carbide spheroid and the surrounding tensile zone together remain less than the critical size for fatigue crack initiation. Network carbides in quenched surfaces are thought to be in tension in two of three directions, with the adjacent martensite in compression. Therefore, the weakest link would be expected to be associated with the carbide itself, and evidence suggests that fatigue cracks do initiate at or within grain boundary carbides (Ref 24). The development of carbides can lead to a local depletion of carbide-forming elements in the adjacent matrix material so that in quenching it is possible that HTTP will be associated with the carbides. Such products will be in tension because they have not expanded as much as the martensite to one side of them; their presence also relieves some of the tension in the carbide film at their other side. Thus, because these high-temperature products are in tension and have less strength than the martensite, they will offer less resistance to cyclic tensile stressing.

Once a fatigue crack has initiated and grown to critical size, crack propagation will likely be assisted by the presence of carbides. Using 20CrNi2Mo and 16CrNi4Mo steels in which the volume fractions of fine carbides were respectively ~6 and 9%, Gu et al. found that fatigue-crack initiation life was mainly dependent on the surface residual stresses (see Fig. 3.17) and that these, in turn, depended on the retained austenite content and the alloy composition (nickel having a positive effect) (Ref 25). Crack propagation appeared to be hastened by fine carbides in excess of ~6% but was slowed by retained austenite (25 to 35%). Network carbides might encourage a propagating crack to follow the grain boundaries, particularly if the maximum austenite-to-martensite transformation has taken place through refrigeration, for example. In such an instance, the microstresses associated with the carbides will be of a high order. However, fatigue cracks do not necessarily follow the grain boundaries, even when those boundaries are filled with carbide.

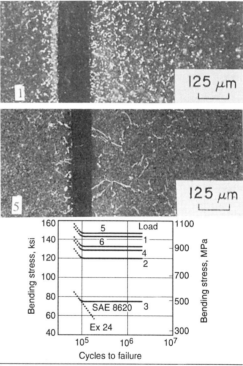

Heat treatment sequence	Edge carbides
1. Carburize 930 °C, 132 min; diffuse 48 min; cool to 845 °C; oil quench 70 °C	Massive carbides
2. Carburize 1050 °C, 35 min; diffuse 13 min; cool to 845 °C; oil quench	Network carbides
3. Carburize 1050 °C, 35 min; diffuse 13 min; cool to 845 °C; oil quench. (With subzero liquid nitrogen)	Network carbides
4. Carburize 1050 °C, 35 min; diffuse 13 min; cool to 845 °C; oil quench; reheat 790 °C; oil quench	Network carbides
5. Carburize 1050 °C, 35 min; diffuse 13 min; cool to 845 °C; oil quench	Network/spheroidal carbides
6. Carburize 1050 °C, 6 min; diffuse 25 min; cool to 595 °C; reheat to 870 °C; oil quench	Network/spheroidal carbides

All tempered at 175 °C

Fig. 3.16 The effect of heat treatment and edge carbides on the bending fatigue endurance of case-hardened SAE 8620 and Ex 24. Source: Ref 14

Influence on Contact Fatigue. Contact fatigue tests, in which surfaces with and without carbide networks were compared, indicate that the carbide-containing structures were somewhat better in high-stress/low-cycle regime (Ref 26). In the low-stress/high-cycle regime, heavy carbide networks had a negative effect on the contact-fatigue resistance if they had an effect at all (Fig. 3.18). Tests conducted by Wang et al. confirmed that under high-stress/low-cycle rolling contact conditions, a surface containing network carbides had a better contact fatigue life than a surface in which there were few carbides, though the longest life was achieved with a surface containing fairly large spheroidal carbides (Fig. 3.19) (Ref 20). Geller and Kozhushnik concluded that a surface with a carbon content of 0.93% with a fine dispersion of carbides was better than a 0.83%C surface with no free carbides (Fig. 3.20) (Ref 27). Vinokur's results were in agreement with these findings, but only up to a point (Ref 13). It was shown that a further improvement of contact-fatigue resistance could be obtained if the spheroidized carbides in a 0.9 to 0.95% C surface were just dissolved by reheating to the Ac_{cm} temperature prior to quenching (Fig. 3.21). This optimum coincided with a retained austenite value of 30%, with no carbides. Vinokur worked with a Cr-Mn-Ni-Mo-V lean-alloy steel and clearly showed that, for contact-fatigue conditions, surfaces with either a low carbon content (~0.7%) or a high carbon content (~1.1%) produced inferior results to surfaces containing 0.9 to 0.95% C. While this carbon range is probably about right for most

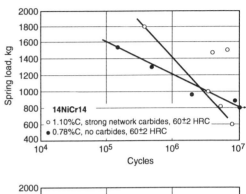

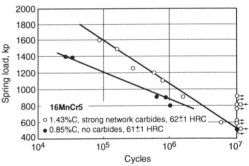

Fig. 3.18 Effect of carbide networks on the contact-fatigue strength of case-hardened steels. Case depths, 1.1 ± 0.1 mm (also see Fig. 4.21). Source: Ref 26

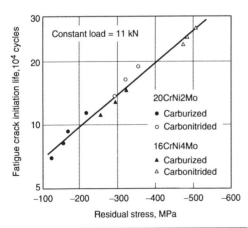

Steel	Carbides			Retained austenite	
	Vol%	Diameter, μm	Spacing, μm	Surface	Maximum at −4 to −5 mm
20CrNi2Mo	12.5	0.56	0.7	~16	~21
16CrNi4Mo	10.5	0.40	0.6	~20	~28

Fig. 3.17 Variation of fatigue-crack initiation lives with residual stress at the notch of tested steels. Surface carbon: 0.95 to 1.05%. Hardness 750 to 780 HV. Crack initiation, 5 μm crack at the notch. Source: Ref 25

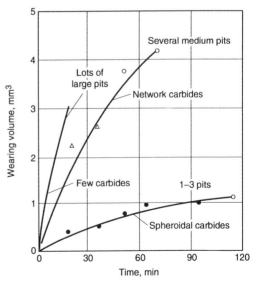

Fig. 3.19 The contact-fatigue wear resistance of different samples. Source: Ref 20

case-hardening steels, the problem in commercial reheat practice is in hitting the Ac_{cm} temperature—each steel grade will have its own.

The idea that carbides are brittle and will fracture readily under load, followed quickly by failure, is not altogether supported by the previously mentioned test nor by certain observations. For example, at a gear tooth surface where heavy sliding and rolling has taken place, network carbides were observed to have bent and flowed with the martensite and retained austenite, together forming an intensely worked surface layer. Cracking of, or at, the carbides was not evident. Beneath the surface of contact-fatigue test discs, in the zone where the Hertzian shear stresses developed during loading, shearing through the network carbides and the matrix was observed (Ref 26). Whether or not such rupturing of the carbide films contributed to either pitting or spalling is not certain, but the test results were favorable, suggesting that in compression due to rolling contact, a cellular (network) structure is strong (Fig. 3.18).

In 52100 steel bearing races, and at stresses above a certain threshold value, the very small carbides contained in the shear bands were seen to have partially sheared via intrusion and extrusion processes or to have sheared through (Ref 28). Cracking was not associated with these features; however, some of the contact damage in races appeared to be associated with lenticular carbides that developed alongside the ferrite shear bands after millions of load cycles (Ref 29).

Influence on Toughness. Intuitively, heavy network carbides in case-hardened surfaces are to be avoided. Surface microstructures in which partial networks of, for instance, less than 50% continuity or those where the carbides exist as small spheroids would be expected to be tougher and, therefore, more acceptable. Surprisingly, Nakazawa and Krauss, working with a quenched and tempered bearing steel, found that microstructures with grain boundary carbides were significantly tougher than those containing a dispersion of small spheroidized carbides (Ref 30). This difference was attributed to the fact that the length of the transgranular fracture path between carbides was much greater in the case of grain boundary carbides. Additional work by Shen and Krauss also implied that network carbides were not especially damaging and, if anything, were

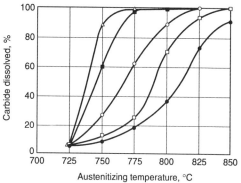

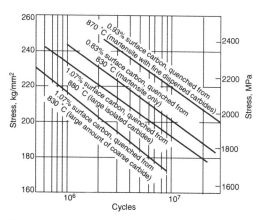

Fig. 3.20 The contact-fatigue strength of carburized 25Kh2GHTA steel (tempered at 180 to 200 °C). Source: Ref 27

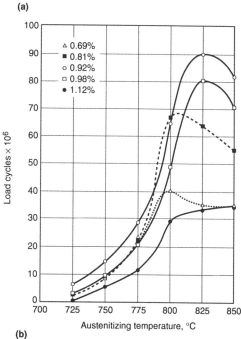

Fig. 3.21 The effect of carbon content and austenitizing temperature on the contact-fatigue life. See Table 3.2 for more detail. Source: Ref 13

not especially damaging and, if anything, slightly beneficial to fracture toughness (Table 3.6) (Ref 6). Despite these observations, heat treaters consider that network carbides make case-hardened parts more susceptible to handling damage during manufacture and impact damage during service.

The results of Sharma et al. suggest that carbides up to 9% have no obvious influence on fracture toughness and that carbon in solution (hence, retained austenite) and nickel content over about 1.5% were much more significant (Table 3.7) (Ref 31). These observations are essentially in agreement with those of Gu et al. in their study of fatigue (Ref 25). For bearing steels at least, a fine distribution of carbides is important because it relates to the amount of carbon in solution in the martensite. A lower austenitizing temperature leaves a larger quantity of undissolved carbides in the micro- structure. Consequently, there is less carbon in the as-quenched martensite, and toughness improves (Fig. 3.22).

Generally, as the carbon content increases, the fracture toughness proportionately decreases. The carbon content of the martensite and, therefore, the type of martensite (plate or lath) also has a bearing on the toughness.

Table 3.6 K_{Ic} values of specimens having different proeuctectoid carbide distributions

Carbide network conditions	Specimens	K_{Ic} values ksi√in.	K_{Ic} values MPa√m
No carbide network	...	14.0–14.4	15.4–16.0
Very few scattered networks of thin discontinuous carbides	4F1T	14.5–15.4	15.9–16.9
	1F1T	15.8–16.8	17.4–18.4
Well-developed networks of thick discontinuous carbides	3F2T	15.6–15.8	17.1–17.4
	2F2T	16.8–16.9	18.5–18.6
Well-developed networks of thin discontinuous carbides	2F1T	17.6–19.0	19.4–20.9
	3F1T	18.4–19.4	20.2–21.3
	...	17.5–17.7	19.4–19.6

Source: Ref 6

Table 3.7 Critical crack size and load carrying capability of various steels

Steel	Composition, % C	Composition, % Ni	K_{Ic}, MPa√m	Critical crack size (Ac), mm	Load carrying capacity (σC)(a), MPa	R(b)	γR, %	FeC %
PS-15	0.99	...	16.6	0.363	690	1.00	39	1
PS-15	0.86	...	22.4	0.660	930	1.35	23	...
PS-15	0.72	...	21.7	0.615	900	1.30	16	...
8697	0.97	0.6	16.6	0.363	690	1.00	33	1
IH-50	0.97	...	20.3	0.543	840	1.22	...	...
4895	0.95	3.4	24.5	0.787	1015	1.47	40	...
4870	0.70	3.5	34.5	1.520	1410	2.04	21	...
ERCH-1	1.00	...	20.7	0.566	860	1.25	42	9
ER-8	0.95	1.5	22.4	0.660	930	1.35	47	1.5
9399	0.99	3.2	27.1	0.965	1125	1.63	49	8.5
5195	0.95	...	2038	0.571	870	1.26	34	1
PS-15 + B	1.00	...	21.2	0.589	885	12.8	31	1

(a) Critical applied stress for a crack size of 0.363 mm. (b) R is σC/σ for PS-15. Source: Ref 31

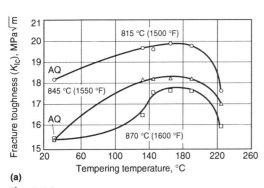
(a)

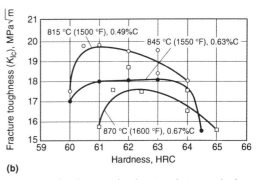
(b)

Fig. 3.22 The effect of austenitizing temperature, tempering temperature, hardness, and carbon in solution on the fracture toughness of an AISI 52100 steel. Source: Ref 18

Globular Carbide Dispersions and Film Carbides

Globular Carbides and Heavy Dispersions

The slow heating of a steel surface through the Ac_1 to Ac_3 temperature range in the presence of a carburizing atmosphere is likely to encourage the formation of globular carbides. This was demonstrated by Dawes and Cooksey, who showed that no carbides were produced when a 4.5%Ni-Cr-Mo steel was heated to the carburizing temperature in a neutral atmosphere then carburized in an active medium (Ref 32). When the carburizing medium was present during the heating period, carbides were produced; when a two-hour soak at 800 °C was included, again with an active atmosphere present, many large carbides formed. Breitbart also established that samples heated in the carburizing atmosphere produced large carbides, whereas those not exposed to the carburizing atmosphere until the carburizing temperature had been reached contained only martensite and austenite (Ref 33). Note that in the temperature range from 750 °C (minimum for safety reasons) to the Ac_3 (for example, 825 °C) an unboosted endothermic carrier gas is decidedly carburizing; however, at 925 °C its carbon potential is only about 0.4% (Fig. 3.23) (Ref 34). If the atmosphere were to be enriched to have a carbon potential of 1% at 925 °C, then in the Ac_1 to Ac_3 temperature range its carbon potential would be exceedingly high.

Such work explains how an undesirable condition for many case-hardened parts could occur (Ref 32, 33). However, carbides are hard and wear resistant, so for certain applications involving heavy wear, the use of case-hardened parts in which heavy deposits of carbides are produced would not seem unreasonable. Consequently, a number of methods have been developed to produce carburized surfaces that contain high densities of carbides, either globular, granular, or spheroidal. The main property of these carbide-containing surfaces is their ability to resist wear. Kern used the term "super carburizing" to describe processes wherein parts were carburized at 930 °C in a highly carburizing atmosphere, cooled to just below the Ac_1 temperature (subcritical anneal), repeatedly carburized and subcritically annealed, and finally quenched and tempered (Ref 35). Surface carbon contents of 1.8% to over 3% were reported along with carbide contents of over 25% and hardnesses of approximately 65 to 67 HRC. Wang et al., aiming to produce a surface containing a fairly dense dispersion of small spheroidal carbides, first precarburized for one to several hours with a carbon potential of 1.1 to 1.6% to produce "seed" carbides in the surface of the part. They then applied a standard carburizing operation lasting several hours to grow the carbides from the seeds (Ref 20). Naito and Kibayashi precarburized to a high carbon content and slow cooled to produce a microstructure containing HTTP (the carbides of which would provide "seed carbides" for further growth during the following carburization process) (Ref 36). The second carburizing cycle consisted of heating slowly from 730 °C to the carburizing temperature, 930 °C, in an atmosphere with a high carbon potential. The product was a surface containing a good proportion of massive spheroidal carbides. Chen (Ref 37) also carburized between the Ac_1 and the Ac_3 temperatures to obtain surfaces containing over 2% C and more than 50% carbides in the microstructure.

Steels used for such processes must contain an adequate amount of one or more of the carbide-forming elements (e.g., chromium, molybdenum, tungsten); therefore, many of the medium-

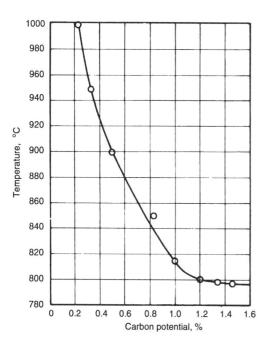

Fig. 3.23 The carbon potential of an endothermic gas of fixed composition as a function of temperature. Derived from Ref 34

alloyed and more highly alloyed case-hardening steels should respond favorably to these processes. However, it has been reported that similar treatments have been successfully applied to 12% Cr steels, wherein chromium carbides (Cr_7C_3) of 0.5 μm mean diameter were developed; more than 30% of the microstructure was taken up with these dispersed carbides (Ref 4, 38).

The development of a structure containing large amounts of carbides leads to a depletion of carbide formers in the matrix. Even so, a reasonable amount of these alloying elements can be retained by the matrix material. For example, Table 3.1 indicates that, depending on the initial chromium content, some chromium will reside in the matrix. Additionally, manganese and molybdenum (assuming their presence) are shared by the carbides and the matrix, and, therefore, it is possible to have sufficient alloying elements in the matrix to produce a eutectoid carbon martensite in the as-quenched item. The actual choice of steel will depend on the ruling section of the part and how quickly it can be cooled after carburizing. Examples of globular carbide deposits are presented in Fig. 3.24.

Film Carbides

Surface film carbides (or flake carbides) form as a continuous or discontinuous film over the surface of the carburized part, possibly with a little penetration (Fig. 3.25). In conventional gas carburizing, this carbide is thought to form due to cooling from the carburizing temperature in an atmosphere of high carbon potential.

In a layer with a depth of ~70 μm, such carbides have been observed at the surfaces of high-speed steel parts subsequent to oil quenching under vacuum (Ref 39). The high-temperature vacuum process produces a clean, active surface, and this reacts with the vapor phase (due to the decomposition of the quenching oil), which quickly develops around the part during the quench. If the quenching oil contains organic compounds, the vapor phase will have a carburizing potential, and consequently, instantaneous carburizing will take place during the quench. The austenitizing temperature and the holding time (within the ranges studied) had no effect on layer formation, whereas quenching temperature, quenchant temperature, atmospheric pressure on the oil surface, and section size all had an influence (Table 3.8).

The Effect of Globular and Film Carbides on Properties

The first choice material for many wear applications must be a type of white cast iron. However, there are wear parts that, for example, require more body toughness than is attainable with white irons, or special machining is needed

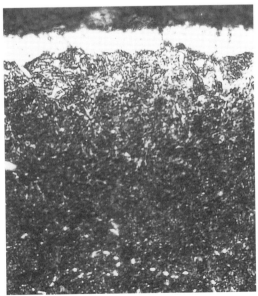

Fig. 3.24 Globular carbides at the surface of a carburized 1%Cr-Mo steel (reheat quenched). 960×

Fig. 3.25 Surface film carbide (1%Cr-Mo steel). 960×

as a final operation, or the wear conditions are not severe enough to justify using special wear-resistant irons. In such cases, a carburized surface with a high carbide content and surface hardness can provide a satisfactory solution.

Influence on Hardness. Carbides themselves are hard; for example, the microhardness of the carbide in a case-hardened plain carbon steel was measured at more than 1000 HV, whereas that in a 2%Ni-Cr steel was found to be ~880 HV (Ref 40). Therefore, a carburized and hardened surface containing 25 to 50% dispersed carbides should have a hardness significantly higher than that of a case-hardened surface devoid of carbides: 65 to 67 HRC (830 to 900 HV) according to Ref 35.

Residual Stresses. The influence of large particles of carbides on the residual stress distribution of a carburized surface could be quite significant. When a carbide particle has developed at the carburizing temperature, it is in a stress-free state with respect to the surrounding material. Once cooled, there will be differences of thermal contraction and transformation behavior between the carbide and the matrix, which in particular will determine the signs and magnitudes of the short range stresses (microstresses) in each phase. The sequence of matrix transformation and the transformation products will have an influence on the long range stresses (macrostresses). If the matrix of the carbide-bearing layer transforms to pearlite or bainite, the surface residual macrostresses will likely be tensile. If the matrix of the carbide-bearing layer transforms to martensite, the macrostresses will probably be compressive; the amount of compression depends on how much austenite is retained and whether the martensite is the plate or lath type.

Kern reported that in super carburized surfaces, compressive stresses exceeding −100 ksi (−690 MPa) could be achieved (Ref 35). Note that by slowly forming large carbides, the matrix will have a eutectoid carbon content that, for the more alloyed steels, is relatively low. This favors the formation of a greater amount of lath martensite in the matrix during the quench resulting in a greater matrix toughness and compressive macrostresses.

Figure 3.26 considers a surface containing numerous carbides developing less compression than the carbide-free surface (−60 MPa versus −500 MPa). Whether or not the reduced compression was due to nonmartensitic microstructures (austenite or bainite) is not known. Table 3.3 provides another example of how different carbides have influenced the surface residual stresses (Ref 20). The influence of film carbides on residual stress distribution is very localized, as Fig. 3.27 illustrates. Again, carbides have caused a loss of compression.

Influence on Bending Fatigue. It has been reported that the "Super Carb" process, when done correctly, produces parts with bending fatigue strengths up to 15% better than those achieved with regular carburized parts (Ref 35). Incorrectly done, the bending fatigue strength could be reduced by 25 to 30% (Ref 43). The difference in fatigue results between properly and improperly processed parts can be related to the matrix microstructure and, therefore, the residual

Table 3.8 Variables affecting the surface carbide film formation of a high-speed steel oil quenched under vacuum

Variable range	Quenching temperature, °C	Quenchant temperature, °C	Layer thickness, μm
Austenitizing temperature, 1100–1210 °C	920	30	35
Holding time, 30–120 min	920	30	35
Quenching temperature, 880–1060 °C	880–1060	20	0–60
Quenchant temperature, 10–30 °C	890	10–30	0–18
Section size, 18–50 mm diam	900	20	11–18
Atmosphere pressure on oil surface, 15–95,000 Pa	920	30	0–40

Source: Ref 39

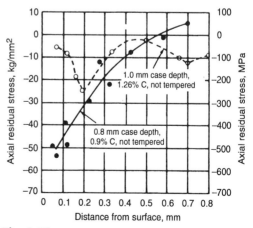

Fig. 3.26 The loss of surface compressive residual stresses due to the presence of a highly developed carbide zone in 20KhNV4MF steel. Source: Ref 41

stresses. The fatigue tests of Wang et al. indicated that for low-cycle fatigue life, a surface containing only a few carbides was superior to one containing large quantities of spheroidal carbides (Table 3.3) (Ref 20); more significantly, fatigue life increased as the compressive residual stresses increased.

No information has been found to indicate the influence of film carbides on fatigue resistance; however, film carbides have depth and penetration characteristics similar to those of internal oxidation (and may even be mistaken for internal oxidation) and can similarly induce the local formation of HTTP. Consequently, in terms of its effect on properties, film carbides should be considered as internal oxidation.

Influence on Contact Fatigue. The contact fatigue lives for four different case-hardened surface conditions are presented in Fig. 3.20. The figure reports the result of slide-roll tests on carburized and tempered 2%Cr-Mn steel. It suggests that coarse carbides can have an adverse effect on contact fatigue and that the quantity of coarse carbides may also be significant (the coarseness of the martensite could also have had an influence). Using a four-ball test, Wang et al. considered three surface conditions: one with few carbides, one with network carbides, and one with large quantities of spheroidal carbides (Ref 20). Figure 3.19 shows that for the test conditions used, the surface containing many spheroidal carbides (deliberately produced for wear resistance) gave the best result. Przylecka found that the optimum contact-fatigue resistance for a carburized and hardened 1.3 to 1.6% Cr bearing steel was achieved when the carbon in solution in the matrix was 0.9 to 0.97%, the retained austenite was 20%, and the carbide content was about 19% (Ref 44). Using SAE 8620 steel, Sheehan and Howes observed that with a surface carbon content of 1.22% (containing some globular carbides), the contact fatigue strength was very similar to that of a steel with 0.91% surface carbon (no free carbides) (Ref 45). Both steels, however, were significantly superior to one with only 0.72% surface carbon. Again, a fairly high matrix carbon content is important for contact-fatigue resistance. (In Ref 13, Vinokur favored a 0.9 to 0.95% matrix carbon content, but no carbides.)

Influence on Toughness. A case-hardened surface of martensite with, for example, 25% massive carbides can be compared with a white iron having the same microstructural features (Ref 31). In white iron there is essentially no change of fracture toughness when the carbide content varies between about 9 and 30% (Ref 46). Above a 30% carbide content, the fracture toughness progressively declines—though never reaches undesired levels—as the carbide content increases. The length of the fracture path between carbides will have a bearing on the toughness, as will the carbon content of the martensite.

With static bending and notched impact testing, carbides at the notch surface can lower the energy for crack initiation, depending on the type of carbide. However, the case matrix structure and the core properties determine the energy

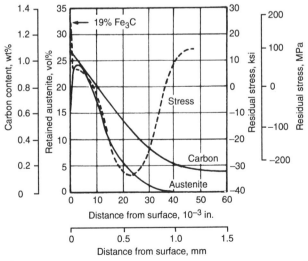

Fig. 3.27 Surface tensile stresses in the outer layer of a carburized SAE 1018 steel caused by the presence of carbides. Source: Ref 42

requirement for crack propagation. The effect of the type of carbide (at the surface) on toughness is indicated in Table 3.3.

Influence on Wear. Carbides are hard and intermetallic. Therefore, sufficient quantities would be expected to provide abrasive wear resistance (due to their hardness) and adhesive wear resistance (due to their intermetallic nature) to surfaces containing them. Generally, the more carbide a surface contains, the more it improves wear resistance. This appears to be the case when globular carbides reside in a nonmartensitic matrix. When the matrix structure is composed of martensite and austenite in varying amounts, the wear resistance does not depend greatly on the amount of carbide present (Ref 47). Przylecka stated that the microstructures that provide the best contact-fatigue resistance also give the best frictional wear resistance, though both contact-fatigue resistance and wear processes are influenced by the quality and condition of any lubrication within the system (Ref 44).

Summary

Fine, dispersed carbide particles are not regarded as detrimental and may even be unavoidable in reheat-quenched carburized alloy steels. Massive carbides at grain boundaries in the outer 0.05 mm (0.002 in.) at corners and edges could be detrimental. Network carbides can be troublesome during posthardening manufacturing processes and also can be detrimental to properties.

- *Preprocess considerations:* Edge rounding deters carbon buildup and massive carbides. A fine grain size might reduce the amount of carbon deposited at the grain boundaries. Lean-alloy grades with high manganese contents might be more prone to developing network carbides.
- *In-process considerations:* Avoid excessively high carbon potentials. Slow cooling from carburizing is likely to promote network carbides. Reheat temperature and time effect how much free carbide is dissolved.
- *Postprocess considerations:* To correct, consider annealing and requenching. A high reheat temperature is feasible but might create other problems, for example, grain growth, retained austenite, and distortion.
- *Effect on properties:* Network carbides are detrimental to bending fatigue. They may be beneficial to contact loading at very high contact pressures, though probably not so at about the fatigue limit. The continuity of the network, the carbide thickness, and depth of penetration likely have a bearing on the strength reduction. Carbides enhance wear resistance.
- *Standards:* ANSI/AGMA 2001 refers to their standard 6033. It accepts semicontinuous networks for grade 1, light discontinuous carbides for grade 3, and somewhere in between for grade 2. See reference photomicrographs in the relevant standard. ISO standard 6336 has no specification for grade ML. For the MQ grade, discontinuous carbides are permitted, provided the length of any carbide does not exceed 0.02 mm (0.0008 in.). Grade ME permits dispersed carbides only.

REFERENCES

1. M. Ichinose, F. Togashi, K. Ishida, T. Nishizawa, Morphological Stability of γ/α Interface Formed by Carburisation in Fe-C-X Alloys, *Metall. Mater. Trans. A,* Vol 25, March 1994, p 531–537
2. K. Kuo, Carbides in Chromium, Molybdenum at Tungsten Steels, *J. Iron Steel Inst.,* April 1953, p 363–375
3. J.T. Sponzilli, G.H. Walter, and D.H. Breen, Complex Carbide Network Lowers Carburised Hardenability, *Met. Eng. Q.,* Nov 1976, p 57–61
4. H.S. Ming, T. Takayama, and T. Nishizawa, Carbide Dispersion Carburizing of a 12% Chromium Steel, *J. Jpn. Inst. Met.,* Vol 45 (No. 11), Nov 1981, p 1195–1201
5. T. Ando and G. Krauss, Effect of Phosphorus Content on Grain Boundary Cementite Formation in AISI 52100 Steel, *Metall. Trans. A,* Vol 12, July 1981, p 1282–1290
6. F.S. Shen and G. Krauss, The Effect of Phosphorus Content and Proeutectoid Carbide Distribution on the Fracture Behaviour of 52100 Steel, *J. Heat Treat.,* Vol 12 (No. 3), June 1982, p 238–249
7. D.E. Diesburg and G.T. Eldis, "Fracture Toughness and Fracture Behaviour of Carburising Steels," 2-193-87/88, Climax Molybdenum Co., July 1975
8. V.M. Pevervezev and V.I. Kolmykov, The Influence of Alloying Elements on Carbide Formation in Iron and Steel During Carburising, *Met. Sci. Heat Treat. (USSR)* (No. 8), Aug 1982, p 11–14

9. E.M. Taleff, K. S. Chol, D.R. Lesuer, and O.D. Sherby, Pearlite in Ultrahigh Carbon Steels: Heat Treatment and Mechanical Properties, *Metall. Mater. Trans. A,* Vol 27, Jan 1966, p 111–118
10. I.S. Kozlovskii, Steels for Carburising and Carbonitriding, *Met. Sci. Heat Treat. (USSR),* Vol 14 (No. 4), April 1972, p 287–289
11. E.C. Rollason (quoting E.C. Bain), "Fundamental Aspects of Molybdenum on Transformation of Steels," Climax Molybdenum Co., London
12. J. Woolman and R.A. Mottram, "The Mechanical and Physical Properties of the British Standard En Steels," Parts II and III, BS 970–1955, BISRA, Pergamon Press, 1966
13. B. Vinokur, The Composition of the Solid Solution, Structure and Contact Fatigue of Case Hardened Layer, *Metall. Trans. A.,* Vol 24, May 1993, p 1163–1168
14. K.D. Jones and G. Krauss, Effects of High-Carbon Specimen Corners on Microstructure and Fatigue of Partial Pressure Carburised Steels, *Heat Treatment 1979 Conf. Proc.,* Metals Society, 1980, p 188–193
15. A.G. Haynes, contribution to discussion of paper by G. Mayer, *Heat Treatment Conference 1960,* (Harrogate, UK), BISRA, 1960, p 32
16. A. Rose and G.P. Hougardy, "Transformation Characteristics and Hardenability in Steels," Climax Molybdenum Co. of Michigan, 1967, p 155–167
17. I. Jenkins, *Controlled Atmospheres for the Heat Treatment of Metals,* Chapman and Hall Ltd., London, 1951
18. B.L. Averbach, Fracture of Bearing Steel, *Met. Prog.,* Vol 118 (No. 7), Dec 1980, p 19–24
19. R. Ruxanda and E. Florian, Microscopic Observations of the Carbide/Matrix Interphasic Interface in a Low-Alloyed Hypercarburised Steel, *Proc. 2nd International Conf. on Carburizing and Nitriding with Atmospheres,* Dec 1995 (Cleveland), ASM International, p 117–121
20. J. Wang, Z. Qin, and J. Zhou, Formation and Properties of Carburised Case with Spheroidal Carbides, *5th International Congress on Heat Treatment of Materials* (Budapest), International Federation for the Heat Treatment of Materials (Scientific Society of Mechanical Engineers) 1986, p 1212–1219
21. H. Stuart and N. Ridley, Thermal Expansion of Some Carbides and Tessellated Stresses in Steels, *J. Iron Steel Inst.,* Dec 1970, p 1087–1092
22. K.Z. Shepelyakovskii, V.D. Kal'der, and V.F. Nikonov, Technology of Heat Treating Steel with Induction Heating, *Met. Sci. Heat Treat. (USSR)* (No. 11), Nov 1970, p 902–908
23. G.H. Robinson, The Effect of Surface Condition on the Fatigue Resistance of Hardened Steel, *Fatigue Durability of Carburised Steel,* American Society for Metals, 1957, p 11–46
24. K. Evanson, G. Krauss, D. Medlin, and M.J. Patel, Bending Fatigue Behaviour of Vacuum Carburised AISI 8620 Steel, *Proc. 2nd Inernational. Conf. on Carburizing and Nitriding with Atmospheres,* Dec 1995, (Cleveland), ASM International, p 61–69
25. C. Gu, B. Lou, X. Jing, and F. Shen, Mechanical Properties of Carburised Cr-Ni-Mo Steels with Added Nitrogen, *J. Heat Treat.,* Vol 7 (No. 2), 1989, p 87–94
26. C. Razim, "Effect of Residual Austenite and Recticular Carbides on the Tendency to Pitting of Case Hardened Steel," thesis, Techn. Hochschule Stuttgart, 1967
27. A.L. Geller and L.G. Lozhushnik, Contact Fatigue Limit of Carburised 25Kh2GNTA Steel, *Met. Sci. Heat Treat. (USSR)* (No. 6), Plenum Publishing Corp., June 1968, p 474
28. J. Bush, W.C. Grube, and G.H. Robinson, Microstructural and Residual Stress Changes in Hardened Steel Due to Rolling Contact, *Trans. ASM,* Vol 54, 1961, p 390–412
29. J.A. Martin, S.F. Borgese, and A.D. Eberhart, Microstructural Alterations of Rolling Bearing Steel Under Cyclic Stressing, *J. Basic Eng. (Trans. ASME),* Sept 1966, p 555–567
30. K. Nakazawa and G. Krauss, Microstructure and Fracture of 52100 Steel, *Metall. Trans. A,* Vol 9, 1978, p 681–689
31. V.K. Sharma, G.H. Walter, and D.H. Breen, The Effect of Alloying Elements on Case Toughness of Automotive Gear Steels, *Heat Treating 1987 Conf. Proc.,* p 291–301
32. C. Dawes and R.J. Cooksey, "Surface Treatment of Engineering Components, Heat Treatment of Metals," Special Report 95, Iron and Steel Institute, 1966, p 77–92
33. S. Breitbart, Maximum Carbon in Carburised Cases, *Met. Prog.,* Vol 47 (No. 6), June 1945, p 1121–1127
34. J. Wünning, *Gasaufkohlungsverfahren Zeitschrift fur Wirtschaftliche Fertigung,* Vol 64 (No. 9), Sept 1969, p 456–464
35. R.F. Kern, Supercarburising, *Heat Treat.,* Oct 1986, p 36–38

36. T. Naito and Y. Kibayashi, Carburising of Ferrous Alloys, *Kabushiki, Kaisha Komatsu Scisakusho Offgas,* 13 May 1980
37. E. Chen, Study of the Carburising and Carbonitriding of Steels with Non-Uniform Austenite, *Heat Treat. Met. (China),* Vol 11, 1982, p 7–13
38. Z. Janchu, W. Jiegan, and C. Yixin, Improving the Morphology of Carbides in Carburised or Carbonitrided Layers by Means of Precarburising Method, *Trans. Met. Heat Treat. (China),* Dec 1980, Vol 1 (No. 2), p 64–71
39. H. Du and G. Zhou, The White Layer of W6Mo5Cr4V2 High Speed Steel after Being Oil Quenched in Vacuum, *5th International Congress on Heat-Treatment of Materials,* International Federation for the Heat Treatment of Materials (Scientific Society of Mechanical Engineers) 1986, (Budapest), p 949–958
40. V.P. Akhant'ev and V.I. Ivlev, The Nature of Troostite in Surface Layers of Steels After Carburising, *Met. Sci. Heat Treat. (USSR),* Vol 17 (No. 2), Feb 1975, p 180–181
41. E.L. Gyulikhandanov and V.G. Khoroshailov, Carburising of Heat Resistant Steels in a Controlled Endothermic Atmosphere, *Met. Sci. Heat Treat.(USSR),* Vol 13 (No. 8), Aug 1971, p 650–654
42. D.P. Koistinen, The Distribution of Residual Stresses in Carburised Cases and Their Origin, *Trans. ASM,* 1958, Vol 50, p 227–241
43. I.S. Kozlovskii, A. Ya. Novikova, A.T. Kalinin, and V.F. Nikonov, Increase in the Strength of Tractor Gears by Case Hardening and Nitrocementation, *Met. Sci. Heat Treat. (USSR)* (No. 5), May 1966, p 388–390
44. M. Przylecka, The Effect of Carbon on Utility Properties of Cemented Bearings, presented at *5th International Congress on Heat Treatment of Materials* (Budapest), International Federation for the Heat Treatment of Materials (Scientific Society of Mechanical Engineers) 1986, p 1268–1275
45. J.P. Sheenan and M.A.H. Howes, "The Effect of Case Carbon Content and Heat Treatment on the Pitting Fatigue of 8620 Steel," Paper 720268, Society of Automotive Engineers, 1972
46. K.H. Gahrzum and W.G. Scholz, Fracture Toughness of White Cast Irons, *J. Met.,* Oct 1980, p 38
47. V.S. Sagaradze, Effect of Heat Treatment on the Properties of High Carbon Alloyed Steels, *Met. Sci. Heat Treat. (USSR)* (No. 12), Dec 1964, p 720–724

SELECTED REFERENCES

- J.Z. Chen, L. Shu, and J.G. Yong, Carbide Formation and Counter-Transformation Behaviors in the Deep-Carburization Process, *Conf: Heat & Surface '92,* 17–20 Nov 1992, (Kyoto, Japan), Japan Technical Information Service, 1992, p 349–352
- C. Dumitrescu, E. Florian, and D. Bojin, Research Studies Regarding the Influence of Carburizing Regimes on the Formation of Carbides, *Bul. Inst. Politeh. 'Gheorghe Gheorghiu-Dej', Metal.,* Vol 46/47, 1984/1985, p 105–115
- H. Sun, C. Chen, H. Wang, and Z. Liu, The Quality Control of the Carburized Layer of Large Heavy-Duty Gears, *Conf.: Heat Treating: Equipment and Processes,* 18–20 April 1994 (Schaumburg, Illinois), ASM International, 1994, p 421–427
- J.-G. Wang, Z. Feng, and J.-C. Zhou, Effect of the Morphology of Carbides in the Carburizing Case on the Wear Resistance and the Contact Fatigue of Gears (Retroactive Coverage), *J. Mater. Eng. (China),* Vol 6, Dec 1989, p 35–40

Chapter 4

Retained Austenite

In steels, austenite is stable at temperatures above the Ac_3 and Ac_{cm} phase boundaries (see Fig. 3.2). On cooling from such temperatures, it becomes unstable and decomposes to some new constituent depending on the chemical composition of the steel and the rate of cooling. When the transformation involves diffusion processes (i.e., processes to form ferrite, pearlite, or bainite), the reaction is essentially complete and no austenite survives. These resulting products are referred to as high-temperature transformation products because they form at relatively high temperatures. For example, for low-carbon steels, these transformations take place at temperatures between the Ac_3 and about 400 °C. Martensite, on the other hand, is a low-temperature transformation product. For a typical low-carbon lean-alloy steel, the martensite transformation range is from 450 to 200 °C, whereas a high-carbon material typical of a case-hardened surface has a martensite transformation range from 200 °C down to about –100 °C. If, on quenching, part of the martensite transformation range lies below the temperature of the quenchant, the transformation of austenite to martensite will remain incomplete; austenite will be retained in the final microstructure. Figure 4.1 represents a high-carbon surface where a part of the martensite transformation range (M_s to M_f) lies below 20 °C. Therefore, the presence of retained austenite is to be expected and indeed is shown in Fig. 4.1(b). However, small amounts of retained austenite have been detected in quenched steels even when their M_f temperatures are above ambient (Fig. 4.2). Such austenite tends to reside at interlath boundaries rather than as volumes typical of a higher-carbon plate martensite.

Austenite Formation

Austenite Stabilization. Retained austenite can become stable. If a part with a high-carbon surface layer is quenched into a refrigerant (i.e., it is cooled straight through the M_s to M_f range), virtually all the austenite will be transformed to martensite. Conversely, if a part is quenched to about room temperature, held there for some time, and then refrigerated to below the Mf, some of the austenite will transform isothermally to martensite and some will survive. This surviving austenite is referred to as thermally stabilized austenite, and it requires a fair amount of energy to destabilize it.

Thermal stabilization involves a strain aging process (Ref 3), where the strain is provided by the accommodation of the martensite and the presence of any tensile residual stresses. Stabilization requires the presence of interstitial atoms (e.g., carbon) and sufficient time for these atoms to segregate to dislocations or to the martensite embryo/austenite interfaces, thereby pinning them (Ref 4). Once the segregation and dislocation pinning have taken place, the austenite is stiffened somewhat, and the growth of martensite is inhibited. In high-carbon austenite typical of a carburized surface, interstitial atom segregation takes place rapidly. Whereas stabilization becomes more complete due to an isothermal hold, a slow cool through the M_s may, nevertheless, be adequate for some stabilization to occur. In Fig. 4.1, the most rapid cool produced an austenite content of 50%, whereas a slower cool produced 60%.

The permanence of stabilization depends on the extent to which the strain aging process has

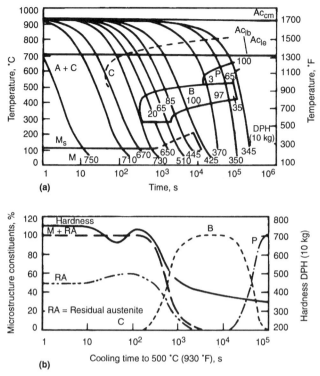

Fig. 4.1 CCT diagram with related hardness and percentage of microstructural constituents for a 20Ni-Mo-6Cr steel carburized to 1.14% C. Austenitized at 930 °C for 30 min. Source: Ref 1

been allowed to proceed and on what subsequent thermal (heating or cooling) or mechanical actions are brought to bear. The clusters of interstitial atoms can either be dispersed or encouraged to precipitate (overaging).

Mechanical working can cause some destabilization by inducing further transformation to martensite, and tempering at temperatures above about 150 °C can also destabilize austenite by transforming it to bainite. Room temperature aging and tempering below about 150 °C favors a more complete degree of stabilization. Refrigeration, while transforming some austenite to martensite, also makes any remaining austenite more ready to transform into bainite during low-temperature tempering.

The Relationship between M_s and Retained Austenite. A consequence of developing a carbon gradient in the surface of a steel component is that the M_s falls as the carbon content increases. An indication of the efficiency of carbon to modify the M_s can be obtained from several sources, but here the Steven and Haynes formula (Ref 5), which determines the M_s from the chemical composition of a steel, will be used:

$$M_s \ (°C) = 561 - 474C - 33Mn - 17Ni - 17Cr - 21Mo$$

This formula is reasonably accurate for steels containing up to 0.5% C. In higher carbon contents, the efficiency of carbon to lower the M_s is reduced, and the correction curves of Fig. 4.3 are required to obtain a truer M_s value. If this complication is ignored for the moment, the elements in the formula do lower the M_s according to their quantity in the

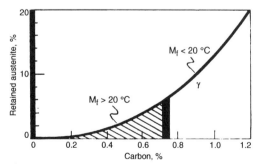

Fig. 4.2 Influence of carbon content on retained austenite content. Source: Ref 2

steel and the respective factor. For example, the factor for carbon is 474; therefore, a 0.1% C increase lowers the M_s by 47 °C (progressively less as the carbon level increases above 0.5%). For alloying elements to have the same effect requires an increase of 1.5% Mn, nearly 3% Ni or Cr, or slightly over 2% Mo. Therefore, carbon up to about 1% is much more influential for lowering the M_s through the case than other alloying elements typical of the carburizing grades of steel. Figure 4.4 shows the effect of carbon for a carburized 3%Ni-Cr steel where two quenching temperatures are considered.

The M_f temperature, at which the martensite reaction ceases, lies approximately 215 °C below the M_s (Ref 5). However, the amount of transformation between M_s and M_f is not linear, and about 90% of the austenite to martensite transformation takes place during the first 110 °C below the M_s. Nevertheless, at high carbon levels and assuming that the difference between M_s and M_f remains essentially constant, incomplete transformation results when some part of the transformation range lies below the temperature of the quenchant. The amount of transformation or, alternatively, the volume of untransformed austenite (V_γ) is therefore related to both the M_s and the quenchant temperature (T_q). This relationship, defined by Koistinen and Marburger (Ref 6), is shown in Fig. 4.5:

$$V_\gamma = e^{-1.10 \times 10^{-2}(M_s - T_q)}$$

Thus, methods are available whereby the M_s can be determined from chemical composition and the quantity of untransformed austenite may be approximated from the M_s. Together they provide the ability to estimate the retained austenite content from chemical composition. This estimation permits the assessment, in terms of retained austenite, of actual and hypothetical steels when subjected to different quenching situations. The approach has been tested against experimentally obtained and published data, which show that 90% of the calculated points fall within ±6% of the measured austenite values and that 80% fall within ±5% (Fig. 4.6). Figure 4.7 further demonstrates how much the retained austenite might vary within one specified composition range and just how significant is the quenching temperature.

An appraisal of the several empirical formulas for determining M_s showed the Steven and Haynes formula to be the most accurate (Ref 7).

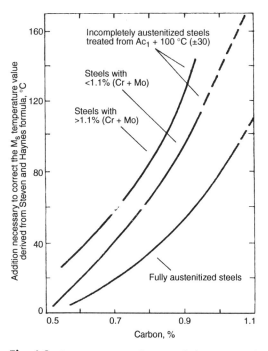

Fig. 4.3 Correction curves for use with the Steven and Haynes formula with author's modifications to original curves shown by dashed lines. When carbon content is less than 0.9%, an 830 °C soak of over 2 h should produce a fully austenitic structure. Source: Ref 5

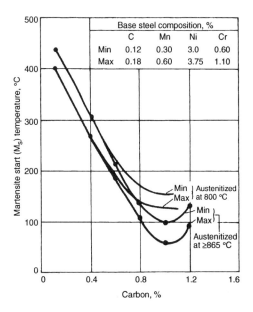

Fig. 4.4 Effect of carbon on the M_s temperature. Calculated for the upper and lower extremes of the composional specification for a carburizing steel at two quenching temperatures

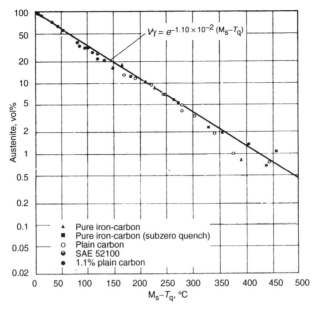

Fig. 4.5 Austenite content as a function of the difference between M_s and the quenchant temperature. Source: Ref 6

Austenite Layering. In case-hardened surfaces containing retained austenite, the quantity of retained austenite is often lower at the surface than at some greater depth beneath the surface. Figure 4.8 shows that the calculated austenite contents agree with those measured away from the surface. The lower surface austenite content is generally attributed to either changes in the surface chemistry, caused by decarburization or internal oxidation (Ref 8), or to a reduced matrix carbon content brought about by the precipitation of carbides (Ref 9). However, even when there is apparently neither a reduction of surface carbon nor of free carbides at or near the surface, the surface retained austenite content may still be lower than at some small distance below the surface. Plastic deformation of the surface during

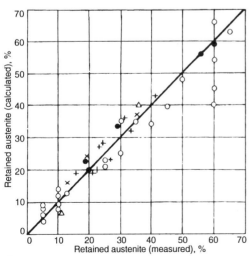

Fig. 4.6 Comparison of calculated and measured retained austenite contents for lean-alloy steels (mainly case-hardening grades).

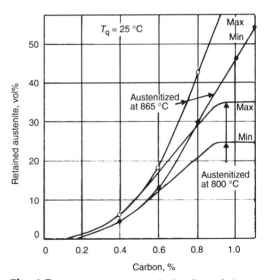

Fig. 4.7 Retained austenite (calculated) in relation to carbon content for a Ni-Cr carburizing steel. See Fig. 4.4 for composition.

prior machining may be influential, although a more likely explanation relates to macrostraining during the transformation stage and/or, perhaps, a degree of austenite stabilization.

Macrostraining can occur during cooling from the austenitic condition. During the quench of a carburized part, transformation takes place in the low-carbon core material while the high-carbon surface is still austenitic. Therefore, at temperatures approximating M_s for the surface material, tensile residual stresses develop in the still austenitic surface; the greatest amount of tension is at the surface. At about the M_s of the surface, plastic deformation can take place at stresses well below the yield stress of the core material, and the martensite reaction can be stimulated. For example, Ankara and West (Ref 10), using a homogeneous 4%Ni-Cr steel, showed that, with free cooling between the M_s and ($M_s - 2$ °C), 7% martensite transformed from the austenite. When stresses of 70, 140, and 1035 MPa (10, 20, and 150 ksi) were applied, the amount of martensite produced increased to 10, 20, and 40%, respectively. Therefore, tensile residual stresses developed at the surface of a case-hardened part near the M_s could lead to more austenite being transformed than a little deeper in the case where the tensile stresses are lower. At still lower temperatures below the Ms, after a fair amount of martensite transformation occurs and the surface residual stresses have reversed and become compressive, further martensite transformation is inhibited. The remaining austenite is protected to some extent by the presence of compressive residual stresses.

The potential contribution of thermal stabilization to austenite layering is implied in Fig. 4.1. The steel has a fixed carbon content and, therefore, a fixed M_s to M_f range, and yet the amount of retained austenite varies (e.g., 50 to 60%). The major difference between samples producing 50 and 60% is the rate at which they cooled; a slower rate, in this instance, gives the higher retained austenite value.

This layering aspect of austenite is discussed because it occurs in case-hardened surfaces, though perhaps not in every instance; and it could have a bearing on both crack initiation and crack propagation under load.

Austenite in the Microstructure

In microscopic examination of as-quenched carburized surfaces, retained austenite is a white etching constituent, as are any free carbides, and even martensite is a light etching. Consequently, differentiation of these structural features can be difficult without special etchants. Fortunately, tempering, which is applied to most case-hardened parts, causes fine carbides to precipitate within the martensite enabling it to etch more rapidly. Thereby, a much greater contrast between the austenite, which remains white etching, and the martensite is produced (Fig. 4.9). Usually the dark etching martensite makes it easier to see any carbides, especially the network carbides.

The white etching austenite volumes are angular; their shape is determined by the plates of martensite that subdivide each austenite grain. The size, or coarseness, of austenite volume relates to the prior austenite grain size and the amount of austenite in the structure, which in turn are mainly determined by the carbon content, the alloying element content, and the quenching temperature.

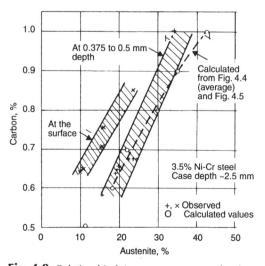

Fig. 4.8 Relationship between percentages of carbon and austenite for carburized components where measurements of each were made at the surface and at a depth of 0.375 to 0.5 mm (0.015 to 0.020 in.) from the surface. Calculated values are also shown.

Effect on Material Properties

Influence on Hardness. Retained austenite is relatively soft, although it is saturated with carbon. Its coexistence with hard martensite reduces

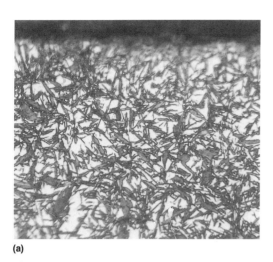
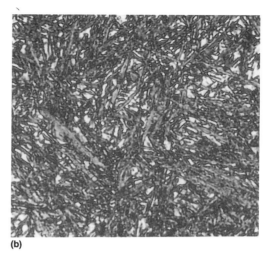

Fig. 4.9 Retained austenite (white) and martensite in the surfaces of carburized and hardened Ni-Cr steel test pieces. (a) ~40% austenite. (b) ~15% austenite. Both 550 ×

the overall macrohardness of a structure to below that of a structure containing only martensite and related to the proportions of the two constituents (Fig. 4.10). Figure 4.10 shows carburized, hardened, and tempered production samples, made from nickel-chromium steels with surface carbon contents of about 0.8%. With a wider coverage of steels and conditions, however, it would not be unusual to obtain hardness values above the upper limit of the band as shown.

Austenite is retained in small amounts at quite low carbon levels (see Fig. 4.2), and as the carbon content increases so too does the retained austenite content, everything else being equal. By relating the hardness to carbon content (Fig. 4.11a), the approximate carbon level needed for austenite to influence the hardness of an as-quenched steel can be determined. The first deviation from an essentially straight line (at ~0.35 to 0.4% C) indicates where retained austenite begins to affect hardness. For direct quenching, the maximum hardness (800 to 880 HV) of the martensite/austenite mix is attained at between 0.6 and 0.75% C, depending on the steel grade. However, the potential hardness is higher still, though probably at around 0.75% C. At yet higher carbon levels, austenite has a marked effect on hardness, especially with direct quenching, as is shown by the sharp decline of hardness (Fig. 4.11).

Influence on Tensile Properties. The room temperature tensile strength and the yield strength in tension decrease as the retained austenite content increases (Table 4.1, Fig. 4.12). However, high levels of strain can induce some austenite to martensite transformation, and the amount of austenite reduction due to 1% strain was observed to be ~7.5 to 10% (Ref 14). Strain induced transformation raises the mean compressive stress, and the martensite produced is more ductile than that from thermally induced transformation (Ref 13). Conversely, Yen et al. (Ref 15) claim that strain induced martensites increase brittleness, and Franklin et al. (Ref 16) provide a reminder that it is untempered, highly strained, and potentially harmful.

Influence on Residual Stresses. The failure of any austenite to transform during quenching means that the volume expansion that should

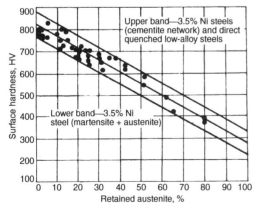

Fig. 4.10 Influence of retained austenite on the surface hardness of carburized alloy steels. Reheat quenched and tempered at 150 to 185 °C

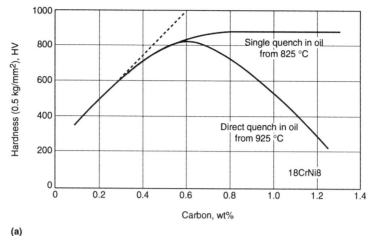

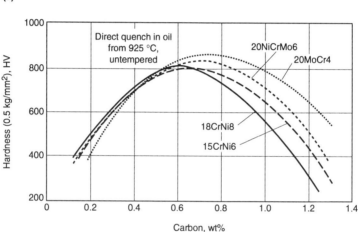

Fig. 4.11 Hardness/carbon relationship for untempered martensite in four case-hardened steels. Source: Ref 11

have accompanied the austenite to martensite reaction did not take place. Therefore, in carburized and hardened surfaces, the development of residual stresses is in some way related to the amount of austenite not transformed. Koistinen (Ref 17) made this point and states that both the distribution and magnitude of the residual stresses are determined by the extent and sequence of the martensite transformation. Maximum compression, therefore, occurs at some distance from the surface where the proportion of martensite to austenite is very high, but lower values of residual compression occur at the surface when the proportion of martensite to aus-

Table 4.1 Tensile test results on through carburized, hardened, and 180 °C tempered 4320 steel

		Offset yield strength, MPa (ksi)				
Condition	Retained austenite, %	0.2%	0.02%	0.01%	0.001%	Ultimate tensile strength, MPa (ksi)
4320 (core)	...	796 (115.5)	523 (75.8)	473 (68.6)	366 (53.0)	...
Carburized	32.6	604 (87.6)	522 (75.6)	456 (66.2)	435 (63.0)	1304 (189.1)
Liquid nitrogen quench	15.0	(a)	1293 (187.5)	1085 (157.4)	577 (83.7)	1503 (218.0)

Carbon content, 1.1%. (a) Strain, <0.002. Source: Ref 12

tenite is low (Fig. 4.13). In Fig. 4.13(b), the peak compression coincides with a carbon content of ~0.55%. However, Salonen (Ref 18) determined a value of 0.6% C at peak compression, whereas others (Ref 19) agreed with Koistinen.

The magnitude of the residual stresses and the residual stress distribution are influenced by chemical composition (including carbon content) and quenching method (including cooling rate). In Fig. 4.14, two steels are compared. One steel has a martensite/austenite outer case, and the other steel also contains some bainite.

Influence on Fatigue Resistance. Fatigue crack initiation and early propagation at and in a case-hardened surface are strongly influenced by the inherent strength of the material and the prevailing residual stresses. High values of compressive residual stresses are favored to negate applied tensile stresses. The presence of retained austenite, however, reduces both the strength (as implied by hardness) and the compressive residual stresses. Therefore, it would be expected to lower the fatigue resistance to a degree dependent on the relative proportions of martensite and retained austenite. This tendency is established by Wiegand and Tolasch (Ref 21), who state that the bending fatigue limit of unnotched case-hardened test pieces decreases as hardness falls below about 680 HV. With notched samples, the fatigue limit progressively falls as the hardness decreases (Fig. 4.15). In terms of actual austenite contents, Razim (Ref 22) obtained a 25% reduction of bending fatigue strength from specimens with a surface austenite content of 80%, and Brugger (Ref 23) observed that, when the austenite contents of a case-hardened 15Cr-16Ni steel were 40 and 20%, the bending fatigue limits were respectively 16 and 4% below the essentially austenite-free condition. Pacheco

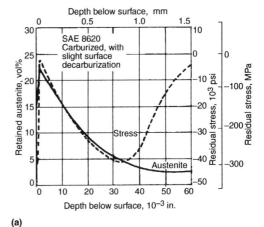

(a)

(b)

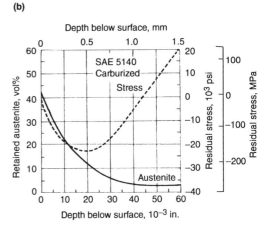

(c)

Fig. 4.13 Retained austenite and residual stress distributions in case-hardened test pieces. Source: Ref 17

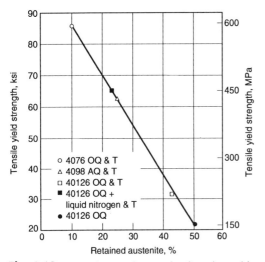

Fig. 4.12 Dependence of stress for first detectable plastic strain (~0.0001) on retained austenite content. OQ, oil quenched; T, tempered; AQ, air quenched. Source: Ref 13

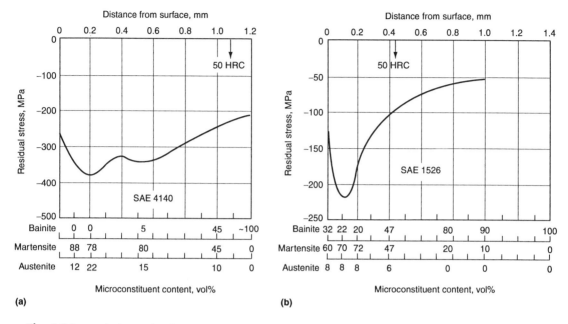

Fig. 4.14 Residual stress distributions in two oil-hardened carburized gears. Source: Ref 20

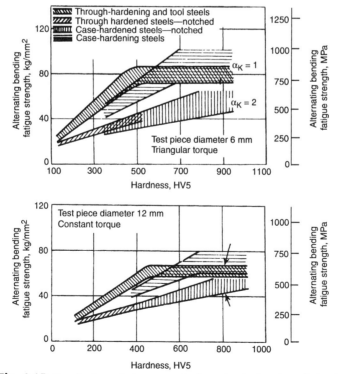

Fig. 4.15 Rotating beam fatigue strength of case-hardening, through-hardening, and tool steels as a function of surface hardness. Source: Ref 21

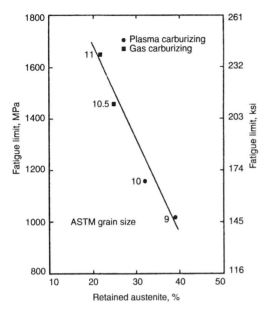

Fig. 4.16 Fatigue limits of plasma and gas-carburized specimens as a function of retained austenite content.

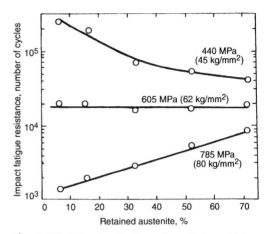

Fig. 4.17 Effect of retained austenite on the impact-fatigue resistance of a 1.45C-11.5Cr steel. Source: Ref 26

and Krauss (Ref 24) confirmed that not only should the austenite be minimal but it should be fine and evenly distributed (Fig. 4.16). It is, therefore, generally accepted that retained austenite is detrimental to the bending fatigue limit (low-stress, high-cycle fatigue) and probably also to the torsional fatigue limit.

At high applied stress levels (high in the finite life part of the S-N curve), on the other hand, it is possible that retained austenite is beneficial by slowing down the crack growth rate. The rate of crack growth slows when the strain ahead of a crack tip (propagating in an austenite containing surface) induces the austenite to martensite reaction, thereby raising the strength and increasing the compressive residual stresses (Ref 25).

At relatively low stresses close to the fatigue limit, there is insufficient strain to induce the austenite to martensite transformation. It is only high applied cyclic stresses that bring about that transformation, which equates to fatigue lives of less than about 1×10^4 load cycles. These points appear to be confirmed by Kozyrev and Toporov (Ref 26), who isolated the effect of austenite on the impact fatigue resistance using a high-carbon 12% Cr alloy steel. The quenching temperature was varied to produce different austenite contents without influencing the grain size. They established that at high levels of applied stress the impact fatigue resistance increases with increasing austenite content. At intermediate stress levels, the austenite content has no effect. At low values of stress, an increased austenite lowers the fatigue resistance (Fig. 4.17).

The results of Panhans and Fournelle (Ref 27) do not altogether agree with the findings of Kozyrev and Toporov. Whereas Panhans and Fournelle (Ref 27) did find austenite to be beneficial at less than 1×10^4 cycles, they also found it to be beneficial at more than 1×10^6 cycles and marginally inferior between the two (Fig. 4.18). This contradiction is difficult to explain on the basis of either plastic deformation of the austenite or the formation of strain-induced martensite. Brugger (Ref 23) did not find austen-

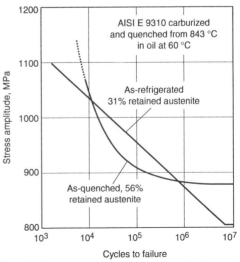

Fig. 4.18 S-N curves for case-hardened 9310 steel (untempered). Source: Ref 27

ite to be beneficial at any stress level, nor did Pacheco and Krauss (Ref 24).

To confuse the issue a little more, Szpunar and Bielanik (Ref 8), studying the crack propagation characteristics of case-hardened steels, report that the two steels studied had different crack propagation behaviors. Steel 20HNMh (SAE 8620) exhibited a maximum crack propagation rate when the retained austenite content was 23%, after which the crack propagation rate declined (Fig. 4.19a). Figure 4.19(a) indicates that over 40% retained austenite appeared to be effective in dulling the crack or even preventing its growth. The 18HGT (Mn-1Cr-Ti) steel, on the other hand, showed little change of the crack propagation rate for retained austenite contents up to about 30% (Fig. 4.19b). Above about 30%, the behavior depended on which of the three load amplitudes was used. At a higher load amplitude, the crack propagation rate increased with increasing austenite content. Whereas at a lower load amplitude, the crack growth rate decreased as the austenite increased with no growth when the austenite content was about 80%. Therefore, the energy absorbed to induce the austenite to martensite reaction could not be available for crack propagation.

Influence on Contact Fatigue. In bending fatigue situations, any benefits of retained austenite appear because of the austenite-martensite reaction. However, in rolling contact situations, the cold working property of the material is more likely to be important, that is, its ability to plastically deform under rolling contact pressures. In rolling contact disc tests, normally the more highly loaded surfaces have longer lives than those tested at intermediate load levels. This result means that the plotted test data have a "C" shape rather than the more familiar S-N plot (Fig. 4.20). The effect has been observed in case-hardened steels with and without retained austenite and in high-strength nitrided steels with no retained austenite. Having said that, an austenite containing case-hardened surface deforms more readily than a wholly martensitic surface or a martensite/bainite surface, thereby contributing more toward improved durability at higher stress levels.

Regarding the influence of contact stressing on surface hardness, case-hardened surfaces with up to about 20% retained austenite and tempered at less than 150 °C harden by rolling contact in a range of 85 to 120 HV, whereas those surfaces tempered at 150 to 250 °C harden by only ~35 HV. Razim (Ref 28) noted that in surfaces containing ~50% austenite hardness increases, in general, from 500 to 1000 HV due to rolling, whereas surfaces containing no austenite are hardly, if at all, affected. The increased hardness in this case is primarily the result of working; both slip lines and induced carbide precipitation were observed. These data provide a rule of thumb measurement of the effect of austenite content (in case-hardened surfaces) on the hardness increases caused by fairly heavy rolling

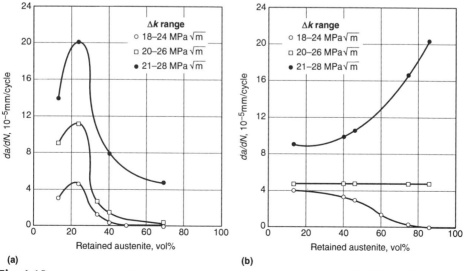

Fig. 4.19 Average propagation rate of fatigue cracks in carburized cases of (a) 20 HNMh and (b) 18 HGT steel depending on percentage of retained austenite and on load amplitude. Source: Ref 8

contact (i.e., for each 20% austenite, the hardness increases by about 130 HV).

The contact fatigue tests carried out by Razim (Ref 28) showed that the surface fatigue resistance increased as the retained austenite content increased (up to about 55%). Further, in comparison with other microstructures likely to be developed during case hardening (e.g., carbides), those containing austenite gave the most favorable results (Fig. 4.21) at all stress levels. The whole of the fatigue curve was raised as austenite increases. Balter and Turovskii (Ref 29) and Diament et al. (Ref 30) agreed that high austenite containing surfaces are superior to those containing only small amounts. Diament also showed that as the contact fatigue tests progress the retained austenite content within the case was altered and the residual stress distribution was modified.

Vinokur (Ref 31, 32) examined the effect of quenching a 0.96% C alloy steel from temperatures within the range 730 to 930 °C (in 20 °C increments) and found that as the quenching temperature increases, the amount of carbide in the final microstructure decreases, and the retained austenite content increases. The best contact fatigue resistance is achieved by quenching from the Ac_m (810 °C for the steel in test), followed by tempering at 150 °C for 2 h (Fig. 4.22). The maximum resistance to contact fatigue relates not only to the austenite content but also to the amount of martensite (65 to 70%) and the near absence of carbides in the structure (matrix carbon 0.8%). The retained austenite was 32% before testing and ~24% after testing, and the optimum structure for fatigue resistance more or less coincided with the initial peak hardness.

The surface residual stresses developed during case hardening can play a part in the contact fatigue life of a component. Barczy and Takacs (Ref 33), after quenching carburized planet pins into a hot quenchant, found a sharp tensile peak residual stress 0.2 mm beneath the surface, and this near-surface stress peak was responsible for surface spalling after a very short life. By raising the surface carbon content and the austenite content (15%), the tensile stress peak

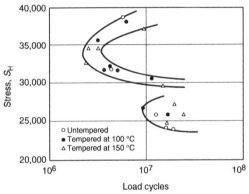

Fig. 4.20 Rolling contact fatigue plots for carburized and hardened 3Ni-Cr steel discs. S_H = (lb/in. of face width)/(relative radius of curvature)

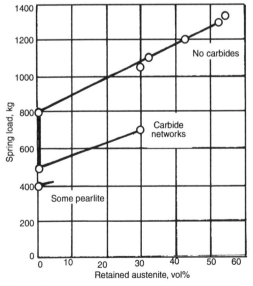

Fig. 4.21 Effect of austenite on the pitting resistance of carburized gears. Taken at 10^7 cycles, which approximates to the "knee" of the S-N curve. Source: Ref 28

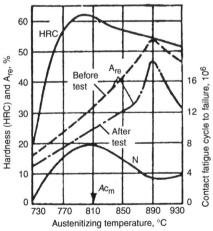

Fig. 4.22 Effect of austenitizing temperature on hardness, amount of retained austenite, and contact fatigue strength of 90KhGNMFL steel under a 3.43 GPa (350 kgf/mm²) load. Source: Ref 32

Table 4.2 Effect of quenching temperature on bend and impact strength of case-hardened 15Cr-6Ni steel

Quench temperature, °C	Retained austenite, %	Hardness, HRC	Bend strength, MPa	Impact strength, MPa	Impact fatigue strength, MPa
900	40	57	1500	2500	670
860	20	59	1420	2500	770
820	0	61	1390	2500	790
790	0	61	1250	2800	800

Source: Ref 23

shifted to 0.4 mm below the surface, and the service life improved. Further increase of the carbon content raised the austenite content to 30% and flattened the residual stress distribution to virtually zero. The outcome was that the service life was appreciably improved. In this instance, compressive residual stresses resulting from heat treatment did not help, and the adverse effect of the sharp tensile peak dominated events. Compressive residual stresses, whether caused by heat treatment or plastic deformation and strain-induced martensite, are thought to be beneficial because they can "squeeze" a crack, which increases the crack-face friction that, in turn, slows down the crack propagation rate (Ref 34).

It seems, therefore, that retained austenite is beneficial under rolling and rolling with sliding conditions, assuming the contacting surfaces remain reasonably separated throughout the working temperature range by adequate lubrication.

Influence on Bending and Impact Fracture Strength. Whereas Brugger's fatigue tests (Ref 23) found no merit in having retained austenite in the carburized case, the accompanying bend tests showed that samples with the highest austenite content produced better results (Table 4.2). This result might be a reflection on the ability of austenite to yield at high surface stresses caused by bending. In these tests, however, the different austenite contents were achieved by using different quenching temperatures. As a consequence, the results might be influenced more by the condition of the material at the core rather than by the presence of austenite at the surface. The impact strength did not seem to be affected by the presence of surface retained austenite, which is in agreement with other works (Ref 19).

Razim considered that case toughness is indicated by the initial crack strength of notched test pieces (Ref 35) and showed that as the retained austenite (and carbon content) increased, the initial crack strength decreased (Fig. 4.23). On the other hand, Thoden and Grosch (Ref 36), working with samples carburized to a surface carbon content of 0.65%, showed that as the retained austenite (and nickel content) increases, the initial crack strength also increases (Fig. 4.24). The bend and impact test results for direct quenched specimens are shown in Fig. 4.25. The results for the double quenched condition (with a finer grain size) generally produced even better values, particularly when the nickel content exceeded about 2 to 3%. Tempering was especially beneficial with respect to the bending strength.

In terms of fracture toughness tests (Ref 37), the K_{Ic} value, which decreases with increasing carbon content, tends to increase with nickel content and with retained austenite content (Fig. 4.26). Here, the critical crack size increases with the nickel content. Consequently, a steel with more than 3% Ni is regarded as having a high initial crack resistance, in keeping with Ref 36.

Influence on Wear Resistance. For straightforward abrasive wear situations, a high surface hardness is the main property requirement. This surface hardness entails developing a surface microstructure of high-carbon martensite with, perhaps, some spheroidized carbides and minimum retained austenite. The influence of retained austenite on the adhesive wear of case-hardened surfaces is complicated by the re-

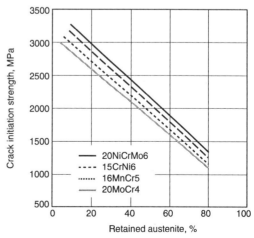

Fig. 4.23 Effect of increasing retained austenite on crack initiation strength. Source: Ref 35

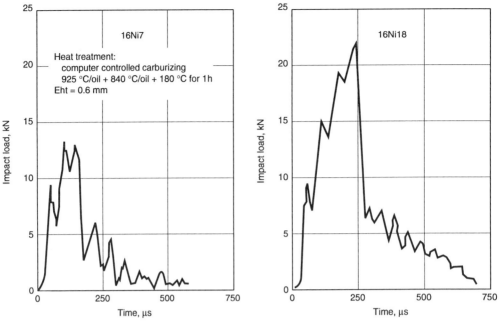

Fig. 4.24 Load-time curves for carburized DVM/DIN 50115 specimens. The initial crack occurs at maximum load. Heat treatment: computer controlled carburizing at 925 °C in oil + 840 °C in oil + 180 °C for 1 h, Eht = 0.6 mm. Source: Ref 36

spective instabilities of austenite and martensite in the microstructure.

Under contact loading, which may involve both rolling and sliding, austenite can be plastically deformed and strengthened, resulting in a greater resistance to wear. Conversely, heat generated by plastic deformation and friction encourages carbide precipitation (primarily from within the martensite) and softening, which can reduce wear resistance. Heat generated at a working surface can also impair the efficiency of the lubricant, favoring the adhesive wear process.

Surface roughness is regarded as detrimental to the efficiency of the lubricant, because surface asperities can penetrate the lubricant film and

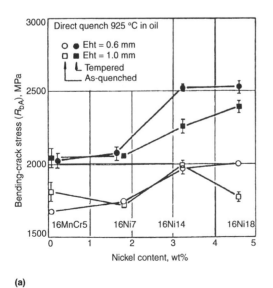

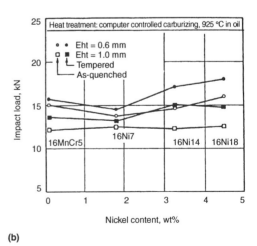

Fig. 4.25 Bending (a) and impact (b) strengths of four carburized and direct quenched steels. Source: Ref 36

Table 4.3 Effect of retained austenite on the scuffing tendency of steel

Steel	Retained austenite, %	Rating(a)	AGMA 2001-C95 welding factor, X_w
Stainless steel	100	0.32	0.45
Case-hardened nickel steel	>20	0.80	0.85 (>30% γ_r)
Through-hardened steel and case-hardened chromium steel	~20	1.00	1.00 (20–30% γ_r)
Case-hardened nickel or chromium steel	<20	1.20	1.15

(a) The higher the rating is, the greater the scuffing resistance will be. Source: Ref 38

make contact with similar asperities on the mating surface. The removal of surface asperities caused by wear or deformation (probably assisted by the presence of retained austenite), therefore, is considered to be a favorable if not crucial happening. It redistributes the load and reduces the frictional effects, provided reasonable lubrication is maintained. Further, with the surface asperities removed, the local contact pressure peaks within the mating surfaces are also removed as well as the risk of microflaking. There are instances, however, where highly polished surfaces can be difficult to lubricate, and a loss of lubricant in such circumstances could lead to adhesive wear.

In their approach to predicting the scuffing tendencies of gears, Niemann and Seitzinger (Ref 38) introduce an X_w factor to account for the potential adverse influence of austenite (Table 4.3). However, the low X_w factor given to stainless steel and attributed to 100% austenite may, in fact, be caused by the nickel content and the absence of carbide precipitates, as shown in Fig. 4.27 (Ref 39). Roberts (Ref 40) points out that both the nickel and the chromium austenitic stainless steels are notoriously difficult to lubricate. Roberts (Ref 40) goes on to assert that there is no direct relationship between scuffing resistance and retained austenite content.

Grew and Cameron (Ref 41) suggest that austenite has a lower affinity than martensite for the surface-active compounds contained in lubricating oils; therefore, it will be more difficult to lubricate. In their tests, which relate the coefficient of friction to the frictional temperature, a carburized and hardened 4%Ni-Cr-Mo steel with 5% retained austenite remained fairly stable, whereas with 25% austenite some instability occurred at 150 °C. Following a 180 °C temper, the instability of the surface originally containing 25% austenite did not occur until a 180 °C friction temperature was reached (Fig. 4.28). The observed instabilities relate to the addition of surface active compounds, and yet, coincidence or not, the observed instabilities seem to relate to the already known thermal instability of austenite. The second stage of tempering commences at about 150 °C, and for an already tempered surface, the second stage of tempering recommences at temperatures above the original tempering temperature. As a reminder, the second stage of tempering involves the transformation of retained austenite to bainite (ferrite with carbide precipitates).

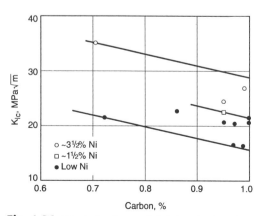

Fig. 4.26 Effect of carbon and nickel contents on fracture toughness. Source: Ref 37

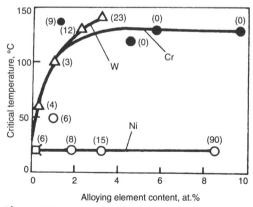

Fig. 4.27 Effect of alloy content in Fe-1%C materials on the critical temperature of a mineral oil. Data in parentheses indicate percentage retained austenite content. Source: Ref 39

Laboratory tests to determine the effect of austenite on scoring resistance of case-hardened surfaces give conflicting information. For example, Kozlovskii et al., using a low- to medium-speed roller test, concluded that a high retained austenite content increases the scoring resistance by virtue of the high capacity of austenite to work harden. On the other hand, hard low-ductility surfaces containing only small amounts of retained austenite are more likely to suffer scoring damage at lower pressures (Fig. 4.29) (Ref 42). Supporting that view to some extent, Manevskii and Sokolov produced results from case-hardened and carbonitrided test pieces suggesting that the hardness for best seizure resistance is approximately 580 to 600 HV; harder surfaces tended to score at lower pressures (Ref 43). Terauchi and Takehara stated that with surfaces of less than 500 HV, the surface hardness increases with repeated rolling, whereas with martensite- only microstructures, the hardness diminishes during testing if it changes at all (Ref 44). They concluded, however, that martensitic surfaces have the highest scoring resistance, and that the scoring resistance decreases as the austenite content increases (Fig. 4.30). It was observed that when there are small quantities of retained austenite in the microstructure, initial seizure occurs by shearing at an austenite volume and spreads from there as the load increases. When the surface has a high retained austenite content, the surface plastically deforms to spread the load; destructive seizure, when it occurs, then takes place across the whole contact area, that is, there is no initial seizure point.

Retained austenite and its contribution to the adhesive wear process could relate to how the mating surfaces are "run-in" and to the response of the material to the running-in process. For example, experience with aluminum bronzes for worm-wheel applications indicated that a material initially in a soft condition work hardened during roll/slide tests to a final hardness level without any signs of scuffing. For such a condition, the load carrying capacity was good. On the other hand, when material was made with an initial hardness equal to the final hardness of the softer alloy, scuffing readily occurred without their being much in the way of work hardening. If the same trend applies to the scuffing behavior of case-hardened sur-

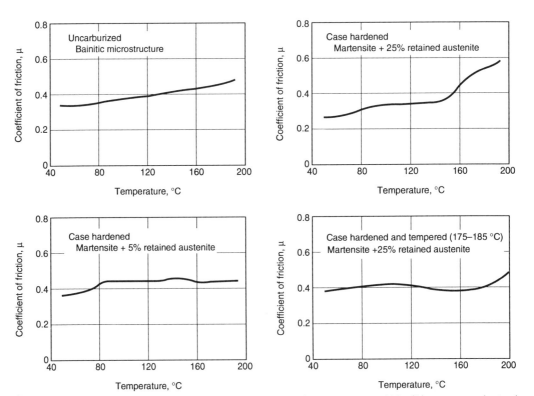

Fig. 4.28 Friction temperature curves from Bowden-Leben machine. Tungsten-carbide slider onto uncarburized, case-hardened, and case-hardened and tempered 4%Ni-Cr-Mo test pieces. Source: Ref 41

faces, then just sufficient retained austenite to assist the running-in process would seem to be in order. The question then is, what is "just sufficient"?

Control of Retained Austenite

The retained austenite content at the surface of a carburized steel is influenced by the alloy content of the steel, the surface carbon content, and the quenching temperature, all of which determine the M_s. The temperature of the quenchant also contributes to the as-quenched retained austenite content. Primary control, therefore, must be by manipulation of these variables.

Considerations Regarding Properties. Whereas certain standards quote maximum retained austenite contents, there is some latitude for the manufacturer to adjust the amount within those limits (i.e., 0 to 35% American Gear Manufacturers Association). Before deciding on the surface carbon content and the quenching temperature for a chosen steel grade, it is necessary to establish what level of retained austenite can be tolerated. Making these decisions requires some knowledge of how the component functions and how it will be loaded in service. In dealing with gears, each type of loading, or action, must be considered. For example, if the safety factor for tooth bending is small compared to surface pitting, the austenite content should be kept to a minimum (preferably without refrigeration). If the converse applies, then a retained austenite content approaching 35% is more appropriate. If occasional tooth bending overloads are expected, then to deter initial cracking it might be prudent to aim for an austenite content of about 25%. If a life of less than 1×10^4 cycles at high loads is required, then much higher austenite contents can be considered. If, on the other hand, adhesive wear (scoring, scuffing) is the more likely failure mode, then it is necessary to lower the austenite content.

Heat Treating. It should be apparent that no single value for retained austenite content will satisfy all requirements of a gear tooth. The foregoing discussion is very nice in theory, but in practical terms it is not altogether realistic. If a designer had the time to determine the best austenite content for a particular gear, could the heat treater comply? In commercial heat treatment, the best an operator can do, given a steel grade and quenching method, is adjust the surface carbon content to obtain a retained austenite content within the recommendation of the standard.

Post case-hardening refrigeration is an effective means of reducing retained austenite, but it can have drawbacks in its influence on material properties. Therefore, carburizing and hardening processes should avoid the formation of excessive amounts of austenite in the first place. If a

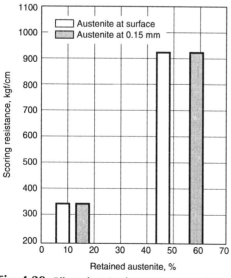

Fig. 4.29 Effect of retained austenite on scoring resistance. Source: Ref 42

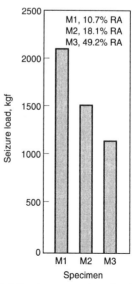

Fig. 4.30 Relation between seizure load and percentage of retained austenite (RA). Source: Ref 44

subzero treatment is unavoidable, then a shallow rather than a deep freeze should be considered, as too should both prior and post tempering operations. The subject of refrigeration is covered in more depth in Chapter 7, "Postcarburizing Thermal Treatments."

Surface Working. The properties of surfaces containing retained austenite can be favorably modified by mechanical methods that induce plastic deformation, such as shot peening or surface rolling. Retained austenite, being relatively soft, is work hardened by these processes with the added benefit that the surface residual stresses are made more compressive. It is important, however, that the balancing tensile residual stresses do not peak just beneath the worked layer. Correctly executed surface working is capable of not only overcoming any adverse effects of retained austenite but also of raising the fatigue strength of the part to well above that obtained solely by thermal means.

Summary

Retained austenite is more or less unavoidable in quenched carburized steels. The amount of retained austenite is determined by the carbon content, quenching, and quenchant temperatures. For most applications some retained austenite is acceptable; if in excess, it can lead to grinding problems.

- *Preprocess considerations:* It is normal to select steel for a component based on size and eventual duty. Therefore, the possibility of producing high retained austenite contents at the surface can be anticipated.
- *In-process considerations:* Carbon-potential control and quenching methods are means of controlling the austenite content. Generally, lean grades of steel are direct quenched, and the more highly alloyed grades are reheat quenched.
- *Postprocess considerations:* If retained austenite is unacceptably high, consider requenching from a lower temperature (watch distortion aspect). Alternatively, use shallow refrigeration if the adverse effects of the process can be tolerated.
- *Effect on properties:* Retained austenite reduces hardness, abrasive wear resistance, and bending-fatigue strength, but it is thought to benefit contact fatigue. Conflicting data exist regarding adhesive wear. In all cases, the austenite and accompanying martensite should be fine and evenly distributed.
- *Standards:* For the lower quality grade neither ANSI/AGMA nor ISO provides a specification. For the middle and the highest grades, ANSI/AGMA quotes 30% maximum retained austenite, whereas ISO calls for 25% maximum.

REFERENCES

1. A. Rose and H.P. Hougardy, Transformation Characteristics and Hardenability of Carburising Steels, Transformation and Hardenability in Steels, Climax Molybdenum Co. of Michigan, 1967, p 155–167
2. E.C. Rollason, "Fundamental Aspects of Molybdenum on Transformation of Steel," Climax Molybdenum Co. of Europe Ltd.
3. R. Priestner and S.G. Glover, Stabilisation Effect in a High Carbon Nickel Steel, *Physical Properties of Martensite and Bainite,* Special report 93, The Iron and Steel Institute, 1965, p 38–42
4. W.S. Owen, Theory of Heat Treatment, *Heat Treatment of Metals,* Hatte Books Ltd., 1963, p 1–28
5. A.G. Haynes, Interrelation of Isothermal and Continuous-Cooling Heat Treatments of Low-Alloy Steels and Their Practical Significance, *Heat Treatment of Metals,* Special report 95, The Iron and Steel Institute, 1966, p 13–23
6. D.P. Koistinen and R.E. Marburger, A General Equation Prescribing the Extent of the Austenite-Martensite Transformation in Pure Iron-Carbon Alloys and Plain Carbon Steels, *Acta Metall.,* Vol 7, 1959, p 59–60
7. C.Y. Kung and J.J. Rayment, An Examination of the Validity of Existing Empirical Formulae for the Calculation of M_s Temperature, *Metall. Trans. A,* Vol 13, Feb 1982, p 328–331
8. E. Szpunar and J. Beilanik, Influence of Retained Austenite on Propagation of Fatigue Cracks in Carburised Cases of Toothed Elements, *Heat Treatment '84,* The Metals Society, p 39.1–39.9
9. L.A. Vasil'ev, Retained Austenite and Hardness of the Carburised Case on Steel 18Kh2N4VA after Quenching in Two Different Media, *Met. Sci. Heat Treat.,* Vol 14

(No. 2), Feb 1972, p 58–59
10. O.A. Ankara and D.R.F. West, Investigation of Transformation Plasticity during Martensite Formation in a Medium Alloy Steel, *Physical Properties of Martensite and Bainite,* Special report 93, The Iron and Steel Institute, 1965, p 183–192
11. K. Bungardt, E. Kunze, and H. Brandis, Betrachtungen zur Direkthärtung von Einsatzstählen, *DEW-Technische Berichte,* Vol 5 (No. 1), 1965, p 1–12
12. R.W. Neu and S. Huseyin, Low Temperature Creep of a Carburised Steel, *Metall. Trans. A,* Vol 23, Sept 1992, p 2619–2624
13. R.H. Richman and R.W. Landgraf, Some Effects of Retained Austenite on the Fatigue Resistance of Carburized Steel, *Metall. Trans. A,* Vol 6, May 1975, p 955–964
14. R.J. Johnson, "The Role of Nickel in Carburising Steels," Publication A 1205, International Nickel Company, Inc.
15. X. Yen, D. Zhu, and D. Shi, The Stress Induced Phase Transformation of Carburising and Carbonitrided Layers and Their Contact Fatigue Behaviour, *Heat Treat. Met. (China),* Vol 1, 1984, p 31–37
16. J. Franklin, P. Hill, and C. Allen, The Effect of Heat Treatment on Retained Austenite in a 1% Carbon-Chromium Bearing Steel, *Heat Treat. Met.,* Vol 2, 1979, p 46–50
17. D.P. Koistinen, The Distribution of Residual Stresses in Carburised Cases and Their Origin, *Trans. ASM,* Vol 50, 1958, p 227–241
18. L. Salonen, The Residual Stresses in the Carburised Layers in the Case of an Unalloyed and a Mo-Cr Alloyed Case-Hardened Steel after Various Heat Treatments, *Acta Polytech. Scand.,* Vol 109, 1972, p 7–26
19. D.E. Diesburg, C. Kim, and W. Fairhurst, Microstructural and Residual Stress Effects on the Fracture of Case-Hardened Steels, paper 23, *Heat Treatment '81* (University of Aston), The Metals Society, Sept 1981
20. M.M. Shea, Residual Stress and Microstructure in Quenched and Tempered and Hot Oil Quenched Carburised Gears, *J. Heat Treat.,* Vol 1 (No. 4), Dec 1980, p 29–36
21. H. Wiegand and G. Tolasch, The Combined Effect of Individual Factors on Raising the Alternating Bending Fatigue Strength of Case-Hardened Test Pieces, *Härterei-Technische Mitteilungen* (BISI 6081), Vol 22, Oct 1967, p 213–220
22. C. Razim, Influence of Residual Austenite on the Strength Properties of Case-Hardened Test Pieces during Fatiguing, *Härterei-Technische Mitteilungen* (BISI 6448), Vol 23, April 1968, p 1–8
23. H. Brugger, Werkstoff und Wärmebehandlungseinflüsse auf die Zahnfuβtragfahigkeit, *VDI-Berichte,* (No. 195), 1973, p 135–144
24. J.L. Pacheco and G. Krauss, Microstructure and High Bending Fatigue Strength of Carburised Steel, *J. Heat Treat.,* Vol 7 (No. 2), 1989, p 77–86
25. M.A. Zaconne, J.B. Kelley, and G. Krauss, Strain Hardening and Fatigue of Simulated Case Microstructures in Carburised Steel, *Conf. Proc. Processing and Performance* (Lakewood, CO), ASM International, July 1989
26. G.V. Kozyrev and G.V. Toporov, Effect of Retained Austenite on the Impact-Fatigue Strength of Steel, *Metal Sci. Heat Treat.,* Vol 15 (No. 12), Dec 1973, p 1064–1066
27. M.A. Panhans and R.A. Fournelle, High Cycle Fatigue Resistance of AISI E 9310 Carburised Steel with Two Different Levels of Retained Austenite and Surface Residual Stress, *J. Heat Treat.,* Vol 2 (No. 1), 1981, p 54–61
28. C. Razim, Effects of Residual Austenite and Recticular Carbides on the Tendency to Pitting of Case Hardened Steels, thesis, Techn. Hochschule Stuttgart, 1967
29. M.A. Balter and M.L. Turovskii, Resistance of Case-Hardened Steel to Contact Fatigue, *Met. Sci. Heat Treat.,* (No. 3), March 1966, p 177–180
30. A. Diament, R. El Haik, R. Lafont, and R. Wyss, "Surface Fatigue Behaviour of the Carbonitrided and Case-Hardened Layers in Relation to the Distribution of the Residual Stresses and the Modifications of the Crystal Lattice Occurring during Fatigue" (BISI 12455), paper presented at 25th Colloque Int. (Caen, France), 29–31 May 1974, International Federation for the Heat Treatment of Materials, 1974
31. B.B. Vinokur, S.E. Kondratyuk, L.I. Markovskaya, R.A. Khrunik, A.A. Gurmaza, and V.B. Vainerman, Effect of Retained Austenite on the Contact Fatigue Strength of Carburised Steel, *Met. Sci. Heat Treat.,* (No. 11), Nov 1978, p 47–49
32. B. Vinokur, The Composition of the Solid Solution Structure and Contact Fatigue of the Case-Hardened Layer, *Metall. Trans. A,* Vol 24, May 1993, p 1163–1168

33. P. Bárczy and J. Takács, Endurance Improvements of Planet Gear Pins by Heat Treatment, paper 24, *Heat Treatment '84*, The Metals Society, 1984
34. G.T. Hahn, V. Bhargava, C.A. Rubin, and X. Leng, Analysis of the Effects of Hardened Layers on Rolling and Sliding Contact Performance, *Conf. Proc. Processing and Performance* (Lakewood, CO), ASM International, 1989
35. C. Razim, Survey of Basic Facts of Designing Case Hardened Components, unpublished private print, 1980
36. B. Thoden and J. Grosch, Crack Resistance of Carburised Steel under Bend Stress, *Conf. Proc. Processing and Performance* (Lakewood, CO), ASM International, 1983
37. V.K. Sharma, G.H. Walter, and D.H. Breen, The Effect of Alloying Elements on Case Toughness of Automotive Gear Steels, *Heat Treatment '87*, The Metals Society, 1987
38. G. Niemann and A. Seitzinger, "Calculation of Scoring Resistance for Spur and Helical Gears," ISO 6336, International Standards Organization, 1996
39. R.M. Matveevsky, V.M. Sinaisky, and I.A. Buyanovsky, Contributions to the Influence of Retained Austenite Content in Steels on the Temperature Stability of Boundary Lubricant Layers in Friction, *J. Lubr. Technol. (Trans. ASME)*, July 1975, p 512–515
40. A.G. Roberts, discussion appended to Ref 39
41. W. Grew and A. Cameron, Role of Austenite and Mineral Oil in Lubricant Failure, *Nature*, Vol 217 (No. 5127), 3 Feb 1968, p 481–482
42. I.I. Kozlovskii, S.E. Manevskii, and I.I. Sokolov, Effect of Retained Austenite on the Resistance to Scoring of Steel 20Kh2N4A, *Met. Sci. Heat Treat.*, (No. 1), 1978, p 70–80
43. S.E. Manevskii and I.I. Sokolov, Resistance to Seizing of Carburised and Carbon Nitrided Steels, *Metalloved. Term. Obrab. Met.*, (No. 4), April 1977, p 66–68
44. J. Terauchi and J.I. Takehara, On the Effect of Metal Structure on Scoring Limit, *Bull. Jpn. Soc. Mech. Eng.*, Vol 21 (No. 152), Feb 1978, p 324–332

SELECTED REFERENCES

- J. Grosch and O. Schwarz, Retained Austenite and Residual Stress Distribution in Deep Cooled Carburized Microstructures, *Conf.: 1995 Carburizing and Nitriding With Atmospheres*, 6–8 Dec 1995, (Cleveland, Ohio), ASM International, 1995 p 71–76
- A. Inada, H. Yaguchi, and T. Inoue, Effects of Retained Austenite on the Fatigue Properties of Carburized Steels, *Kobelco Technol. Rev.*, Vol 17, April 1994, p 49–51
- G. Krauss, Advanced Performance of Steel Surfaces Modified by Carburizing, *Conf.: Heat & Surface '92*, 17–20 Nov 1992 (Kyoto, Japan), Japan Technical Information Service, 1992, p 7–12
- G. Krauss, Microstructures and Properties of Carburized Steels, *Heat Treating*, Vol 4, *ASM Handbook*, ASM International, 1991, p 363–375
- F. Li and C. Li, The Influences of Heat Treatment After Carburizing and Retained Austenite in the Carburized Layer on the Strength and Toughness of Steel, *Trans. Met. Heat Treat. (China)*, Vol 6 (No. 2), Dec 1985, p 59–68
- J.D. Makinson, W.N. Weins, and R.J. de Angelis, The Substructure of Austenite and Martensite Through a Carburized Surface, *Conf.: Advances in X-Ray Analysis*, Vol 34, 30 July to 3 Aug 1990 (Steamboat Springs, Colorado), Plenum Publishing Corporation, p 483–491
- D.L. Milam, Effect of Interrupted Cooling on Retention of Austenite and Development of Case Hardness in a Carburizing Process, *Conf.: 1995 Carburizing and Nitriding With Atmospheres*, 6–8 Dec 1995 (Cleveland, Ohio), ASM International, p 111–116
- R.W. Neu and H. Sehitoglu, Transformation of Retained Austenite in Carburized 4320 Steel, *Metall. Trans. A*, Vol 22 (No. 7), July 1991, p 1491–1500
- M. Przylecka, M. Kulka, and W. Gestwa, Carburizing and Carbonitriding Bearing Steel (LH15-52100), *Conf.: Heat Treating: Equipment and Processes*, 18–20 April 1994 (Schaumburg, Illinois), ASM International, 1994, p 233–238
- E. Shao, C. Wang, and C.-S. Zheng, Improving Rolling Contact Fatigue Strength for Carburizing Gears With Retained Austenite and Carbide in Their Case and Its Application, *Trans. Met. Heat Treat. (China)*, Vol 10 (No. 1), June 1989, p 31–43

- J. Siepak, Effect of Retained Austenite in Carburized Layers on Rolling Wear, *Neue Hütte,* Vol 29 (No. 12), Dec 1984, p 452–455
- B.B. Vinokur and A.L. Geller, The Effect of Retained Austenite on Contact Fatigue in Cr-Ni-W Carburized Steel, *J. Met.,* Vol 49 (No. 9), Sept 1997, p 69–71
- D. Zhu, F.-X. Wang, Q.-G. Cai, M.-X. Zheng, and Y.Q. Cheng, Effect of Retained Austenite on Rolling Element Fatigue and Its Mechanism, *Wear,* Vol 105 (No. 3), 1 Oct 1985, p 223–234

Chapter 5

Influential Microstructural Features

Microstructural features observed within case-hardened steels include grain size, microcracking, microsegregation, and nonmetallic inclusions. Although these microstructural constituents are not considered in the same detail as those presented in the first four chapters, their significant influence on properties justifies their review.

Grain Size

For optimal properties, it is essential that the grain size of a carburized and hardened component is both uniform and fine. Generally, the starting point is a grain-refined steel having an ASTM grain size from No. 5 to 8, though subsequent mechanical and thermal processing can change the final microstructure to be either more coarse or fine.

At one time, grain size control was very much in the hands of the heat treater who, supplied with coarse-grained steels, employed double quenching treatments to ensure that both case and core of the carburized item were refined. However, the advent of grain refinement by alloying obviated the need for carburized components to be double quenched. Now, the majority of heat treaters either employ direct quenching or single reheat quenching to harden their products. Figure 5.1, coupled with Table 5.1, illustrates different thermal cycles used to effect hardening and generally ensure a fine-grained product, provided the carburizing temperature for all programs except D is not excessive. Much of the flexibility for process cycles must be attributed to the availability of grain-refined steels.

In addition to the process cycles shown in Fig. 5.1, some cycles include a subcritical annealing, or a high-temperature tempering operation, between the carburizing and hardening stages. By slow cooling from the carburizing temperature then annealing or tempering from 600 to 650 °C, the part is rendered suitable for any intermediate machining needed. Another use of the high-temperature tempering involves a version of the double quenching program (D in Fig. 5.1). The first quenching operation is into a medium

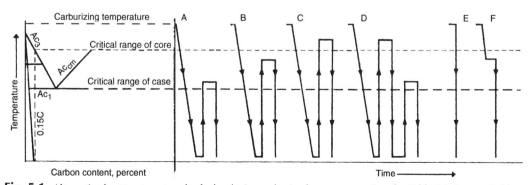

Fig. 5.1 Alternative heat treatment cycles for hardening carburized components. See also Table 5.1. Source: Ref 1

Table 5.1 Case and core characteristics resulting from the various heat treatments shown in Fig. 5.1

Treatment	Case	Core
A: Best adapted to fine-grained steels	Refined; excess carbide not dissolved; distortion minimized	Unrefined; soft and machinable
B: Best adapted to fine-grained steels	Slightly coarsened; some solution of excess carbide	Partially refined; stronger and tougher than A
C: Best adapted to fine-grained steels	Somewhat coarsened; solution of excess carbide favored; austenite retention promoted in highly alloyed steels	Refined; maximum core strength and hardness; better combination of strength and ductility than B
D: Best treatment for coarse-grained steels	Refined; solution of excess carbide favored; austenite retention minimized	Refined; soft and machinable; high degree of toughness and resistance to impact
E: Suitable for fine-grained steels only	Unrefined with excess carbide dissolved; austenite retained; distortion minimized	Unrefined but hardened
F: Suitable for fine-grained steels only	Unrefined; excess carbide avoided if combined with diffusion treatment; austenite retention reduced; distortion minimized	Unrefined but hardened

Source: Ref 1

held at a temperature not too far below the Ac_1 from which the workpieces are then reheated for the second quench. The aim is to ensure grain refinement (Ref 2) and reduce distortion and the risk of cracking.

Evaluation of Grain Size

Grain Growth. Vinograd et al. (Ref 3), who studied the behavior of twelve different steels heated from 850 to 1250 °C, concluded that grain growth takes place by three different mechanisms:

- The resorption of grains, which occurs at 50 to 100 °C above the Ac_3
- The formation of new boundaries and grains between 250 and 300 °C above the Ac_3
- Boundary migration, which occurs at all temperatures but affects grain growth only at temperatures above 1100 °C

The first two mechanisms involve the decomposition of old boundaries, whereas the third involves the movement of boundaries.

Grain Size Control. Grain refinement by alloying is accomplished by adding certain elements (e.g., aluminum or vanadium) to the molten steel in the ladle after a thorough deoxidation treatment, usually with silicon. Silicon in adequate quantities, although harmful in terms of internal oxidation during carburizing, is important as a deoxidizer and, therefore, for grain size control (Fig. 5.2). Additions of aluminum and/or vanadium encourage the formation of compounds (e.g., AlN or V_4C_3) that, because of their extreme fineness and relative stability, are able to mechanically restrain grain boundary movement during subsequent austenitizing treatments. Of the grain refining agents, AlN is more important for fine-grain stability. The effect of grain refining treatments is limited, because above a certain temperature within the austenite range, the precipitated compounds coalesce and then dissolve. Consequently, they can no longer prevent grain boundary movement. However, the grain coarsening temperature is further increased by adding both aluminum and titanium. This addition is important when any high-temperature carburizing process is being used, if indeed grain coarsening is a problem. The efficiency of aluminum to inhibit austenite grain growth is illustrated in Fig. 5.3, which compares an untreated steel with an aluminum treated steel. Figure 5.3(a) shows that the untreated steel coarsens progressively as the temperature increases, whereas the treated steel

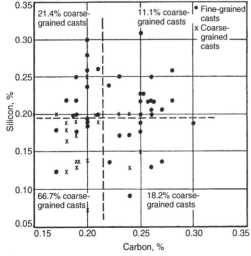

Fig. 5.2 Effect of carbon and silicon on grain size control of 665M17 steel made by basic electric-arc process with aluminum addition of 16 oz/ton. Source: Ref 4

exhibits no serious grain growth until it reaches about 925 °C. Thereafter, the grains of the aluminum treated steel coarsen rapidly. Grain refining agents other than aluminum and vanadium include boron, titanium, niobium, cerium, and sometimes chromium.

The duration of the austenitizing treatment has some significance with regard to grain coarsening. For example, the steel referred to in Fig. 5.3(b) was fine grained when treated at 925 °C for 1 h, whereas a 10 h treatment at that temperature causes the average grain size to rise.

Each steel has its own coarsening tendencies whether grain refined with aluminum or vanadium or left untreated, and steels containing alloying elements, such as nickel and/or molybdenum, have a greater resistance to coarsening at conventional carburizing temperatures than plain-carbon steels.

The mechanical and thermal histories of a part can influence the grain-coarsening temperature. For example, Fig. 5.4 shows that prior normalizing lowers the grain coarsening temperature more than annealing does. As early as 1934, Grossman (Ref 6) showed that repeated heating to slightly above the Ac_3 temperature also lowers the coarsening temperature, and although controlled hot working reduces the initial grain size, it nevertheless lowers the coarsening temperature. Others (Ref 7) have suggested that an incorrect hot-working temperature contributes to the occasional occurrence of coarse-grained aluminum treated case-hardening steels.

With respect to heat treatment, Kukareko (Ref 8), working with an 18KhNVA steel, related that an increased heating rate to 930 °C results in the formation of a metastable fine-grained structure with a tendency to rapid grain growth via grain merging during subsequent isothermal heat treatment. On the other hand, a preliminary tempering operation and slow heating to the austenitizing temperature produces a stabilized grain structure.

In terms of modern carburizing practice, it is unlikely that conventional treatments at 925 °C using satisfactorily grain-refined steels give rise to coarse austenite grain sizes. However, the trend to higher carburizing temperatures, sometimes in excess of 1000 °C, requires consideration of all aspects of grain refinement (e.g., alloying additions, hot working, and subsequent thermal treatments) to end up with a fine-grained

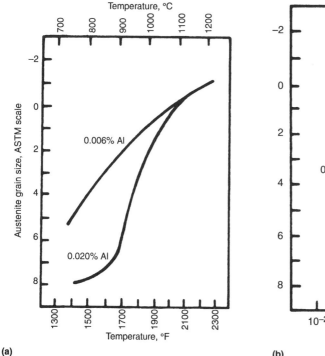
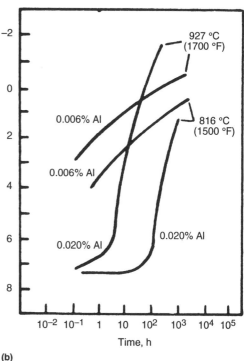

Fig. 5.3 Relationship between austenitizing parameters and grain size for grain-refined and non-grain-refined AISI 1060 steel. (a) Effect of austenitizing temperature (2 h soak). (b) Effect of austenitizing time. Source: Ref 5

product. Child (Ref 9) claimed that grain growth is not a problem with gas carburizing temperatures up to 1000 °C. Nevertheless, Fig. 5.5 suggests that coarsening, at least in the leaner grades of carburizing steels, becomes likely at 1000 °C. This observation is supported by Benito and Mahrus (Ref 11), who encountered an excessive increase in grain size when vacuum carburizing steels SAE 5115, 4320, and 8620 at 1000 to 1100 °C. When four German steels were carburized from 920 to 1100 °C (Ref 12), grain growth was slight, although there was some local coarsening at above 1020 °C in some samples. The chromium-manganese steels were found to be more susceptible than the molybdenum-chromium steels. Thus, grain size is influenced by the chemical composition, the mechanical and thermal history, and the carburizing temperature. However, the reheat temperature is of particular importance. Effective grain refinement of the case is achieved by reheat quenching from about the Ac_{cm} temperature.

Metallography. The austenite grain size of a carburized and hardened steel is generally not very obvious unless the carbon content is so high and the postcarburizing heat treatments such that network carbides are visible during the metallographic examination (Fig. 5.6). More often than not, there are no network carbides, because they are regarded as detrimental and therefore avoided. Assessment of such a structure, at best, is qualitative (i.e., coarse, normal, or fine) and made from the features of the austenite decomposition products. If it is necessary to determine the austenite grain size, then the standard McQuaid-Ehn test is employed. In the McQuaid-Ehn test, a steel sample is carburized in a medium of sufficient carbon potential to decorate the grain boundaries with continuous films of cementite. The sample is then prepared for viewing at 100× magnification, and the grain size graded according to a table, such as Table 5.2, or by comparison with examples, such as Fig. 5.7. For case-hardening grades of steel, a grain size of ASTM No. 5 or finer (ASTM E 112) is specified.

The austenite grain size influences the size of the austenite decomposition products. The grain size of the product is smaller than the austenite grain in which it grew. In the case of the high-temperature transformation products (ferrite, pearlite, or bainite), many new grains nucleate simultaneously along each austenite grain boundary, and those reaching critical size first continue to grow at the expense of the others. Thus, each austenite grain is replaced by several grains of the transformation product. With respect to the diffusionless transformation that produces martensite, many faults develop within each austenite grain during the quench. Each fault is a potential nucleation site from which a martensite plate can grow; growth, when it does occur, is rapid. The first formed plates are likely to be the largest and no longer than the austenite grain diameter. Subsequently formed plates then subdivide the remaining volumes of the austenite.

Therefore, thinking in terms of carburized surfaces, the austenite grain size at the onset of the quench influences martensite plate size (hence, the microcrack content) and the size of the austenite volumes in the as-quenched product. Also,

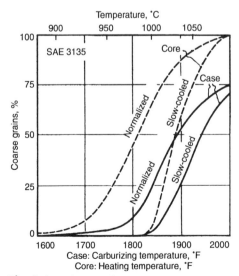

Fig. 5.4 Reduction of grain-coarsening temperature due to normalizing. Source: Ref 6

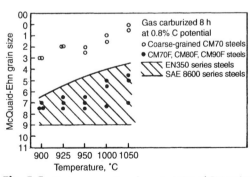

Fig. 5.5 Grain-coarsening characteristics of CM series steels compared with conventional carburizing grades. Source: Ref 10

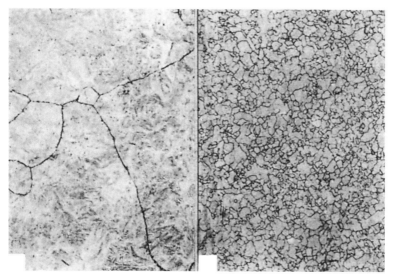

Fig. 5.6 Prior austenite grain boundaries. (a) Coarse-grained SAE 1015 carbon steel, carburized. 100×. (b) Fine-grained SAE 4615 nickel-molybdenum steel, carburized. 100×

grain size affects the frequency and depth of internal oxidation and, therefore, the grain size of any high-temperature transformation products (HTTP) associated with it.

The austenitic grain size data for incoming materials may be provided by the steel supplier or determined by the customer as part of routine acceptance testing. Unfortunately, grain size determinations made on a small diameter forged test bar may not truly reflect on those of the parts that the test bar represents. For example, the amount of hot working can be different, as too the heating and cooling rates during any simulated carburizing treatments. Dietrich et al. (Ref 13) warned how normalizing prior to grain size testing could produce a false result and that slow heating (3 °C/min) to the testing temperature can lower the content of coarse grains. Subcritical annealing (710 °C) has a similar effect. Even so, grain size control and assessment by steelmakers are sufficiently adequate to assure the customer that fine-grained steels are being provided.

The hardenability of a fine-grained steel is less than that of a coarse-grained steel having an identical chemical composition. The same applies to the case hardenability. However, whereas this distinction is especially meaningful for plain carbon and very lean-alloy case-hardening grades and may lead to problems during processing, it is of less concern for the more alloyed steels (Ref 14).

Francia (Ref 15), who provides a reminder that hardenability is critical in gear steel selection, makes the point that hardenability should not be sought by the use of coarse-grained materials. Apart from having the potential to produce inferior properties, coarse-grained steels distort more than their fine-grained counterparts.

Table 5.2 ASTM grain size

ASTM grain number	Section viewed at 100×, grains/in.^{2}	Grain surface(a), in.^{2}/in.^{3}	Interfacial area(b), in.^{2}/in.^{3}	Grain diameter(c), in.
1	1	4.0	270	0.0113
2	2	5.6	340	0.0080
3	4	8.0	480	0.00567
4	8	11.3	679	0.00400
5	16	16.8	961	0.00283
6	32	22.6	1360	0.00200
7	64	32.0	1920	0.00142
8	128	45.3	2720	0.00100

(a) Calculated for cubical grains at 100×. (b) Average calculated for a 14-sided solid of maximum ability for close packing. (c) Equivalent spherical grain, not magnified

Effect of Grain Size on Properties

Most properties are modified by changes of grain size, and some properties are adversely affected by an increase of grain size. However, information dealing specifically with the effects of grain size on the properties of case-hardened parts is somewhat limited. Therefore, where direct information is absent, other studies (of non-case-hardened steels) are referenced to obtain reasonable trends and guidance.

Hardness. Austenitic grain size does not greatly affect the surface hardness of carburized and hardened surfaces and only marginally influences the core hardness. For high-carbon martensite, Kelly and Nutting (Ref 16) considered that grain size has a negligible influence on the increase of hardness, and other factors, such as carbon in solid solution within the twinned martensite or carbide precipitation, are far more significant. However, Zaccone et al. (Ref 17), using steels of varying chromium content and grain size, showed with direct quenched samples that for a given grain size the hardness decreases as the retained austenite increases. With reheat quenched samples in which both grain size and retained austenite vary, there is a different trend (Fig. 5.8). This trend difference might be attributed to grain size. The reheat quenched samples were, as expected, finer grained than their direct quenched counterparts.

For low-carbon ferritic steels (i.e., ferrite, ferrite/pearlite, and the quenched and high-temperature tempered microstructures), an approximately linear relationship between hardness and grain size might be expected. The results for three plain-carbon steels in the worked, quenched, and high-temperature tempered condition are shown in Fig. 5.9. The trend is for hardness to increase as grain size decreases. However, over the range of grain sizes typical for case-hardened steels, the change in hardness is less than 20 HV.

Tensile Strength Properties. In heat-treated steels, the tensile yield strength is influenced by a number of factors including grain size (prior austenite grain size or as-heat-treated grain size). The yield strength is inversely proportional to the square root of the austenite grain size (Ref 16). This is illustrated in Fig. 5.10 for two martensitic steels where the yield strength increases as the grain size decreases by about 21 MPa per grain size ($d^{-1/2}$, mm$^{-1/2}$) change. Other factors that influence yield strength include carbon segregation, precipitates, and the substructure of the martensite (dislocations, internal twinning).

Schane (Ref 20), employing fine- and coarse-grained SAE 1040 steels, showed that an initially coarse-grained steel began to further coarsen at a relatively low austenitizing temperature (810 to 840 °C), while an initially fine-grained material resisted coarsening until temperatures above about 1000 °C were used. The

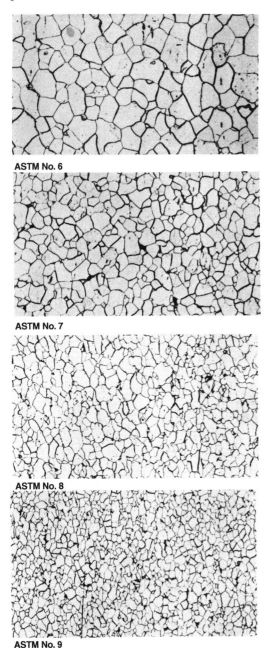

Fig. 5.7 Comparison of nominal ASTM 6 to 9 grain size (with calculated grain size numbers of 6.08, 7.13, 8.03, and 8.97, respectively). Nital etch, 100×

properties obtained after normalizing at temperatures up to 1093 °C are shown in Fig. 5.11. In terms of ductility, the fine-grained steel was the better of the two. The inflections in the curves coincide with the onset of grain coarsening.

Residual Stresses. Acknowledging that grain size differences might influence the proportions of the transformation products in a case-hardened surface, particularly the plain-carbon grades, then it can be reasoned that the residual stress distribution within the carburized case will be modified accordingly. Also, coarse-grained steels are more prone to distortion (which is a residual stress effect) than fine-grained steels, which could be a reflection on the influence of grain size on hardenability. Coarse-grained steels also tend to be more prone to cracking and microcracking during quenching or grinding; again, macrostresses and microstresses are involved.

Fatigue Strength. The grain size of a steel affects its response to cyclic loading. Macherauch (Ref 21) showed that, for a low-carbon steel, coarse-grained structures are inferior to fine-grained structures when tested under bending fatigue conditions (Fig. 5.12). The same

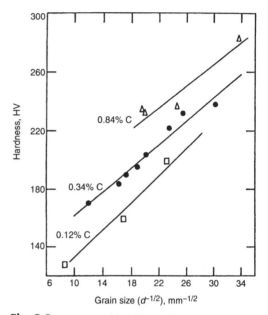

Fig. 5.9 Variation of hardness with ferrite grain size. Source: Ref 18

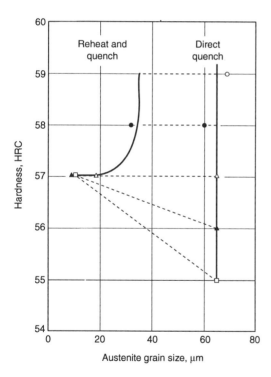

Fig. 5.8 Hardness vs. grain size. Steel composition: 0.82C, 0.9Mn, 0.31Si, 1.76Ni, 0.72Mo, and Cr (per table). Source: Ref 17

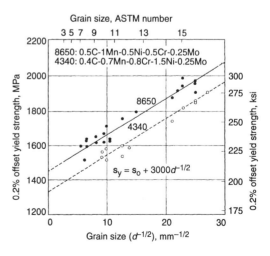

Fig. 5.10 Effect of prior-austenite grain size on the strength of martensite. Source: Ref 19

106 / Carburizing: Microstructures and Properties

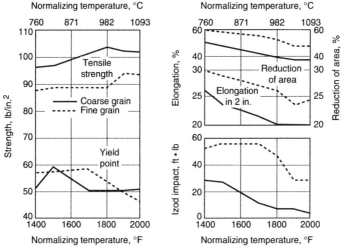

Fig. 5.11 Effect of austenitizing temperature (and grain size) on room temperature properties of a 0.4% C steel. Solid line, coarse grain; dashed line, fine grain. Source: Ref 20

trend has been observed with carburized steels (Ref 22). When reducing the grain diameter from 100 to 4 μm, more than doubled fatigue strength (Fig. 5.13). However, it is not clear what other factors are involved in achieving such an improvement. Pacheco and Krauss (Ref 23) illustrated how grain size and retained austenite work together to influence the fatigue strength (Fig. 5.14), although the individual contribution of each could not be isolated. For high bending fatigue strengths in carburized and hardened parts, the martensite of the case should be fine and any retained austenite should be minimal and finely dispersed. Such benefits are somewhat dependent on the condition of the actual surface being cyclically stressed, as Fig. 5.15 shows (Ref 24).

With coarser grained case-hardened surfaces, such as developed by direct quenching, the tendency is for crack initiation to be intergranular. In fine-grained surfaces (e.g., developed by reheat quenching), crack initiation tends to be transgranular (Ref 25). Propagation is often transgranular, but the overload fracture can either be intergranular or transgranular depending on the grain size or method of quenching (Ref 26).

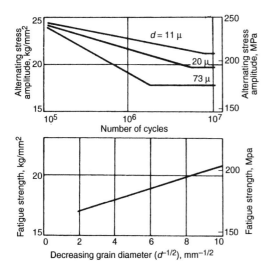

Fig. 5.12 Effect of grain size on the fatigue strength of a 0.11% C mild steel. Source: Ref 21

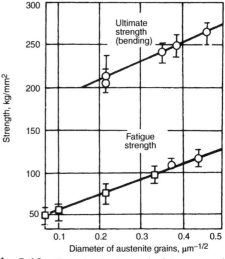

Fig. 5.13 Effect of grain size on the fatigue strength of case-hardened test pieces. Source: Ref 22

For instance, quenching from above the Ac_{cm} produces intergranular overload fractures.

Bending and Impact Fracture Strength. According to Wang and Chen (Ref 27), in precracked fracture toughness specimens, crack propagation occurs from a microcrack the size of a second-phase particle (e.g., a carbide precipitate). In notched specimens, propagation occurs from a crack the size of a ferrite grain. Therefore, in impact testing, grain size is more directly significant than it is for fracture toughness testing with precracked specimens.

In impact and fracture toughness testing, the ratio of intergranular to transgranular fracture surfaces relates to the toughness of the test piece: as the intergranular fracture ratio falls, the toughness property increases (Ref 28). Further, a small grain size appears to equate to a reduced intergranular fracture ratio. Therefore, the toughness of a case-hardened surface is dictated by the intergranular strength. The finer the grain size is, the lower the intergranular fracture ratio is, hence the higher the toughness is. When the transgranular mode of fracture takes over, the presence of retained austenite in the structure becomes more significant with respect to toughness. A point to bear in mind is that as the grain size increases, the impact transition temperature increases.

The effect of grain size in relation to the impact strength of a plain-carbon steel in the normalized condition is shown in Fig. 5.11. The same trends apply to case-hardening steels in the blank carburized condition (see Table 5.3), where the coarse-grained samples have higher tensile strengths, lower reductions of area, and lower impact strengths than those of the fine-grained samples. Double quenching treatment makes for an improvement in each of these properties. Table 5.4 considers a 0.45% C steel where bending and impact strengths are highest when the grain size is at its smallest.

Microcracking

A structural feature sometimes observed in quenched bearing steels that can develop in carburized and hardened surfaces is microcracking associated with the martensite plates. Such cracks run either across the martensite plate or along the side of the plate (Fig. 5.16), that is, along the interface separating the martensite and the adjacent austenite. Microcracks have also been observed at the prior austenite grain boundaries (Ref 30). Detection requires careful preparation of the metallographic sample and only a light 2% nital etch immediately prior to viewing at 500 to 1000× magnification.

Microcracks are formed when the strains generated at the tip of a growing martensite plate are sufficient to induce cracking in any plate or boundary with which the growing plate im-

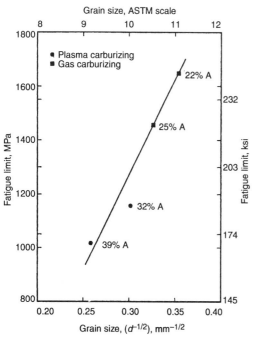

Fig. 5.14 Fatigue limits of plasma and gas-carburized specimens as a function of austenitic grain size. A, retained austenite. Source: Ref 23

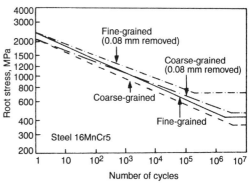

Fig. 5.15 Effect of grain size on fatigue strength of case-hardened gears with and without removal of a 0.08 mm surface layer. Coarse-grained steel, ASTM 1–4; fine-grained steel, ASTM 6–8. Source: Ref 24

108 / Carburizing: Microstructures and Properties

Table 5.3 Effect of grain size on the core properties of case-hardened steels

Steel	Heat treatment	Grain	McQuaid-Ehn test, ASTM grain size	Izod impact ft·lb	Izod impact kg/cm²	Reduction of area, %	Maximum tensile stress tons/in.²	Maximum tensile stress kg/mm²
Carbon	Pot quench	Fine	8	42	7.2	54	46	72
		Coarse	2–3	16	2.8	39	49	77
	Single quench	Fine	8	58	10.0	62	37	58
		Coarse	2–3	13	2.2	45	41	65
	Double quench	Fine	8	99	17.1	69	37	58
		Coarse	2–3	52	9.0	59	38	60
Nickel-molybdenum	Pot quench	Fine	7–8	96	16.6	64	51	80
		Coarse	3	31	5.4	54	60	94
	Single quench	Fine	7–8	87	15.0	62	51	80
		Coarse	3	40	6.9	61	56	88
	Double quench	Fine	7–8	93	16.1	64	48	76
		Coarse	3	72	12.4	56	49	77

Source: Ref 1

Table 5.4 Deterioration of properties with increasing grain size

Solution temperature, °C	Bend strength, kg/mm²	Deflection, mm	Hardness, HRC	Work to fracture, kg Smooth	Work to fracture, kg Notched	Impact (notched), kg/cm²	Bend angle of fractured sample	Length of martensite needles, (Gost scale), grade	Maximum length of martensite, μm	Austenite grain size (Gost scale), grade
850(a)	361	7.12	53	4.2	0.6	2	20°30′	7–8	...	...
950	414	7.97	57	12	0.7	2.6	50°30′	8–9	12–16	9–10
1050	397	5.57	57	11	0.6	2	48°30′	9–10	16–20	8–9
1150	385	5.46	56	6	0.6	1.9	26°00′	...	>20	8–6

Steel 45, induction hardened and tempered, 160–170 °C. (a) Some ferrite with martensite. Source: Ref 29

pinges. The growing plate itself may be cracked by the impingement (Ref 31). A crack must always involve an impingement, although not all impingements cause cracks.

Fig. 5.16 Microcracking in a Ni-Cr steel that also exhibits microsegregation. 1000×

Factors Influencing Microcracking

Carbon Content of the Steel. Microcracks form only in plate, not in lath, martensite. For iron-carbon alloys, plate martensite starts to appear when the carbon content is above about 0.6%, as shown in Fig. 5.17 (Ref 26). Although microcracking is not observed in steels with carbon levels of less than the eutectoid carbon (0.78% C) (Ref 32), above that value the microcracking incidence increases with carbon content. Consequently, in case-hardened surfaces, microcracking may be unavoidable, because surface carbon contents often exceed the eutectoid value. Further, as the alloy content increases, the eutectoid carbon content decreases and could fall below 0.6% for a 3%Ni-Cr steel.

Provided that quenching of the austenitized test pieces is adequate, the final microstructure consists of martensite and austenite. The quantity of austenite largely depends on the M_s and M_f temperatures, which are strongly influenced by the carbon content. The retained austenite content is unlikely to have much bearing on the

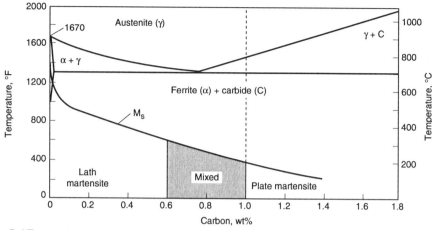

Fig. 5.17 Morphological classification of martensite in Fe-C alloys. Source: Ref 26

microcrack count, which has more to do with how much martensite is present.

Most of the studies on the subject of microcracking have used steel with a carbon content in excess of 1.0% and an inherently coarse grain size. In contrast, carburized steels generally have surface carbon contents of less than 1.0% and are grain refined.

Carbon in the Martensite. The temperature and duration of austenitizing determines how much of the carbon is dissolved and how much of it remains combined as carbides. With respect to microcracking, it is the carbon in solution in the martensite that is important (Ref 33). Figure 5.17 shows the Ac_{cm} for a Fe-1.39%C alloy is ~920 °C. On heating from the Ac_1 (~710 °C) up to 920 °C, the amount of carbon going into solution increases progressively. At and above 920 °C, following a reasonable soak, all the carbon is in solution. Quenching the steel from any temperature within this range determines the degree of solution and, hence, the extent of the microcracking (Fig. 5.18). Once full solution is attained, there is little change to the microcracking sensitivity.

Increasing the carbon content of the martensite increases the tetragonal distortion and, therefore, the brittleness of the martensite (Ref 33). It also changes the morphology of the martensite from one with a {225} habit plane to a {259} habit plane. Nickel favors the {259} type, whereas chromium and manganese stabilize the {225} habit planes. The {259} martensite forms in bursts by autocatalytic action; therefore, many of the impingements are not collisions. Any microcracks associated with this type of martensite tend to be interface or boundary cracks. In the long, slender {225} martensite plates, the impingements are generally collisions. However, the habit plane is of secondary importance in microcrack formation (Ref 34). The primary factors are the length of the martensite or, perhaps, a high length to width ratio at the instant of impingement (Ref 34). As the martensite plate thickens, there is a tendency for the cracks to be along the interface.

Effect of Alloying Elements. It has been suggested that microcracking is more likely to occur in

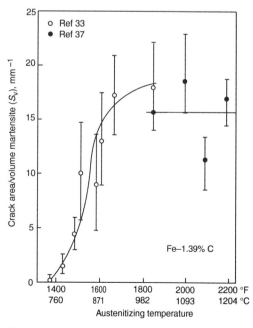

Fig. 5.18 Effect of austenitizing temperature on microcrack sensitivity. Source: Ref 33, 37

steels for which the major alloying elements are the carbide formers (Ref 35). On the other hand, Davies and Magee (Ref 34) do not regard alloying elements (Ni, Cr, Mn) as having a direct influence on microcracking. Indirectly, though, alloying elements do alter the martensite habit plane of high-carbon materials and can thereby affect the microcrack event. In commercially heat treated parts, bearing steel 52100 and case-hardened, coarse-grained, SAE 8620 steels have received the most attention. Neither of these steels is particularly well alloyed; there are many case-hardening grades more heavily alloyed, yet the microcracking incidence seems to be no worse. What these two steels can have in common is the carbon content, and of the elements added to steels, carbon is the most potent with respect to microcracking.

Plate Size and Grain Size. Although microcracks are known to occur in small martensite plates (Ref 32, 36), they are most frequently observed in the larger plates (Ref 32, 34). The highest density is associated with the longest and thinnest plates, presumably because the larger plates are likely to be subjected to more impingements by other large plates striking them at the necessary velocity and tip strain energy (Ref 34). The maximum plate size, or volume, is directly related to the grain size and, therefore, it might be expected that microcrack incidence increases with austenite grain size, up to a point (to 200 μm grain diameter according to Ref 34). Rauch and Thurtle (Ref 32) were in agreement, stating that both the frequency and the size of microcracks increased with grain size (Fig. 5.19). Brobst and Krauss (Ref 30), studying a Fe-1.22%C alloy, also confirmed that there were fewer cracks in the smaller grains, but for those cracks which did form, the proportion of grain boundary cracks had increased.

Quench Severity. Kern (Ref 35) suggested that vigorous quenching contributes to microcrack formation. However, this conclusion is not supported by laboratory tests in which water quenching, oil quenching, or simulated martempering are employed (Ref 37) or where quenching into 3% caustic solution, oil at 60 °C, or a salt bath at 180 °C are utilized (Ref 32). Microcracks formed regardless of the type of quenching used, thereby implying that quench severity is not significant.

Tempering. Microcracking in case-hardened surfaces may be aggravated by the presence of hydrogen, which is absorbed during carburizing and reheating in an endothermic atmosphere. Welding research indicates that the microcracking of martensite in the heat affected zone of a weld is influenced by the presence of hydrogen (Ref 38). For example, the martensite of a weld cooled to and soaked at temperatures above 150 °C is free from microcracking, whereas a comparable weld cooled to below 130 °C contains microcracks even when soaked at that temperature. When microcracks exist in the weld area, major cracks propagate from them due to the action of low applied loads and delayed hydrogen cracking (Ref 39).

Conventionally carburized and freshly quenched parts contain hydrogen, may contain microcracks, and are somewhat brittle. It would seem prudent, therefore, to temper components soon after quenching to drive off the hydrogen and reduce the risk of delayed cracking. Immediate tempering would also induce some measure of microstress relaxation to toughen the martensite/austenite structure. Tempering has an added benefit of causing the smaller microcracks to heal (Ref 40–42), and tempering at temperatures as low as 180 °C for as little as 20 min has been shown to have a marked effect (Fig. 5.20). Healing is attributed to volume changes and associated plastic flow (Ref 42).

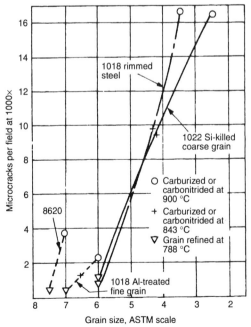

Fig. 5.19 Influence of grain size on microcrack frequency in some carburized steels. Source: Ref 32

Effect of Microcracking on Properties. Microcracking has attracted attention in terms of causes and contributing factors, but few studies, as yet, have been carried out to determine the effect of microcracking on the more important material properties of heat-treated bearing steels or case-hardened steels. Occasionally, premature failures have been attributed to the presence of microcracks, but in failure analyses, there are generally so many variables involved that it is difficult to isolate the primary contributor to failure.

The presence of microcracks implies that structural microstresses have been relieved by microcrack formation. However, now the cracks must be regarded as stress concentrators, and the following questions apply. Are the cracks large enough to be significant as initiators of service cracks? Do they generally reside in locations of maximum potency? The cracks are mostly located within the confines of the prior austenite grain. If a service crack initiates at a grain boundary and starts to propagate along the grain boundaries, most of the microcracks are out of harm's way. When a service crack is, or becomes, transgranular, microcracks would likely hasten its propagation. As yet there is no clear evidence of just how detrimental microcracks are to the service life of a component. Most microcracks are probably not close enough to the surface and not of critical defect proportions to initiate a transgranular service crack. Nevertheless, a crack is a crack, and it should be regarded with some mistrust.

If it is assumed that cracks will be closed by compressive macrostresses and opened by tensile macrostresses, then the influence of microcracks could vary according to situation. Carburized and hardened cases contain residual compressive macrostresses, which reduce any known or unknown adverse traits that microcracks might have. Through-hardened high-carbon steels, on the other hand, might have surfaces in which the residual macrostresses are decidedly tensile after quenching. The macrostresses in this instance add to any adverse influence of the microcracks.

Influence on Fatigue. Apple and Krauss (Ref 43) investigated the influence of microcracking on the fatigue resistance of case-hardened samples, using different quenching treatments to vary the microcrack population. The quenching treatments led to differences in grain size, hardness, retained austenite content, and microcrack density (Fig. 5.21). With so many variables to consider, it is difficult to assess the contribution of any one of them. Even so, Apple and Krauss (Ref 43), through examining the data including fracture characteristics, concluded that the difference between the fatigue strength of the direct quenched and reheat quenched samples may be attributed to the differences in microcrack distribution and size. Also, the absence of microcracks at the immediate surface of the double reheated samples may, in part, account for the relatively superior test results. In viewing these results, one might speculate that grain size and hardness accounted for about 80% of the difference in fatigue results, and microcracking accounts for the rest.

Of the three heat-treated conditions (Ref 43), the largest microcracks were observed in the double quenched samples, although the crack severity was the least. However, the overall maximum crack severity was not at the surface but about 0.1 to 0.2 mm beneath the surface. Panhans and Fournelle (Ref 44) reported the

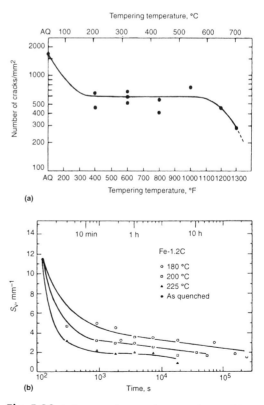

Fig. 5.20 Influence of tempering on microcracking. (a) Effect of tempering temperature on the number of cracks per unit volume. (b) Effect of tempering temperature and time on S_v, microcrack area per unit volume of specimen. Source: Ref 41, 42

same findings. Removal of a surface layer, by whatever means, could actually bring the zone of maximum crack incidence to the surface.

Control of Microcracking. The use of grain-controlled steels and the control of carburizing condition to limit surface carbon content help reduce the likelihood of microcrack formation. The quenching conditions then become important to ensure that the grain size remains fine.

High-temperature carburizing at greater than 1000 °C can produce coarse-grained or mixed coarse-and-fine-grained microstructures, because the grain-coarsening temperature has been exceeded. If grain size and microcracking are related, a high incidence of microcracking might be expected when quenching a coarse-grained material from above the Ac_{cm} or with an increased surface carbon content. When a steel is carburized from 925 to 935 °C and below the conditions necessary to promote grain coarsening, the resultant grain size is likely to be normal for that quench. The extent of the microcracking should be less than it is for the high-temperature treatment. Again, an increase of the carbon content could lead to more microcracks.

Quenching from below the Ac_{cm} produces a finer grain size and forms carbides in high-carbon surfaces (or the lack of carbide solution during reheating for quenching). If some of the carbon is tied up as carbides, then the matrix material has a lower carbon content and, hence, a reduced risk of microcracking during quenching. In reheat quenched materials, microcracking is rare (Ref 26). However, when considering a lower quench-

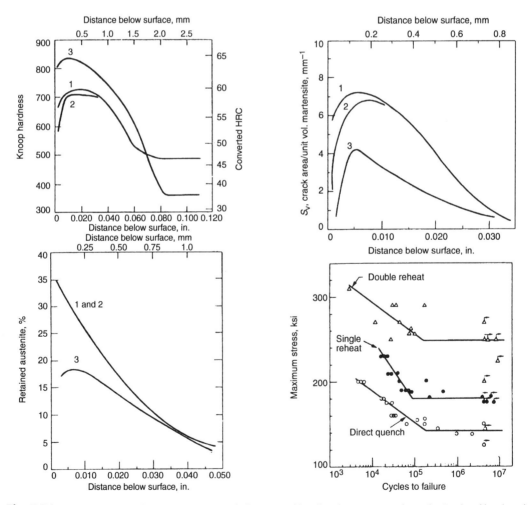

Fig. 5.21 Hardness, retained austenite, microcrack density, and bending fatigue curves for carburized and hardened SAE 8620 steel quenched by three methods: direct quench, ASTM 1–3 grain size; single reheat, ASTM 4–5 grain size; double reheat. Source: Ref 43

ing temperature to control microcracking, it is necessary to be mindful of the effect that low-temperature quenching will have on the core microstructure. If ferrite is unacceptable in the core, then there is a limit to how low the quenching temperature can be reduced.

Krauss (Ref 26) reported that quenching an initially coarse-grained steel from below the Ac_{cm} causes fine microstructural features, few microcracks, and that on testing the overload fractures were transgranular. Quenching from above the Ac_{cm} produces a coarse-grained structure, microcracks, and predominantly intergranular overload fractures.

The rate of quenching has already been discussed; it has little significance on microcrack severity. Interrupted quenching, on the other hand, is very significant. Lyman (Ref 46), investigating microcracking in AISI 52100 steel, quenched the austenitized samples into oil held at a temperature just below the M_s (138 °C) of the steel. After an appropriate dwell, the samples were transferred into a salt bath held at 260 °C and then quenched off. The aim was to temper, stress relieve, and toughen the first formed martensite. The results are as follows:

- Without the 260 °C temper, simply an arrest at just below the M_s before continuing the quench, microcracking was severe.
- When the sub-M_s arrest prior to 260 °C temper produced ~10 to 20% martensite, only a few microcracks formed.
- When the martensite formed prior to tempering was ~30 to 40%, no microcracks formed.

Clearly, there are a number of ways of controlling microcracking, including tempering, which can heal the smaller microcracks.

Microsegregation

For design purposes, it is generally assumed that steels are homogenous. Unfortunately, they are not. Macrochemical analysis surveys of ingots cast from a single ladle of molten steel reveal differences of composition. These surveys also show that composition varies from the bottom to the top and from the center to the outside of each ingot. A batch of forgings made from a single ingot will therefore exhibit part-to-part composition variability, as well as variations within each part. However, such variations do not generally depart from the intended steel specification range. What is of more concern is the variability that occurs on a microscale over very small distances. This variability is referred to as microsegregation.

Formation of Microsegregation

Ingot Solidification. Microsegregation of steels arises during the solidification process. In this section, only microsegregation caused by solidification following casting is discussed. Subsequent segregation of elements during heat treatments, such as the segregation of phosphorus to austenite grain boundaries during austenitizing prior to quenching, is not addressed here.

When a steel is cast into the ingot mold, the first material to solidify is that adjacent to the cooler mold walls. If there is no special preparation applied to the mold walls to slow the cooling rate, then the first metal to solidify, called a "chilled" surface layer, is a thin layer of small equiaxed crystals with the same composition as the liquid metal. At the inner surface of the chilled layer, the situation is energetically favorable for crystals above a critical size to continue growing inward. The growth is cellular initially but soon gives way to dendritic growth parallel to the thermal gradient. The dendrites grow long (see the columnar zone, Fig. 5.22 and 5.23), but such growth slows as the thermal gradient becomes flatter. Meanwhile, in the liquid ahead of the thermal gradient when the temperature has fallen sufficiently, nucleation takes place at many sites (i.e., on suitable substrates, such as nonmetallic inclusions), followed by uniform dendritic growth at each site until solidification terminates

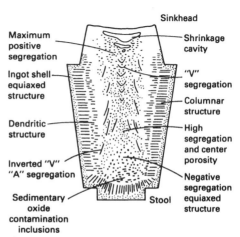

Fig. 5.22 Schematic illustrating macrosegregation in a large steel ingot

when neighboring grains impinge. These grains constitute the equiaxed grained zone at the center of the ingot.

The Development of Microsegregation. An alloy of composition X_0 is presented in the hypothetical constitution diagram of Fig. 5.24. When the temperature of the molten metal falls to the liquidus at X_1, solid material of composition A_1 nucleates within the melt. With further cooling, the composition of any growing crystal changes from A_1 at its center through to A_4 on its surface. Meanwhile, the composition of the surrounding liquid changes from X_1 to X_4, which is the last of the liquid to solidify. Therefore, there is a distinct difference in impurity or solute content between the first and the last materials to solidify. This difference is what is called "microsegregation." Figure 5.25 shows an example of microsegregation in which the composition of a cross section of a dendrite is recorded (Ref 47). It shows a comparatively low solute content, in this case nickel and chromium, at the center of the dendrite.

As the columnar grains grow forward together from the surface (Fig. 5.23), they push the more impure liquid ahead of them toward the central zone. Therefore, the central zone contains more of the impure material than the columnar zone. When the central zone has almost solidified, the least pure liquid, including the lower melting point nonmetallic inclusions, has nowhere to go except to the interfaces between the impinging dendrites. Consequently, on final solidification, the central zone is the more heavily macro- and microsegregated (Fig. 5.26). Laren and Fredriksson (Ref 47) observed that the maximum microsegregation in 1 and 9 ton ingots occurs at three-fourths of the distance from the surface to the center of the ingot.

Steels solidify in the manner described, although the proportion of columnar growth to equiaxed growth varies depending on the length and steepness of the thermal gradient. Thus, the mode of solidification is essentially columnar for continuously cast steels and equiaxed for very large conventionally cast ingots.

Microsegregation Tendencies of Alloying Elements. Steels are complex alloys containing many elements. Some elements are intentional and wanted, and others are residual and possibly unwanted. Each element has its own segregation tendency, although in some instances the segregation tendency of one element can be modified by the presence of others. For example, silicon and manganese influence the segregation of molybdenum, and manganese influences that of sulfur. The generally accepted order of the susceptibility of elements to segregation is, from most prone to least: sulfur, niobium, phosphorus, tin, arsenic, molybdenum, chromium, silicon, manganese, and nickel. Table 5.5 provides a basis of comparison for some of the common elements. Table 5.5 clearly indicates that, when solidification has been directional (columnar growth), the intensity of segregation is less than in the central, equiaxed grain zone.

The Effect of Mechanical and Thermal Treatments. Case-hardening steels are generally supplied in the wrought condition, that is hot rolled or forged to some convenient shape and size. During these hot-working processes, the microsegregated areas are given a directionality related

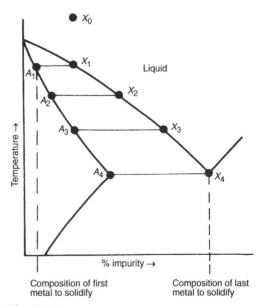

Fig. 5.24 Schematic constitutional diagram of how compositional variations develop during solidification

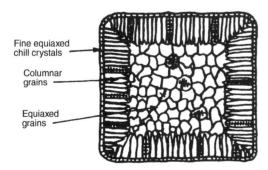

Fig. 5.23 Section through a conventionally cast ingot

to the amount and direction of working. Not only that, the intensity of microsegregation is reduced somewhat, as indicated in Table 5.6 for chromium and nickel segregation.

A high degree of homogenization can be effected thermally by soaking the segregated material at an elevated temperature. However, the soak times to ensure virtual complete homogeneity can be very long, particularly below 1200 °C (Fig. 5.27). Nevertheless, a significant reduction of microsegregation can be achieved by a balanced combination of mechanical and high-temperature heat treatments.

Additional mechanical and/or thermal treatments to remove or reduce microsegregation add to the cost of manufacture. Unless there is a clear-cut reason to have a homogenous material, then perhaps things are best left as they are.

Macrostructures and Microstructures. The distribution of microsegregation in wrought steels depends on how much working has been done to shape the part. In a large forging with little deformation, the cast structure is not eliminated, and much of the dendritic form of segregation persists (Fig. 5.28). In small forgings, on the other hand, the metal is gener-

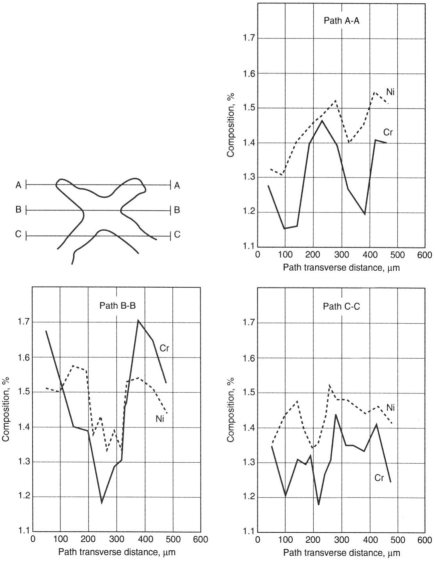

Fig. 5.25 Chromium and nickel segregation around a dendrite cross. Melt composition: 0.36C, 0.35Si, 0.68Mn, 1.48Cr, 1.44Ni, 0.20Mo. Source: Ref 47

ally more severely worked. While segregation, albeit less intense, survives the mechanical treatment, it is redistributed as alloy-rich and alloy-lean bands flowing in directions dictated by the working process (Fig. 5.29, 5.30).

In microsegregated materials, the heat treatment response of the alloy-rich areas is different from that of the alloy-lean areas. Consequently, it is possible for each to transform at different times and to different products on cooling from the austenitizing temperature. In low- and medium-carbon steels, austenitized and slow cooled, the microsegregation is identified as alternating areas of ferrite and pearlite (banding).

Reaustenitizing and fast cooling in the pearlitic range causes the carbon to redistribute fairly evenly to give the impression that the segregation has been removed. Unfortunately, the alloy segregation will still be there, essentially unaffected. It is merely masked by the redistributed carbon. In the surface of a carburized case where the carbon may be evenly distributed, the presence of microsegregation might be detected by observing alternate areas of martensite and bainite or of

Table 5.5 Segregation indices for some elements in steel

Alloying element	Segregation index (IS)(a)	
	Columnar crystals	Equiaxed crystals
Chromium	1.31–1.38	1.47–2.10
Nickel	1.06–1.24	1.10–1.50
Manganese	1.07–1.27	0.40–1.95
Molybdenum	1.70–2.00	1.9–7.3
Niobium	...	>2000
Phosphorus	2.05	>170
Arsenic	1.23	~75

(a) (Concentration at interdendritic site)/(concentration at dendritic core). Source: Ref 48, 49

Table 5.6 Changes in microsegregation intensity due to hot working between 1200 and 1000 °C

Alloying element	Segregation index (IS)					
	Columnar crystals			Equiaxed crystals		
	A	B	C	A	B	C
Chromium (as cast)	1.15	1.16	1.72	1.54	1.35	1.69
Chromium (98% reduction)	1.08	1.12	1.45	1.20	1.14	1.46
Nickel (as cast)	1.12	1.08	1.40	1.30	1.27	1.42
Nickel (98% reduction)	1.02	1.08	1.39	1.29	1.05	1.39

Source: Ref 50

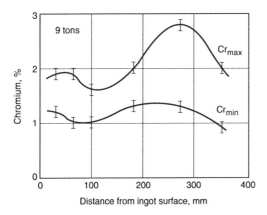

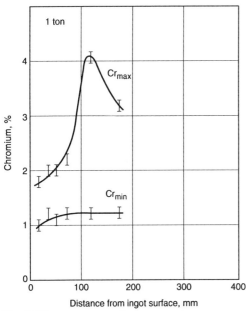

Fig. 5.26 Maximum and minimum chromium concentrations as a function of distance from surface. Source: Ref 47

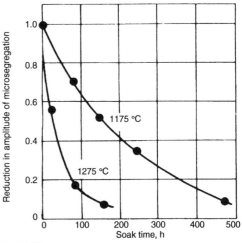

Fig. 5.27 Effect of homogenization temperature and time on the intensity of manganese segregation in an EN39 steel. Source: Ref 51

austenite and martensite, depending on the actual compositions and the cooling rates involved. Where the tempered martensite normally etches dark, a light 2% nital etch should determine if significant microsegregation is present.

Effects of Microsegregation on Properties

Influence on Hardness. Figure 5.31(b–d) shows an example of the variations in microstructure that can result from the presence of microsegregation; the cooling rates are those of an end-quenched hardenability bar. Within this range of cooling rates, the alloy-rich areas have transformed to either martensite or bainite. Meanwhile, the alloy-lean areas have transformed to martensite and bainite with increasing amounts of pearlite and ferrite with the slower cooling rates. Figure 5.31(a) relates these microstructures to the hardness.

Microsegregation influences the hardness of the case, particularly if it leads to local concentrations of retained austenite or bainite in the mainly martensitic structure. Both austenite and bainite are softer than the martensite at a given carbon level.

Influence on Tensile and Toughness Properties. The tensile strength and yield strength are unlikely to be affected by the presence or absence of microsegregation. The ductility indicators, on the other hand, will be affected. Kroneis et al. (Ref 53) showed that the reduction of area of a conventionally melted and forged medium-carbon alloy steel (5:1 to 10:1 reductions) was about

Fig. 5.28 Dendritic microsegregation in a fractured gear tooth. 2×

Fig. 5.29 Section through a forged sliding clutch hub. 0.8×

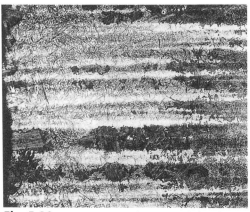

Fig. 5.30 Microstructure of an air-cooled carburized bar end. 50×

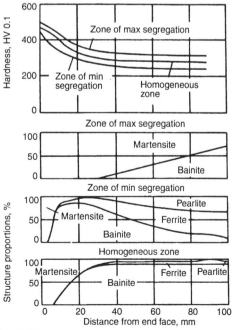

Fig. 5.31 Hardenability curves and corresponding microstructures for a 25CrMo4 steel. Source: Ref 52

doubled by special treatment to remove the microsegregation. An electroslag melted steel was also improved by the same balanced mechanical and thermal treatment. At tensile strengths typical for the core of a carburized part made from a wrought blank, it is unlikely that microsegregation has much influence on ductility and toughness indicators in the longitudinal direction. On the other hand, it does have a pronounced negative affect on those properties in the transverse direction (Fig. 5.32).

Influence on Fatigue. It is difficult to determine the effect of microsegregation alone on the various properties of a steel because of the presence of nonmetallic inclusions. Similarly, it is difficult to isolate the influence of nonmetallic inclusions on properties. However, if it is assumed that electroslag remelted (ESR) steels are essentially cleaner than the conventionally melted steels, then some measure of the respective influences of microsegregation and nonmetallic inclusions can be obtained.

Kroneis et al. (Ref 53, 54) carried out such a comparison using a Cr-Mo-V steel and showed the ESR steel to have superior general properties. In this way, the effect of nonmetallic inclusions in particular could be assessed. With a special treatment to reduce the microsegregation of the steels melted by the two processes (ESR and conventional), the properties of each were further improved. For example, the bending fatigue strength of the ESR material (Fig. 5.33) was superior to that of the conventionally melted steel. Therefore, the conclusion is that nonmetallic inclusions have a significant influence on the fatigue strength of this particular high-strength, through-hardened, and tempered steel. The fatigue strength of the ESR steel is improved even further by applying a special treatment to reduce the severity of the microsegregation. The fatigue limit of the conventionally melted steel is little enhanced by the special treatment, although in the high-stress, low-cycle region, the treatment did lead to an improvement.

What these tests suggest is that for high-cycle fatigue applications, the effect of nonmetallic inclusions far outweigh any adverse influences of microsegregation. Presumably, the nonmetallics provide sites at which fatigue cracks can initiate and from which they can grow. In the low-cycle domain (in this example, $<2 \times 10^5$ cycles), microsegregation appears to be more influential than nonmetallic inclusions.

These test results (Ref 53, 54) refer to a high-strength, through-hardened material, and whether the same trends apply to a carburized and hardened part is not really known. Nonetheless, some speculation can be made. For example, for the core, there will be a similar trend: microsegregation can affect low-cycle fatigue and have little, if any, influence on the high-cycle fatigue strength. For the case, the local transformation behavior might be affected by microsegregation. However, the effect will be detrimental only if it leads to the formation of mixed microstructures, such as martensite plus bainite or martensite plus austenite, depending on the grade of steel. If, for example, the retained austenite at a carburized and hardened surface is patchy, varying appreciably over short distances due to microsegregation (Fig. 5.34a, b), then the microhardness will vary accordingly, as too will the local residual macro- and microstresses. The bending fatigue strength would then approach that of the softer transformation product.

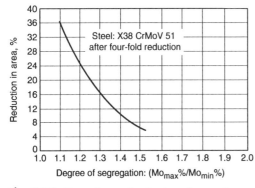

Fig. 5.32 Dependence of reduction of area of transverse specimens on degree of molybdenum segregation. Hardened and tempered; tensile strength, 1550 to 1770 MPa. Source: Ref 54

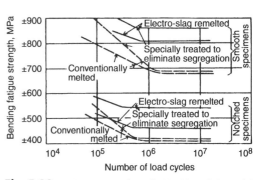

Fig. 5.33 Fatigue curves for a Cr-Mo-V steel. Material reduced four fold and heat treated. Specimen position, longitudinal at 1/2 radius; tensile strength, 1900 MPa. Source: Ref 53

Other Aspects of Microsegregation. A danger period for heavily microsegregated steels, especially the more alloyed steels, is during a slow cool after carburizing. The slow rate of cooling can lead to a carburized layer microstructure that consists of distinct areas of HTTP residing alongside areas of low-temperature transformation products. The short range residual stresses developed during the cool can be very high and can induce cracking.

Another danger period is during the grinding of the carburized and hardened surface. If the structure consists of alternating areas of austenite and martensite, each of these features will react differently to the action of the abrasive wheel. Grinding will deform and possibly tear the austenite volumes in the surface, and also cause the immediate surface temperature to rise more than would a wholly martensitic surface. Therefore, burning and cracking will tend to occur more easily in a surface that contains alternating areas of martensite and austenite (See Chapter 8 on grinding).

With regard to machining during the early stages of manufacture, microsegregation can be responsible for heavy tool wear, although much depends on the actual microstructure. Banded structures, consisting of layers of ferrite and pearlite, each one grain thick, machine better than those where bands are several grains thick that give rise to tearing through the ferrite and its adhesion to the tool tip.

Nonmetallic Inclusions

Origin of Nonmetallic Inclusions

Steels are not commercially made free from nonmetallic inclusions, nor are they likely to be. It is estimated that there are millions of inclusions per tonne of steel. Table 5.7 indicates this estimate for a hypothetical steel in which the oxygen and the sulfur contents are 1 ppm. In actuality, steels contain more oxygen and sulfur (>10 ppm); however, increasing the oxygen and sulfur do not so much increase the number of inclusions, but rather is more likely to increase their average size. Also, inclusions within a piece of steel are not one size, or close to one size, but are within a wide range of sizes. A clean steel, therefore, may have 10^{12} inclusions per tonne where most are less than 0.2 μm in diameter and few exceed 20 μm. An average steel has more inclusions of this size range, and dirty steels contain many inclusions having sizes much larger than 20 μm.

In steels, there are two classes of nonmetallic inclusions: exogenous and indigenous. The former occurs as a result of pieces or particles of

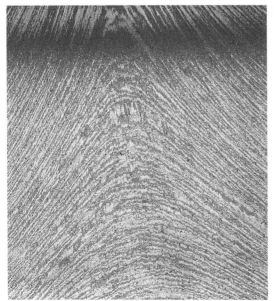

Fig. 5.34(a) Flow of microsegregation caused by forging in a Ni-Mo steel part. Note the surface carburized layer.

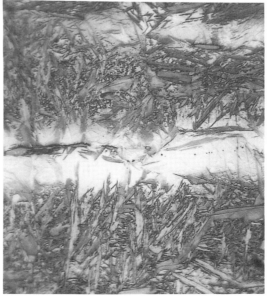

Fig. 5.34(b) Nonmetallic inclusions and banding in a heavily microsegregated 1% C alloy steel sample. Retained austenite, light

refractory material being separated from surfaces of the steelmaking equipment with which the molten steel has made contact (e.g., furnace linings, runners, and ladles). Such inclusions can range from extremely small to quite large lumps. They can provide surfaces onto which the indigenous types of inclusions can nucleate, grow, and with which they can chemically react. The indigenous inclusions form as a result of reactions that take place during steelmaking and the deoxidation processes (deoxidation products). Primary deoxidation takes place in the melt at high temperatures, followed by secondary deoxidation during the liquid cool, and to some extent when the steel has solidified. Much of the nonmetallic material present in deoxidized liquid steel rises to the surface, leaving a fair quantity either trapped in the dendrites growing in from the surface or carried to the bottom of the ingot by falling dendrites. Small inclusions, which rise only slowly in the liquid metal, are more likely to be retained in the steel.

The products of deoxidation are the oxide types of inclusion, and the oxides of most importance include aluminum oxide, Al_2O_3, and silicon dioxide, SiO_2. The main product of the desulfurizing reactions is manganese sulfide, MnS. Without adequate quantities of manganese present in the steel (>0.3%), iron sulfides form, leading to problems of hot-shortness during metalworking operations. These compounds are the pure forms, but they can and do combine with other inclusion-making materials, such as $MgO \cdot Al_2O_3$ or iron and manganese silicates. The oxide, SiO_2 itself can have modifications (cristobalite, tridymite, and quartz). The manganese in MnS can be substituted in varying degrees by other elements present in the system, such as iron and chromium. Furthermore, inclusions of one type can be found attached to or within inclusions of another type. Thus, the subject of nonmetallic composition and identification is quite complex. Nevertheless, to an experienced metallographer, a study of the inclusions observed within a sample of steel can be of value in assessing how that steel was made.

The distribution and the average size of nonmetallic inclusions in a finished ingot is not usually uniform. It depends on the melting and the casting-pit practices and the method and materials of the deoxidation sequence. For example, in a killed steel, the largest population of oxide inclusions is in the bottom part of the ingot. Increased aluminum additions progressively reduce that population. Sulfides tend to be smaller in the columnar solidification zone than in the equiaxial solidification zone, and the central zone of the middle of the ingot contains more sulfide than either the top or the bottom of the ingot.

Anisotropy and Nonmetallic Inclusion Shape Control. One problem encountered by steel users is the difference between longitudinal and transverse properties, called the anisotropy effect (Table 5.8). This effect is primarily related to the presence and behavior of deformable nonmetallic inclusions and, to some extent, microsegregation, which become elongated in the direction of working. Manganese sulfide is a significant contributor to the anisotropy effect; it is soft and deformable under cold and hot working conditions. However, the ability to deform tends to increase with type (I, II, or III) and decrease with substitution of manganese by other elements in the system, such as iron, chromium, and nickel, which raise the hardness of the inclusion (Table 5.9). Calcium aluminate ($CaO \cdot Al_2O_3$) and corundum (Al_2O_3) do not deform, and pure silica inclusions are brittle at metalworking temperatures. Therefore, unless they are present as stringers, they contribute less than MnS to the anisotropy effect. Iron-manganese silicates are deformable above 850 °C, so in hot-worked steels they too contribute to the difference between longitudinal and transverse properties.

Table 5.7 Influence of inclusion size on inclusion numbers in a hypothetical steel with 1 ppm oxygen and 1 ppm sulfur as spherical Al_2O_3 and MnS inclusions

Inclusion diameter, μm	No. of inclusions/t	Volume of steel per inclusion	Mean distance between inclusions
10^3	2.3×10^3	55 cm^3	3.8 cm
10^2	2.3×10^6	55 mm^3	3.8 mm
10	2.3×10^9	55×10^6 μm^3	380 μm
1	2.3×10^{12}	55×10^3 μm^3	38 μm
10^{-1}	2.3×10^{15}	55 μm^3	3.8 μm
10^{-2}	2.3×10^{18}	55×10^{-3} μm^3	3.8 μm
10 Å	2.3×10^{21}	55×10^{-6} μm^3	380 Å

Source: Ref 55

Steelmaking processes, such as electroslag remelting (ESR) or vacuum arc remelting (VAR), reduce the amount of significantly sized nonmetallic materials within a steel (Ref 57). The ESR process produces greater freedom from sulfides, whereas VAR gives the steel a lower oxide content (Ref 55). Electron beam remelting appears to effectively reduce both sulfides and oxides (Ref 58).

Calcium, rare earth (RE), or titanium treatments further decrease the number of visible inclusions and better distribute and modify them. These treatments make the manganese sulfide inclusions less deformable during metalworking processes, thereby significantly reducing the anisotropy effect (Ref 58). A reduced anisotropy is of special interest to plate users and fabricators.

The composition, size, and distribution of nonmetallic inclusions within a steel are initially determined by the steelmaking parameters, such as the boiling time, the refractory type, the deoxidizing materials, the deoxidation practice, and the tapping and teeming methods. Modern steelmaking methods, in particular the remelting technology, coupled with a good understanding of the compositions and properties of nonmetallic materials, make it possible for the steelmaker to adjust the cleanness of a steel according to the eventual application.

Stability. At one time it was thought that nonmetallic inclusions were relatively stable. However, because reactions in such materials take place fairly slowly, inclusions are regarded as being initially metastable (in the as-cast ingot). For example, after cooling, the sulfur within the ingot may be tied into an iron-manganese-sulfur inclusion where the iron content is high. However, during subsequent heat treating, the iron component is replaced by manganese from the steel matrix, thereby producing a truer MnS particle surrounded by manganese depleted steel. High-temperature heat treatments can also bring about shape changes (e.g., rods of type II MnS can spheroidize) (Ref 59), composition and phase changes (Ref 60, 61), and the precipitation of one type of inclusion within another (e.g., CrS within MnS). In some instances, partial or complete dissolution of sulfides can occur (Ref 62), followed by reprecipitation at prior austenite grain boundaries, as in overheating.

Effects of Nonmetallic Inclusions

Influence on Tensile Properties. Nonmetallic inclusions, typical of the commercial grades of wrought steels, do not significantly influence the ultimate tensile strength or the yield related properties of a material in either the longitudinal or transverse directions. On the other hand, the ductility indicators (reduction of area and elongation) are affected, at times appreciably so. Thus, for two steels that are identical apart from their nonmetallic counts, the stress-strain curves are basically the same up to the point of maximum stress. Beyond the maximum stress, when an unstable condition develops and necking occurs, the cleaner steel necks down more (a larger reduction of area) and stretches further (larger elongation) than the less clean steel (Fig. 5.35). For a given steel, the reduction of area tends to fall as the tensile strength increases. The effect, however, becomes more pronounced as the inclusion count increases, as Fig. 5.36 (Ref 58) shows (where values of true strain E_f were determined from the cross-sectional areas at the fractures and, therefore, reflect on the reduction of

Table 5.8 Effect of sulfide inclusions on mechanical properties in a Cr-Ni-Mo steel

Property	Many inclusions	Few inclusions
Ultimate tensile strength, MPa (kg/mm^2)	775 (79)	700 (72)
Reduction of area, %		
Longitudinal	62.8	59.5
Transverse	17.5	29.7
Reduction of area anisotropy, transverse/longitudinal	0.28	0.50
Bending fatigue, MPa (kg/mm^2)		
Longitudinal	390 (39.4)	375 (38.5)
Transverse	320 (32.4)	320 (32.6)
Fatigue anisotropy, transverse/longitudinal	0.82	0.85
Fatigue ratio, longitudinal fatigue/ ultimate tensile strength	0.5	0.54

Source: Ref 56

Table 5.9 Hardness of nonmetallic inclusions

Nonmetallic inclusion	Microhardness, kgf/mm^2
Manganese sulfides(a)	
MnS	170
MnS + Cr	<450
MnS + Ni	<250
MnS + Co	<240
MnS + Fe	<300
MnS + Ti	<215
MnS + V	<340
Al_2O_3	>3000
SiO_2	1600
$MnSiO_2$	750
MnO	400

Nonmetallic inclusions with MgO present, usually from refractories, tend to be hard (1000 to 1200 kgf/mm^2) or very hard (2100 to 2400 kgf/mm^2), such as $MgO \cdot Al_2O_3$. (a) Room temperature, includes manganese sulfides in which manganese has been substituted by other elements. Source: Ref 55

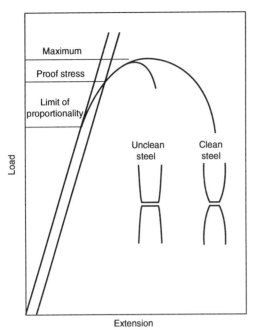

Fig. 5.35 Load/extension curves indicating the trend formed by steel cleanness

area). Ductility indicators are adversely affected by any type of nonmetallic, but the stringer type is more detrimental than the globular type (Ref 63). Also, the more elongated the stringer is, the worse the effect will be.

The anisotropy effect due to metalworking has already been mentioned, though not in any detail. The ductility indicators in conventionally produced steels are lowest in the short transverse direction, intermediate in the long transverse direction, and highest in the longitudinal direction (Ref 64). The variation of reduction of area (ROA) relative to orientation is illustrated in Fig. 5.37. An example of the effect that forging ratio has on the ROA is shown in Fig. 5.38, which shows that anisotropy increases with the amount of metalworking. However, these results are from about 1949; since then, steelmaking and metalworking processes have advanced appreciably so that differences due to orientation can be better controlled (Table 5.10). In the cleaner grades of steel, the difference in ductility between the longitudinal and long transverse directions can be kept small, although that for the

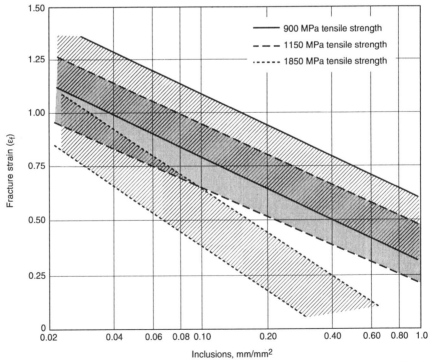

Fig. 5.36 Scatterbands for the tensile true fracture strain of a 0.4% C steel at three strength levels as a function of the total inclusion projected length in the fracture plane. Inclusions, total projected length. Source: Ref 58

short transverse direction lags some way behind. Even so, it might not be necessary to make steels with equal ductility in all directions. The important achievement would be to have adequate ductility in the short-transverse direction, and this can be done by producing a low-sulfur steel along with a RE treatment.

Both microsegregation and nonmetallic inclusions vary depending on where in the forging or bar the samples were taken. Consequently, because they both have an effect on the ductility indicators, the ductility should vary with location. Table 5.11 shows that near surface ductility is superior to that at the mid-radius or center locations. This comparison might be of interest to gear manufacturers who, in general, cut the gear teeth in the near surface region of the forging or bar.

Influence on Fatigue Resistance. Leslie (in Ref 65) considered that in the high-cycle regime ($>10^5$ load cycles) nearly all fatigue cracks initiate at nonmetallic inclusions. In the low-cycle regime (10^3 to 10^5 load cycles) and especially in the 10^1 to 10^3 cycle range, slip band cracks provide the initiation sites. The potency of a nonmetallic inclusion with respect to fatigue stressing depends on the chemistry, size, location, and quantity of inclusions; strength of the steel; and the residual stress state immediately adjacent to the inclusion (Ref 66).

Effect of Inclusion Chemistry. The hard types of inclusions that resist deformation during processes, such as rolling or forging, have a tendency to develop cone-shaped cavities in the direction of metal flow during the working operations (Ref 67). These cavities, which do not weld up again, increase the effective size and the stress-concentrating effect of the inclusion, thereby becoming more potent as a fatigue crack initiation site. These hard, undeformable types of inclusions, such as single-phase alumina (Al_2O_3), spinels (e.g., $MgO \cdot Al_2O_3$), calcium aluminates (e.g., $CaO \cdot Al_2O_3$), titanium nitride, and some silicates, all have lower thermal expansion coefficients than steel (Fig. 5.39). Therefore, a tensile stress field should develop around each inclusion as a result of hardening heat treatments, and the tensile stresses could then add to any stresses applied in service. Thus, with

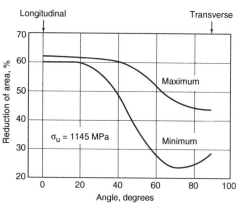

Fig. 5.37 Relationship between reduction of area and angle between the longitudinal direction in forging and the specimen axis. Source: Ref 64

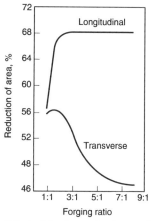

Fig. 5.38 Effect of forging reduction on longitudinal and transverse reduction of area. Tensile strength, 840 MPa (118 ksi). Source: Ref 64

Table 5.10 Mechanical properties of 15NiCrMo 16 5 carburizing steel after melting (after K. Vetter)

Melting type	Sulfur content, %	Degree of deformation	0.2 proof stress, MPa	Tensile stress, MPa	Elongation at fracture(a), %		Reduction of area, %		Notch toughness(b), J/cm²	
					Longitudinal	Transverse	Longitudinal	Transverse	Longitudinal	Transverse
Conventionally melted (electric furnace)	0.010	~6×	1160	1450	10	7	48	27	50	35
Electroslag remelted	0.004	~6×	1160	1450	12	10	55	46	...	69
Electron-beam remelted	0.007–0.009	10 to 30×	1060–1120	1350–1420	13.5	13.5	58	51	83	65

(a) $l_o = 5d_o$. (b) DVM specimen. Source: Ref 57

Table 5.11 Mechanical properties of heat-treated forged bars of X 41 CrMoV 51 steel

Position of specimen in ingot	Ingot type of melting	Elongation at fracture(a) Longitudinal	Elongation at fracture(a) Transverse	Reduction of area, % Longitudinal	Reduction of area, % Transverse	Ratio of transverse/longitudinal specimen Elongation	Ratio of transverse/longitudinal specimen Reduction of area
Edge	Conventionally melted (electric furnace)	12	6	45	12	0.5	0.27
	Electroslag remelted	12	12	45	43	1.0	0.97
½ radius	Conventionally melted (electric furnace)	10	2	34	5	0.2	0.15
	Electroslag remelted	10	10	40	38	1.0	0.95
Center	Conventionally melted (electric furnace)	10	2	30	4	0.2	0.13
	Electroslag remelted	10	9.5	40	36	0.95	0.90

Diameters from 230 to 350 mm; tensile strengths from 1550 to 1700 MPa; averages of ~200 heats. (a) $L_0 = 5d_0$. Source: Ref 57

respect to properties, the hardest nonmetallic inclusions are thought to be the most harmful (Ref 68–70), and the softer inclusions, such as manganese sulfide and manganese oxide, are the least harmful (Ref 70). The softer inclusions are essentially harmless even in high-strength bearing steels (Ref 60), although there could be a little more life variability. A hard inclusion can be rendered less deleterious if it contains or is enveloped by manganese sulfide. This is due in part to the deformability of manganese sulfide and in part to the tendency of hard nonmetallic inclusions to be noncoherent with the ferrous matrix, whereas manganese sulfide is semi-coherent.

In general, the term "nonmetallic inclusion" refers to the exogenous and indigenous types. However, a third type develops by the intentional addition of lead, which is essentially im-

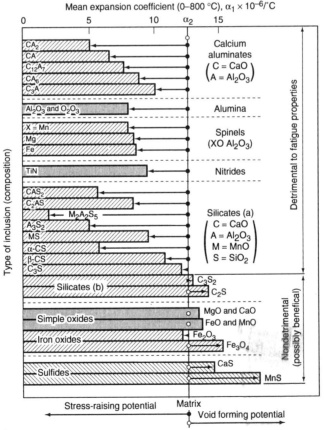

Fig. 5.39 Stress-raising properties of inclusions in 1%C-Cr bearing steel. Source: Ref 55

miscible in steels and which forms a fine dispersion within a lead-treated steel. It is added for the sole purpose of giving the steel free-machining properties. Lead is soft and, therefore, contributes to anisotropy in wrought steels similar to manganese sulfide. Goldstein et al. (Ref 71) found that lead reduced the fatigue limit by about 8% (compared with unleaded steels) but anisotropy accounted for an ~30% difference in the fatigue limits of both leaded and unleaded steels.

Effect of Inclusion Size and Location. Critical inclusion size varies depending on the composition of the inclusion, because composition determines shape (round or angular) and whether or not residual stresses, or cavities, are associated with the inclusion. Critical size increases with distance from the surface; the further an inclusion is from the surface, the larger is its critical size. For alumina inclusions, the critical size for an inclusion close to the surface is 10 μm, but it increases to 30 μm when the inclusion is located 100 μm from the surface (Ref 72). Kawada et al. (Ref 73), studying the bending fatigue strength of hardened bearing steels, agreed with the 10 μm critical defect size for a surface inclusion, but they were not so optimistic regarding the critical size at 100 μm (Fig. 5.40) unless the local data are averaged. When a subsurface inclusion is above the critical size, its strength reducing effect is proportional to the cube root of its diameter (Ref 72).

The critical inclusion or defect size with respect to fatigue loading is not the same for fracture toughness or metalworking situations.

Effect of Inclusion Quantity. For an inclusion to initiate a fatigue crack, it must be in the path of the applied stress. Often that path is quite narrow, such as at the fillet of a gear tooth. The probability of an adequately sized inclusion being critically located within a narrow load path increases as the number of adequately sized nonmetallic inclusions in the steel increases. Figure 5.41 shows how the fatigue limit falls as the inclusion count increases. Crack propagation can be assisted or impeded by nonmetallic inclusion, perhaps depending on orientation.

In design, it is not unusual for a stress concentrator to be added close to another stress concentrator in order to reduce the stress raising effect of the original. Therefore, with respect to nonmetallic inclusions, is it possible that neighboring inclusions could reduce the stress raising potency of each other? In other words, are two 40 μm inclusions that are close to one another less harmful than an isolated inclusion of the same diameter?

Effect of Steel Strength. For a critical location within a steel, the critical inclusion size is smaller the harder the steel is (Fig. 5.42). In high-hardness steels including case-hardening steels, calcium aluminates are the most detrimental of inclusions (Fig. 5.43). Calcium aluminates are far more harmful than, for example, alumina or titanium carbonitride, shown in Fig. 5.44 (Ref 65). Manganese sulfide is not particularly damaging (Ref 56), especially in relatively low-strength materials (Table 5.8).

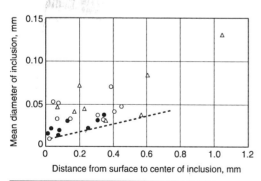

Fig. 5.40 Fatigue fracture of bearing steel caused by inclusions in rotary bending. Distance from steel surface vs. diameter of inclusions that initiated fatigue fracture. Source: Ref 60, 73

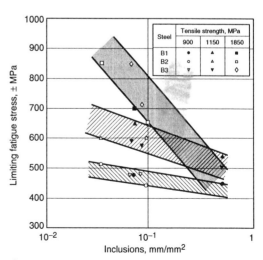

Fig. 5.41 Relationship between the limiting fatigue stress in smooth, rotating bending tests and total inclusion projected length. Results for two test orientations and three strength levels are shown. Source: Ref 58

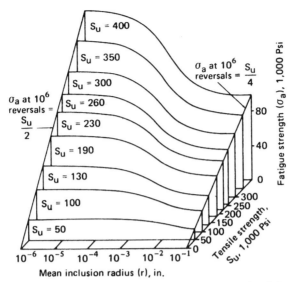

Fig. 5.42 Effect of inclusion size (spherical) on the fatigue strength of steel. Source: Ref 74

Effect of Residual Stresses from Inclusions. As previously mentioned, tensile stress fields can develop around the harder type of nonmetallic inclusion, and cavities can form at hard inclusions during metalworking operations to act as stress raisers. Also, angular inclusions (Ref 75) or strings of fragmented inclusions can act as stress raisers, even without the presence of cavities. In through-hardened and low-temperature tempered steels for which the inclusion critical size is small anyway, such features are detrimental to the fatigue resistance of the steel. In case-hardened parts, the surface layers are hard, which implies that the critical inclusion size is small and that other adverse features are also present. However, a case-hardened surface with its compressive residual stresses can better tolerate such adverse features than can a residual stress-free surface.

Any surface treatment that induces hardening and compressive residual stresses offsets much, if not all, of the adverse influence of nonmetallic inclusions. Cook and Dulieu (Ref 58) and others have demonstrated this with shot peening. Having said that, Fig. 5.43 shows the damaging influence of calcium aluminates

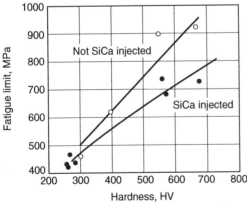

Fig. 5.43 Fatigue limits for SiCa-injected and not SiCa-injected steels at different hardness levels. The SiCa-injected steel with D type inclusions has a lower fatigue limit. Source: Ref 65

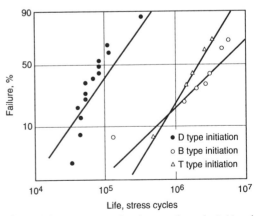

Fig. 5.44 Rotating bending fatigue of samples initiated by B (alumina, irregular), D (calcium aluminate, spherical), and T (titanium carbonitride, cuboid) type inclusions in an SAE 52100 steel. Source: Ref 65

on the fatigue limit of case-hardened test pieces, although it is not clear if the fractures were surface or core initiated. It is not uncommon, in studies related to the fatigue strength of case-hardened test pieces, for failures to initiate at inclusions within the case or in the core just beneath the case where the residual stresses are tensile (Ref 76, 77). Such failures occur in the high-cycle area of the S-N curve, and they are attributed to the presence of hydrogen concentrated at the site of an inclusion (Ref 65).

Influence on Contact Fatigue. As in other cyclic loading situations, nonmetallic inclusions can contribute to failure during rolling or rolling with sliding, such as in bearings and gears. Again, size, hardness, coherency with the matrix, location, and the number of the inclusions are the important factors.

Hard inclusions situated between the working surface and the depth of significant Hertzian stresses are potentially damaging. The more inclusions there are above a critical size and residing in the stressed layer, the more likely they are to reduce the life of the part (Fig. 5.45). Therefore, clean steels with inclusion size control should be more resistant to failure.

The stress associated with a hard inclusion is tensile and at a maximum at the interface between the inclusion and the steel matrix. According to Winter et al. (Ref 78), the stress-raising effect of an oxide inclusion is 2.5. Therefore, some point on the interface, depending on aspects of the inclusion and the direction of the applied stresses, is where damage or cracking eventually initiates. If a cavity exists at an inclusion (due to forging) or a crack develops due to loading, the stress concentrating effect of the inclusion increases. If, on the other hand, shearing (or whatever mechanism is involved) occurs, which gives rise to white etching areas associated with the inclusion called a "butterfly" (Fig. 5.46), then the stress-raising effect of the inclusion is reduced (Ref 79).

Soft inclusions are not regarded as damaging in rolling contact situations. Even so, they could

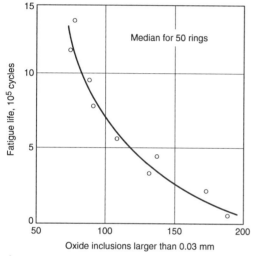

Fig. 5.45 Effect of number of large oxide inclusions on the flaking of rig-tested bearings (1309 outer rings). Through-hardened steel for ball bearings; composition, 1C, 0.5Mn, 1.5Cr. Source: Ref 60

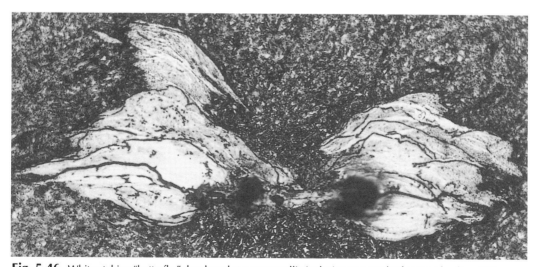

Fig. 5.46 White etching "butterfly," developed at a nonmetallic inclusion as a result of contact loading. 750×

have some effect if they lie in the zone of cyclic stress. Littman and Widner (Ref 79) state that in bearing fatigue tests, cracks and butterflies are almost never associated with sulfide inclusions. Sugino et al. (Ref 80) confirm that MnS is fairly harmless in this respect, but Al_2O_3 (with or without associated MnS) encourages the formation of butterflies (Table 5.12). Winter et al. (Ref 78), referring to gears, observed very small areas of white etching material at sulfide inclusions but did not associate them with contact failures. Consequently, soft inclusions are of less concern than the harder types. Nevertheless, their size, type, and quantities are best controlled to improve other properties.

Influence on Impact Fracture Strength and Toughness. The Charpy toughness (shelf energy) is increased by increasing the cleanness of a steel (Ref 81) or by reducing the average size of inclusions and the number of large inclusions. The transition temperature, on the other hand, is little if at all affected by an increase of cleanness.

The toughness in the longitudinal direction is not necessarily improved by increasing cleanness, but for each of the transverse directions, especially the through-thickness direction, toughness is improved (Ref 58, 81).

Consequences of Producing Clean Steels

Machinability. The improvements to metal shaping machines and tools have not always matched the improvements in steel cleanness. As a result, machining problems are encountered when the sulfur content of a lean-alloy steel falls below about 0.009%. Difficult machining can lead to poor workpiece surface finish and heavy tool wear. Even when the sulfur content is above 0.009%, the machinability can vary somewhat depending on the size of the sulfide particles. Steels with larger particles tend to machine more easily than do those in which the sulfides are small.

Manganese sulfide is well known for its contributions to easy machining and good tool life. It assists in chip breaking of the swarf as it flows from the cutting zone. It is not particularly abrasive to the flank of a tool, and it provides a protective layer on the rake face, which protects against cratering.

The crystalline silicates and alumina (the nondeformable types) tend to give heavy tool wear and are not effective as chip breakers. Those silicates that become deformable at about 850 °C assist machinability when the cutting speed is sufficiently high to generate temperatures in excess of 850 °C in the swarf metal as it shears away from the rake face of the cutting tool.

The ceramic and the coated tools developed for clean steel machining might not be durable for machining the less clean steels. Sulfide inclusions have the potential to chemically react with the cutting surfaces of such materials, thereby increasing tool wear.

Overheating. The appearance of matte facets on the fracture faces of impact test pieces indicates that the steel has been "overheated" during the preheating and hot-working processes. The dissolution of manganese sulfide and its reprecipitation as very fine particles on the austenite grain boundaries during hot working are regarded as the cause (Ref 58). Decreasing the sulfur content lowers the onset temperature for overheating where the onset temperature is at its lowest when the sulfur content is between 0.002 and 0.004%. The steel composition and the cooling rate from the working temperature have an effect on overheating. For example, nickel lowers the overheating temperature, and cooling rates between 10 and 300 °C/min seem to produce more matte facets on the impact test fracture surfaces than other cooling rates (Ref 82). A reheat temperature of 1150 °C is considered safe, and in the event of overheating, a reheat to 1200 °C reduces, if not eliminates, the effect. Thus, overheating is reversible. According to Ref 59, RE or calcium modifications to a low sulfur steel can prevent overheating by raising the overheating temperature.

McBride (Ref 82) asserts that severe overheating can reduce the fatigue resistance by 20 to

Table 5.12 List of inclusion type and occurrence associated with butterfly formation

Type of nonmetallic inclusion	Appearance of nonmetallic inclusion	Frequency of butterflies
MnS	Elongated (<3 µm)	Rare
	Elongated (<3 µm)	Few
	Very thin and long	None
Al_2O_3	Finely dispersed stringers	Many
TiN	...	None
MnS + Al_2O_3	...	Many

25%. Gardiner (Ref 83) determined the best combination of preheating temperature and percent metalworking reduction to maximize the overheating effect of two VAR S82 case-hardening steels. Using these conditions to induce overheating, test pieces were prepared (carburized and uncarburized) and tested. Overheating had no detrimental effect on:

- Rotating bending fatigue strength of either carburized and uncarburized samples
- Rate of crack growth in uncarburized samples
- Fatigue crack initiation sites
- Tensile strength
- Fracture toughness

It did, however, reduce the tensile ductility, the impact toughness, and the K_{ISCC}.

Burning takes place at temperatures above ~1450 °C and is irreversible. It is caused by liquation at grain boundaries, encouraged by the segregation to the boundaries of solute elements, such as phosphorus and sulfur. When the sulfur content is high, the temperature for burning is not much above that for overheating. When the sulfur content is low, the overheating temperature is lowered, thereby increasing the range between the overheating and burning temperatures.

Welding. It is not good practice to weld parts before or after carburizing and hardening. However, because cleanness affects the response of a steel to welding, the following information is included.

The use of cleaner steels significantly improves weldability. The risk of heat affected zone (HAZ) liquation cracking diminishes as the sulfur content is reduced, and lamellar tearing becomes less of a problem as the through-thickness ductility is improved as a result of steel cleanness.

However, clean steels are prone to HAZ cracking during and following the welding operation. This cracking occurs due to a high HAZ hardenability, which results from the relative absence of sulfide and oxide inclusions. These inclusions, had they been present in quantity, would have stimulated the formation of HTTP (Ref 84). Instead, there is a greater tendency for martensite to form in the HAZ, which, coupled with the intake of hydrogen into the zone (from the steel and from the welding), can result in hydrogen-induced HAZ cracking. Hewitt (Ref 85) reported a definite trend between hydrogen cracking and sulfur content. Therefore, for clean steels, a low-hydrogen welding practice is recommended.

Summary

Grain size

Grain size affects both case and core properties; small grain sizes are required.

- *Preprocess considerations:* Purchase steel of appropriate quality; adhere to the grain-size requirement. A normalizing heat treatment favors an initially uniform and small grain size.
- *In-process considerations:* High-temperature carburizing (>1000 °C) tends to encourage some grain growth. Direct-quenched cases tend to be coarse in terms of grain size, martensite plate size, and the size of austenite volumes. Re-heat quenching favors a refined structure.
- *Postprocess considerations:* Consider reheat quenching, keeping in mind the distortion aspect. For some applications, shot peening might counter some of the adverse effects of a coarse-grained structure (although no data are available to show this).
- *Effect on properties:* A fine grain size can occasionally have an adverse influence on the case and the core hardenability of lean-alloy carburizing steels. However, in general, a fine grain size is preferred. Coarse-grained steels are thought to distort more during heat treatment. A coarse austenite grain will, when quenched, produce large martensite plates and large austenite volumes. Large martensite plates are more prone to microcracks than are small plates. Altogether, these adverse effects contribute to reduced fatigue and impact strengths.
- *Standards:* Specifications on grain size call for a grain size of predominantly five or finer.

Microcracks

Microcracks refer to cracks that develop during the formation of plate martensite. They are confined to the high-carbon regions of the case. Large high-carbon martensite plates are more likely to have microcracks.

- *Preprocess considerations:* Plain-carbon and very lean-alloy grades of steel are more prone to microcracking during case hardening.

Therefore, the steel grade needs to be considered, as does grain size.
- *In-process considerations:* High carburizing temperatures favor grain growth and, hence, coarse plate martensite. Microcracking tendency increases with carbon content; therefore, consider case carbon control. Direct quenching retains more carbon in solution in the martensite; therefore quenching method and temperature are important.
- *Postprocess considerations:* Tempering is regarded as beneficial, and refrigeration might have a negative effect.
- *Effect on properties:* This is difficult to assess, though it is thought by some that fatigue strength is reduced by as much as 20%.
- *Standards:* ANSI/AGMA has no specification for grades 1 and 2. For the grade 3 quality, 10 microcracks in a 0.0001 in.^{2} (0.06 mm^2) field at 400 is specified.

Microsegregation

Microsegregation is unavoidable; it is influenced by the modes of solidification during casting. The degree of microsegregation is affected by the alloy content of the steel, including tramp elements. The distribution and directionality of microsegregation are affected by metalworking processes, and its intensity can be reduced to some degree during the mechanical and thermal treatments. Generally speaking, manufacturers and heat treaters have to live with microsegregation.

- *Preprocess considerations:* Microsegregation is difficult to assess during acceptance testing; it is often masked by interstitial elements, for example, by carbon in normalized products. It can create machining problems.
- *In-process considerations:* Microsegregation can cause severe cracking in the more alloyed carburizing grades during slow cooling from the carburizing temperature. In carburized and quenched surfaces, microsegregation can cause alternating zones of martensite and austenite, which are regarded as unsatisfactory. It can also influence the growth and distortion behavior of a heat-treated part.
- *Postprocess considerations:* Where actual case-hardened parts are not sectioned for quality assessment, the presence of microsegregation and its effect on structure will go undetected. Its presence should be considered during service failure analyses.
- *Effect on properties:* See comments for "In-process factors" in this list. Microsegregation, along with nonmetallic inclusions, appreciably influences the transverse properties of wrought steels, particularly the toughness and ductility indicators. Austenite/martensite or martensite/bainite banding, if either occur, is expected to have a negative influence on fatigue resistance, especially in those surfaces containing bands of bainite.
- *Standards:* No specifications

Nonmetallic Inclusions

All steels contain numerous nonmetallic inclusions, but the cleaner grades have fewer large or significant inclusions than do the conventional grades. Clean steels, however, are more prone to overheating problems during forging.

- *Preprocess considerations:* Steel quality requirement, and the steel purchased, is determined by the product quality sought, for example, ANSI/AGMA 2001 grade 1, 2, 3, or aircraft quality. For special applications, ultrasonic surveys maybe carried out to determine if, and where, significant inclusions or inclusion clusters reside within the workpiece. Actions are taken accordingly. Clean steels can be more difficult to machine than conventionally melted steels.
- *In-process considerations:* No in-process considerations
- *Postprocess considerations:* Shot peening of critically stressed areas will, to some extent, offset any adverse effects of nonmetallics residing at, or close to, the surface of that area.
- *Effect on properties:* Nonmetallic inclusions can have a significant adverse influence on transverse toughness and ductility. In terms of fatigue properties, soft inclusions are less harmful than are hard inclusions, size for size. In case-hardened parts nonmetallics might not seriously affect the fatigue limit but could contribute to greater life variability at stresses above the fatigue limit.
- *Standards:* For AGMA grade 2 and grade 3 qualities, (ultrasonic and magnetic particle nondestructive testing) requirements are specified. Sulfur contents to 0.040% are permitted for grade 2, and to 0.015%S for grade 3.
- *Note:* Magnetic particle examination may be needed for detecting defects at or close to a steel surface; such defects are the more likely to affect service life. In magnetic-particle ex-

amination of gear teeth larger than 10 pitch normal diameter (P_{nd}), indications of less than $\frac{1}{64}$ in. (~0.4 mm) are ignored, yet the critical defect size for case-hardened surfaces could be less than this. Therefore, while such inspections are important, they may not guarantee a freedom from harmful surface breaking inclusions.

REFERENCES

Grain Size

1. *Nickel Low-Alloy Steels Case Hardening,* International Nickel Ltd., 1968
2. M.A. Krishtal and S.N. Tsepov, Steel Properties after High Temperature Vacuum Carburising, *Metalloved. Term. Obrab. Met.,* Vol 6, June 1980, p 2–7
3. M.I. Vinograd, I.Y. Ul'Yanina, and G.A. Faivilevich, Mechanism of Austenite Grain Growth in Structural Steel, *Met. Sci. Heat Treat.,* Vol 17 (No. 1–2), Jan/Feb 1975
4. B.L. Biggs, Austenite Grain-Size Control of Medium-Carbon and Carburising Steels, *J. Iron Steel Inst.,* Vol 192 (No. 4), Aug 1959, p 361–377
5. O.O. Miller, Influence of Austenitizing Time and Temperautre on Austenite Grain Size of Steel, *Trans. ASM,* 1951, Vol 43, p 260–289
6. M.A. Grossman, Grain Size in Metals, with Special Reference to Grain Growth in Austenite, *Grain-Size Symp.,* American Society for Metals, 1934, p 861–878
7. R.G. Dressel, R. Kohimann, K.J. Kremer, and A. Stanz, The Influence of Rolling and Heat Treatment Conditions on Austenitic Grain Growth in Case Hardened Steels, *Härt.-Tech. Mitt.,* Vol 39 (No. 3), May/June 1984, p 112–119
8. V.A. Kukareko, The Regularities of Austenite Grain Growth in Steel 18KhNVA, *Metalloved. Term. Obrab. Met.,* (No. 9), Sept 1981, p 15–17
9. H. Child, Vacuum Carburising, *Heat Treat. Met.,* Vol 3, 1976, p 60–64
10. J.D. Murray, D.T. Llewellyn, and B. Butler, Development of Case Carburizing C-Mn Steels, *Low-Alloy Steels,* ISI publication 114, The Iron and Steel Institute, 1968, p 37–53
11. L.N. Benito and D. Mahrus, *35th Annual Congress of Association Brasiliera de Metals,* Vol 1, Association Brasiliera de Metals, July 1980
12. A. Randak and E. Kiderle, The Carburisation of Case-Hardening Steels in the Temperature Range between 920–1100 °C (BISI 5231), *Härt.-Tech. Mitt.,* Vol 21 (No. 3), 1966, 190–198
13. H. Dietrich, S. Engineer, and V. Schüller, Influences on the Austenitic Grain Size of Case Hardening Steels, *Thyssen Edelstahl Tech. Ber.,* Vol 10 (No. 1), July 1984, p 3–13
14. W.T. Cook, The Effect of Aluminum Treatment on the Case Hardening Response of Plain Carbon Steels, *Heat Treat. Met.,* 1984, (No. 1), p 21–23
15. V. Francia et al., Effect of the Principal Metallurgical Parameters of Steel on Carburising and Hardening, *Metall. Ital.,* Vol 74 (No. 12), Dec 1982, p 851–857
16. P.M. Kelly and J. Nutting, Strengthening Mechanisms in Martensite, MG/Conf/85/65, p 166–170
17. M.A. Zaccone, J.B. Kelley, and G. Krauss, Strain Hardening and Fatigue of Simulated Case Microstructures in Carburised Steels, *Conf. Proc. Processing and Performance* (Lakewood, CO), ASM International, July 1989
18. K. Onel and J. Nutting, Structure-Properties Relationship in Quenched and Tempered Carbon Steel, *Met. Sci.,* Oct 1979, p 573–579
19. R.W.K. Honeycombe, Steels—Microstructures and Properties, Edward Arnold, 1982
20. P. Schane, Effects of McQuaid-Ehn Grain-Size on the Structure and Properties of Steel, *Grain-Size Symp.,* American Society for Metals, 1934, p 1038–1050
21. E. Macherauch, Das Ermüdungsverhalten Metallischer Werkstoffe, *Härt.-Tech. Mitt.,* Vol 22 (No. 2), 1967, p 103–115
22. V.D. Kal'ner, V.F. Nikonov, and S.A. Yurasov, Modern Carburizing and Carbonitriding Techniques, *Met. Sci. Heat Treat.,* Vol 15 (No. 9), Sept 1973, p 752–755
23. J.L. Pacheco and G. Krauss, Microstructure and High Bending Fatigue Strength in Carburised Steel, *J. Heat Treat.,* Vol 7 (No. 2), 1989
24. H. Brugger, Werkstoff und Wärmebehandlungseinflüsse auf die Zahnfußtragfähigkeit, *VDI Berichte,* (No. 195), 1973, p 135–144
25. H.-J. Kim and Y.-G. Kweon, High-Cycle Fatigue Behavior of Case-Carburized Medium-Carbon Cr-Mo Steel, *Metall. Mater. Trans. A,* Vol 27, 1996, p 2557–2564
26. G. Krauss, The Microstructure and Fracture of a Carburised Steel, *Metall. Trans. A,* Vol 9, Nov 1978, p 1527–1535
27. G.Z. Wang and J.H. Chen, A Comparison of Fracture Behaviour of Low Alloy Steel with

Different Sizes of Carbide Particles, *Metall. Mater. Trans. A,* Vol 27, July 1996
28. K. Isokawa and K. Namiki, Effect of Alloying Elements on the Toughness of Carburised Steels, *Denki Seiko (Electr. Furn. Steel),* Vol 57 (No. 1), Jan 1986, p 4–12
29. I.I. Prokof'eva, M.V. Taratorina, and I.V. Shermazan, Effect of Martensite Needle Size on the Mechanical Properties of Steel 45 after Induction Hardening, *Metalloved. Term. Obrab. Met.,* (No. 10), Oct 1978, p 10–13

Microcracking

30. R.P. Brobst and G. Krauss, The Effect of Austenite Grain Size on Microcracking in Martensite of an Fe-1.22C Alloy, *Metall. Trans.,* Vol 5, Feb 1974, p 457–462
31. A.R. Marder and A.O. Benscoter, Microcracking in Fe-C Acicular Martensite, *Trans. ASM,* Vol 61, 1968, p 293–299
32. A.H. Rauch and W.R. Thurtle, Microcracks in Case Hardened Steel, *Met. Prog.,* Vol 69 (No. 4), April 1956, p 73–78
33. M.G. Mendiratta, J. Sasser, and G. Krauss, Effect of Dissoved Carbon on Microcracking in Martensite of an Fe-1.39C Alloy, *Metall. Trans.,* Vol 3, Jan 1972, p 351–353
34. R.G. Davies and C.L. McGee, Microcracking in Ferrous Martensites, *Metall. Trans.,* Vol 3, Jan 1972, p 307–313
35. R.F. Kern, Controlling Carburising for Top Quality Gears, *Gear Technol.,* Mar/April 1993, p 16–21
36. L. Jena and P. Heich, Microcracks in Carburised and Hardened Steel, *Metall. Trans.,* Feb 1972, Vol 3, p 588–590
37. A.R. Marder, A.O. Benscoter, and G. Krauss, Microcracking Sensitivity in Fe-C Plate Martensite, *Metall. Trans.,* Vol 1, June 1970, p 1545–1549
38. J. Hewitt, The Study of Hydrogen in Low-Alloy Steels by Internal Friction Techniques, *Hydrogen in Steel,* Special report 73, The Iron and Steel Institute, 1962, p 83–89
39. R.G. Baker and F. Watkinson, The Effect of Hydrogen on the Welding of Low-Alloy Steels, *Hydrogen in Steel,* Special report 73, The Iron and Steel Institute, 1962, p 123–132
40. A.R. Marder and A.O. Benscoter, Microcracking in Plate Martensite of AISI 52100 Steel, *Metall. Trans.,* Vol 1, Nov 1970, p 3234–3237
41. A.R. Marder and A.O. Benscoter, Microcracking in Tempered Plate Martensite, *Metall. Trans.,* Vol 5, Mar 1974, p 778–781
42. T.A. Balliett and G. Krauss, The Effect of the First and Second Stages of Tempering on Microcracking in Martensite of an Fe-1.22C Alloy, *Metall. Trans. A,* Vol 7, Jan 1976, p 81–86
43. C.A. Apple and G. Krauss, Microcracking and Fatigue in a Carburized Steel, *Metall. Trans.,* May 1973, Vol 4, p 1195–1200
44. M.A. Panhans and R.A. Fournelle, High Cycle Fatigue Resistance of AISI 9310 Carburised Steel with Different Levels of Surface Retained Austenite and Surface Residual Stress, *J. Heat Treat.,* Vol 2 (No. 1), June 1981, p 54–61
45. R.F. Kern, Carburize Hardening: Direct Quench vs. Reheat and Quench, *J. Heat Treat.,* March 1988, p 17–21
46. J. Lyman, High Carbon Steel Microcracking Control during Hardening, *J. Eng. Mater. Technol. (Trans. ASME),* Vol 106, July 1984, p 253–256
47. I. Larén and H. Fredriksson, Relations between Ingot Size and Microsegregation, *Scand. J. Metall.,* Vol 1, 1972, p 59–68

Microsegregation

48. T.B. Smith, Microsegregation in Low-Alloy Steels, *Iron Steel,* Vol 37, Nov 1964, p 536–541
49. J. Philibert, E. Weinry, and M. Ancey, A Quantitative Study of Dendritic Segregation in Iron-Base Alloys with the Electron-Probe Microanalyser, *Metallurgia,* Vol 72 (No. 422), Nov 1965, p 203–211
50. T.B. Smith, J.S. Thomas, and R. Goodall, Banding in a 1½% Nickel-Chromium-Molybdenum Steel, *J. Iron Steel Inst.,* Vol 201, July 1963, p 602–609
51. R.G. Ward, Effect of Annealing on the Dendritic Segregation of Manganese in Steel, *J. Iron Steel Inst.,* Vol 203, Sept 1965, p 930–932
52. A. Kulmburg and K. Swoboda, Influence of Micro-Segregations on the Transformation Behaviour and Structural Constitution of Alloy Steels, *Prakt. Metallogr.,* Vol 6, 1969, p 383–400
53. M. Kroneis et al., *Berg Hüttenmann. Montash.,* Vol 113, 1968, p 416–425
54. M. Kroneis et. al., Properties of Large Forgings of Electroslag Remelted Ingots, *5th Int. Forging Conf.* (6–9 May 1970), Terni, p 711–737

Nonmetallic Inclusions

55. R. Kiessling, *Nonmetallic Inclusions in Steel IV,* Metals Society, 1978; R. Kiessling, *Nonmetallic Inclusions in Steel II,* Metals Society,

56. J. Watanabe, Some Observations on the Effect of Inclusions on the Fatigue Properties of Steels, *Proc. 3rd Japanese Cong. of Testing Materials,* Kyoto University, 1960, p 5–8
57. E. Plöckinger, Properties of Special Constructional Steels Manufactured by Special Melting Processes Including Steels for Forging, *Stahl Eisen,* Vol 92 (No. 20), 1972, p 972–981
58. W.T. Cook and D. Dulieu, Effect of Cleaners on Properties of Heat-Treated Medium-Carbon and Low-Alloy Steels, *Heat Treatment '87* (London), Institute of Metals, 1987
59. W.H. McFarland and J.T. Cronn, Spheroidisation of Type II Manganese Sulfide by Heat Treatment, *Metall. Trans. A,* Vol 12 (No. 5), May 1981, p 915–917
60. R. Kiessling, *Nonmetallic Inclusions in Steel III,* Metals Society, 1978
61. S.I. Gubenko, Phase Transformations in Nonmetallic Inclusions during Heat Treatment of Steels, *Izv. V.U.Z. Chernaya Metall.,* (No. 12), 1986, p 67–71
62. S.I. Gubenko, Dissolution of Nonmetallic Inclusions during High Temperature Heating, *Izv. Akad. Nauk SSSR, Met.,* (No. 2), Mar/April 1983, p 103–107
63. W.A. Spitzig and R.J. Sober, Influence of Sulfide Inclusions and Pearlite Content on the Mechanical Properties of Hot-Rolled Carbon Steel, *Metall. Trans. A,* Vol 12, Feb 1981, p 281–291
64. G.E. Dieter, *Mechanical Metallurgy, Metallurgy and Metallurgical Engineering Series,* McGraw-Hill, 1961, p 270–271
65. G. Kiessling, *Nonmetallic Inclusions in Steel V,* Metals Society, 1989
66. D. Brooksbank and K.W. Andrews, Production and Application of Clean Steels, *J. Iron Steel Inst.,* Vol 210, 1972, p 246–255
67. K.B. Grove and J.A. Charles, Further Aspects of Inclusion Deformation, *Met. Technol.,* Vol 1 (No. 9), Sept 1974, p 425–431
68. H.N. Cummings, F.B. Stulen, and W.C. Schulte, Tentative Fatigue Strength Reduction Facors for Silicate-Type Inclusions in High-Strength Steels, *Proc. ASTM,* Vol 58, 1958, p 505–514
69. T. Ohta and K. Okamoto, Effect of Nonmetallic Inclusions on Fatigue Life of Bearing Steels, *Tetsu-to-Haganè (J. Iron Steel Inst.),* Vol 52, 11, Lecture 104, 230, 1966
70. T. Tsunoda, I. Uchiyama, and T. Araki, Fatigue (of Steels) in Relation to the Non-Metallic Inclusion Content, *Tetsu-to- Haganè (J. Iron Steel Inst.),* Vol 52, 11, Lecture 229, 103, 1966
71. Y.E. Goldstein, A.Y. Zaslavskii, and B.S. Starokozhev, Effect of Lead on the Fatigue Characteristics of Structural Steels, *Metal Sci. Heat Treat.,* 5 May 1973, p 397
72. W.E. Duckworth and E. Ineson, *Clean Steels,* Special report 77, The Iron and Steel Institute, 1963, p 87–103
73. Y. Kawada, H. Nakazawa, and S. Kodama, *Mem. Fac. Technol., Tokyo Metrop. Univ.,* (No. 15), 1965, p 1163–1176
74. R.F. Kern, Selecting Steels for Heat-Treated Parts, *Met. Prog.,* Vol 94 (No. 5), Nov 1968, p 60–73
75. L.O. Uhrus, *Clean Steels,* Special report 77, The Iron and Steel Institute, 1963, p 104–109
76. F. Vodopivec and L. Kosec, The Fatigue Strength and the Influence of Nonmetallic Inclusions on the Fatigue Fracture of Case-Hardening Steels (BISI 5906), *Härt.-Tech. Mitt.,* Vol 22, July 1967, p 166–173
77. S. Gunnarson, Structure Anomalies in the Surface Zone of Gas-Carburized, Case-Hardened Steel, *Met. Treat. Drop Forging,* June 1963, p 219–229
78. H. Winter, G. Knauer, and J.J. Gamel, White Etching Areas in Case Hardened Gears, *Gear Technol.,* Sept/Oct 1989, p 18–44
79. W.E. Littman and R.L. Widner, Propagation of Contact Fatigue from Surface and Sub-Surface Origins, *Trans. ASME,* Sept 1966, p 624–636
80. K. Sugino et al., Structural Alterations of Bearing Steels under Rolling Contact, *Trans. ISIJ,* Vol 10, 1970, p 98–111
81. G.R. Speich and W.A. Spitzig, Effect of Volume Fraction and Shape of Sulfide Inclusions on Through-Thickness Ductility and Impact Energy of High-Strength 4340 Plate Steel, *Metall. Trans. A,* Vol 13, Dec 1982, p 2239–2259
82. J.S. McBride, Overheating—Reappraisal of a Reccurring Problem, *Met. Mater.,* Vol 8 (No. 5), May 1974, p 269–270
83. R.W. Gardiner, Effect of Overheating on the Fatigue Strength and Other Mechanical Properties of Carburized and Uncarburized VAR S 82 Steel, *Met. Technol.,* Dec 1977, p 536–547
84. P. Hart, Effects of Steel Inclusions and Residual Elements on Weldability, *Met. Constr.,* Oct 1986
85. J. Hewitt, *Hydrogen in Steel,* B.I.S.R.A./The Iron and Steel Institute, 1961

Chapter 6

Core Properties and Case Depth

When a case-hardened machine part is subjected to normal loading, its failure, should failure occur, usually results from contact damage or damage due to bending stresses, possibly acting on an engineering or metallurgical stress raiser. Contact damage often initiates at the surface as a consequence of frictional effects (predominantly sliding and wear processes), and, therefore, the surface condition and the metallurgical structure of the surface are very much involved. Contact damage can also develop at subsurface locations due to shear stresses, which are generated below the surface when one surface rolls over another. In this situation, the metallurgy of the material beneath the surface (where the maximum shear stresses develop) needs to be considered. When bending conditions prevail, as at a gear tooth fillet, the stresses developed are greatest at the surface and decrease steeply beneath the surface; again the surface and near-surface metallurgy are important. Therefore, if the case depth of a case-hardened part is deep enough, the material of the case is strong enough, and the load-bearing areas are large enough, the notion that the core is "just stuffing" is more or less valid. Unfortunately, it is not that simple. In general, economic and ecological considerations dictate the use of the leanest (or cheapest) steel for the job, as well as the shortest time in the furnace needed to produce just enough carburized case to cope with the applied stresses in a minimum weight-to-power ratio situation. The aim is to not over design (to do so increases the cost); therefore, there is a risk of under designing with respect to material selection and case depth specification. This is designing nearer to the limit, and in such circumstances, the contact and/or bending stresses experienced by the core material beneath the case can be significant. Under these conditions, the core ceases to be "just stuffing."

Core Factors

The core properties of a carburized part are dictated by the chemical composition of the steel and the rate at which the part cools during quenching. The alloy content is mainly responsible for the depth to which a steel will harden (hardenability), whereas the carbon content largely determines the hardness.

Core Hardenability

Case-hardening steels are usually lean alloy with total alloying-element contents ranging from about 1 to 6.5% and carbon contents between about 0.1 and 0.25%. When the total alloy range is 4 to 6.5%, the main alloying element is usually nickel. The available case-hardening steels can produce as-quenched core hardnesses between 20 and 45 HRC (depending upon size), and are suitable for case-hardened parts from a few millimeters to almost a meter in section. Case-hardening steels for special applications, for example, high-temperature service applications, may contain at least 10% alloying elements. Such steels are generally tool steels adapted for carburizing and require special heat treatments to develop desired properties. Therefore, they are not directly discussed in this work.

When selecting a steel for an application, two important requirements must be satisfied. First, the steel must have adequate case hardenability for the job; that is, the carbon-rich layer induced by carburizing will suitably harden when quenched. A number of steels might satisfy this requirement, but each will likely have a different core hardness range. Therefore, the second requirement to be satisfied is core hardness (or strength) of the critical core areas of the component. This primarily relates to the hardening depth (the core-hardenability). Figure 6.1 provides a general view of core hardenability in terms of how average tensile strength varies with section size and alloy content.

Microstructures observed within the cores of case-hardened parts are: ferrite (undissolved or precipitated), bainite (upper and lower), and martensite (low-carbon) (see Fig. 6.2). Pearlite is not found in a quenched core material unless the section size is large and the steel has low hardenability. In such an instance, the case might not satisfactorily harden. One reason ferrite could be present in an as-quenched core material is that it was austenitized (prior to quenching) at a temperature below the Ac_3 temperature, leaving some ferrite undissolved. Another reason, perhaps the most common, is that a steel is employed that has less hardenability than required for the section. This is illustrated in Fig. 6.3 where the upper continuous-cooling transformation (CCT) diagram represents a lean-alloy case-hardening steel, and the lower diagram represents a medium-alloy case-hardening steel. By superimposing near-surface cooling curves (e.g., for ~12 mm and ~50 mm diameter bars),

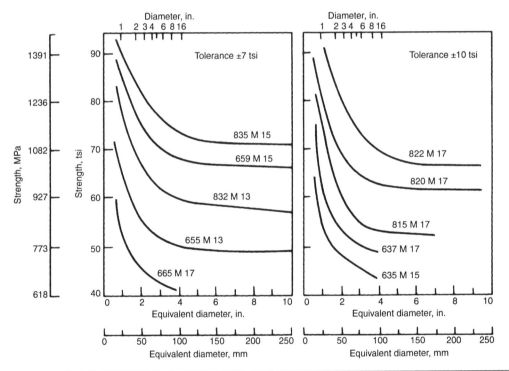

Steel	Nominal composition, %				
	C	Mn	Ni	Cr	Mo
835 M 15	0.15	0.40	4.1	1.1	0.20
659 M 15	0.15	0.40	4.1	1.1	...
832 M 13	0.13	0.40	3.2	0.90	0.20
655 M 13	0.13	0.50	3.2	0.90	...
655 M 17	0.17	0.50	1.8	...	0.25
822 M 17	0.17	0.60	2.0	1.5	0.20
820 M 17	0.17	0.80	1.8	1.0	0.15
815 M 17	0.17	0.80	1.4	1.0	0.15
637 M 17	0.17	0.80	1.0	0.80	0.05
635 M 15	0.15	0.80	0.80	0.60	0.05

Fig. 6.1 Strength versus section diameter for a number of U.K. carburizing steels. Use lower scale to estimate bar center strength; use upper scale to estimate bar surface strength. Source: Ref 1

it is seen that for such conditions, the formation of ferrite is unavoidable in the leaner grade but avoidable in the more alloyed grade. Consequently, if ferrite is not desired in the core of a part, an adequately alloyed steel must be selected with quenching from the fully austenitic condition. Increasing the cooling rate might suppress ferrite formation up to a point. The selection of a borderline grade of steel might, on a batch-to-batch basis, lead to a fair amount of core structure variability. This is because steelmakers cannot work to precise chemical compositions; they must have reasonable working tolerances. In the unlikely situations where each alloying element in a steel is either at the bottom or at the top of its specification range, the transformation characteristics can be extremely variable, as Fig. 6.4(a) and (b) suggest. Even variations of carbon content alone can make a difference in transformation behavior (Fig. 6.4c). This indicates for this steel that if the carbon content is low at 0.08%, the time to the ferrite nose is less than 8 s, which means, for

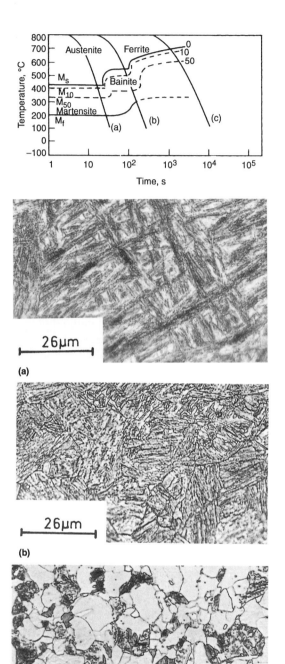

Fig. 6.2 Microstructures obtained by cooling a 0.16%C-3%Ni-Cr steel from 920 °C. (a) Fast cool (920–200 °C in 30 s) giving low-carbon martensitic structure of 1590 MPa UTS. 800×. (b) Intermediate cool (920–250 °C in 200 s) giving bainitic structure of 1360 MPa UTS. 800×. (c) Slow cool (920–250 °C in 10^4 s) giving a ferrite/pearlite structure of 740 MPa UTS. 800×. Source: Ref 1

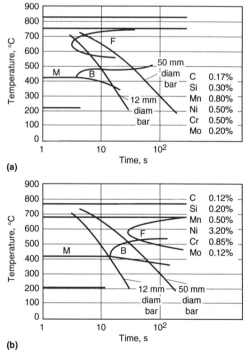

Fig. 6.3 A comparison of the continuous-cooling transformation diagrams for (a) BS 970 805M20 (SAE 8620) (composition: 0.17 C, 0.30 Si, 0.80 Mn, 0.50 Ni, 0.50 Cr, 0.20 Mo) and (b) BS 970 832M13 (composition: 0.12 C, 0.20 Si, 0.50 Mn, 3.20 Ni, 0.85 Cr, 0.12 Mo). F, ferrite; B, bainite, M, martensite. Surface cooling is shown for a 12 mm diam bar and a 50 mm diam bar.

example, that an oil-quenched 12.5 mm (½ in) diameter bar will contain some ferrite at its center. When the carbon content is raised to 0.18% and the other elements remain the same, the time to the ferrite nose is nearer to 200 s. This allows a 125 mm (5 in.) diameter bar to be oil quenched without any ferrite production. Fortunately, steel manufacturers can generally maintain the amount of each element in a steel grade within narrow limits. The purchase of controlled-hardenability steels also makes for a greater degree of consistency from batch to batch. In terms of end-quench hardenability, Fig. 6.5(a) and (b) show the hardenability curves that correspond to the CCT curves of Fig. 6.3; the extremes of the bands of Fig. 6.5(b) correspond to the CCT curves of Fig. 6.4(a) and (b).

There are areas within the core of a carburized and hardened component that experience little, if any, stress during service; for these areas, the microstructure and strength are of little concern. It is those areas just beneath the case where high stresses develop during service for which core structure and strength are important. Therefore, when using CCT diagrams to assess the suitability of a steel for a given component, the near-surface cooling rate is the most meaningful.

Ferrite cores, although not uncommon, are often regarded as unacceptable for critically loaded components. Bainite-type cores are perhaps more common and more desirable than ferrite. The martensitic, or predominantly martensitic, core structures tend to be found in those components with small sections, and these are also desirable if the core carbon and the core hardness are not too high.

In selecting a steel for a particular component, care must be taken to choose one without too much carbon and hardenability for the shape and the section involved, otherwise growth, distortion, and internal cracking (as under the tops of gear teeth) might become a problem. Also, one needs to decide whether to choose a steel of reasonable hardenability and harden it with a mild quench, or to adopt a lower hardenability steel and hard quench it. The choice between a low hardenability steel with a high carbon content, and a higher hardenability steel with a low carbon content often depends on the duty of the finished item. However, a low hardenability steel with a high carbon content is more prone to size variations. One good reason for selecting an alloy case-hardening steel with a low nominal carbon content (~0.13%) is that it should develop an as-quenched core strength of less than about 1000 MPa (145 ksi). With this in mind, one can carburize, slow cool, and subcritical anneal to facilitate the removal of areas of the case where, after quenching, further machining operations can be performed.

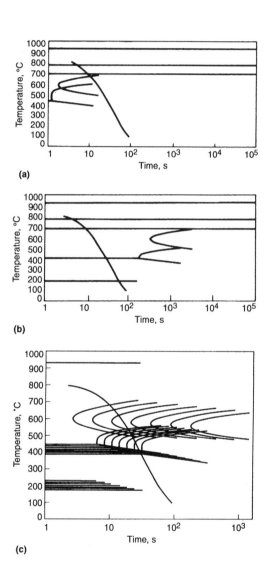

Fig. 6.4 Continuous-cooling transformation diagrams for selected 3%Ni-Cr case-hardening steels. Specification En 36 is now replaced by 655M13 and 831M13. (a) Ni, Cr, and Mo contents all at the bottom of the specification range (En 36). Composition: 0.12 C, 0.20 Si, 0.40 Mn, 3.00 Ni, 0.60 Cr, 0.00 Mo, 0.00 V. (b) Ni, Cr, and Mo contents all at the top of the specification range. Composition: 0.12 C, 0.20 Si, 0.60 Mn, 3.50 Ni, 1.10 Cr, 0.25 Mo, 0.00 V. (c) Effect of carbon content with alloying elements at constant levels. Composition: 0.06–0.18 C, 0.20 Si, 0.54 Mn, 3.18 Ni, 0.91 Cr, 0.04 Mo, 0.00 V. Source: Ref 1

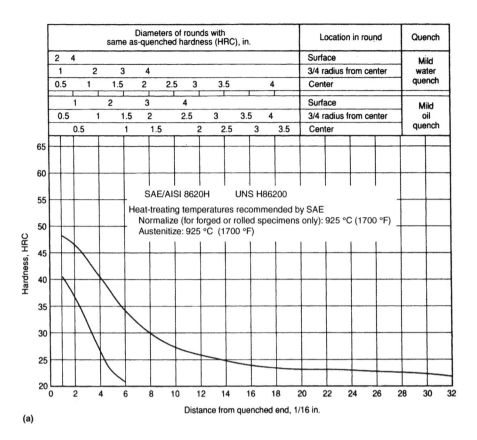

(a)

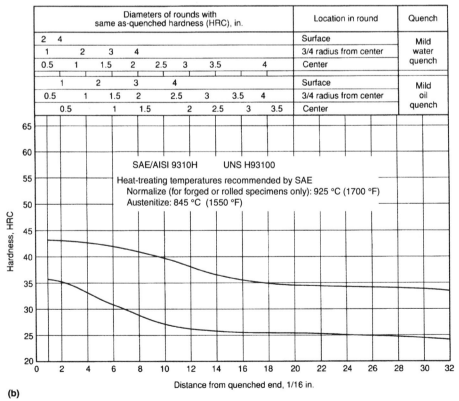

(b)

Fig. 6.5 Hardenability ranges for two case-hardening grades of steel. Source: Ref 2

Core Microstructure and Hardness

Core hardness is sometimes specified on engineering drawings for case-hardened parts, implying that a heat-treated part is to be sacrificed to permit the necessary sectioning and hardness testing. In such instances, the hardness tests are carried out as a standard procedure in which test impressions are made at specified locations. For gear teeth, these might be at the center of the tooth and at the center of the tooth on the root circle diameter. Such tests relate the minimum tooth hardness, which has some value even though, in the locations specified, the service-applied loads are negligible. Perhaps a more meaningful test site would be at, for example, 2 × the total case depth at the midflank and tooth fillet positions. Hardness tests are quick and easy to perform and, as a rule, are reliable.

When it is unacceptable to determine the core hardness of actual parts, the best alternative is to have a simulation test piece that contains appropriate geometric features and cools during quenching at a rate similar to that of the actual part. The third option is to ensure that the material selection is related to the size of the component (Fig. 6.1 or equivalent), that the material supplied is to specification, and that process control is tight. In that way, the attainment of a minimum core hardness of the finished part is virtually ensured.

From a designer's point of view, a hardness test value is only a means of conveying what is required or what has been achieved; strength data are far more meaningful to the designer. Fortunately, the relationship between hardness test values and equivalent tensile strengths is quite good (Fig. 6.6), and it is possible to approximate other properties from the tensile strength.

The connections between strength properties and microstructure tend to be somewhat blurred. This is because core microstructures are often a mixture of different phases; cores with hardnesses of <25 HRC will have high ferrite contents, whereas >40 HRC cores would indicate a predominantly martensitic microstructure. Hardness values of mixed microstructures are between 25 and 40 HRC. The effect of structure variability on the hardness of an alloyed case-hardening steel is illustrated in Fig. 6.7; note that other steels with different core carbon and alloy contents will shift the lines of the diagram to the left or to the right along the hardness scale, depending on the actual chemical composition.

Core Tensile Properties

Core Tensile Strength. The influence of core carbon on the center tensile strength of 3%Ni-Cr steel parts with different sections is shown in Fig. 6.8. This figure suggests that there can be a 310 to 460 MPa wide band on the ultimate tensile strength (UTS) value due simply to a variation of carbon content within a range of 0.08% C. Normally, for case-hardening steels, the tolerance band for carbon content is 0.05 or 0.06% C, and steelmakers can readily achieve this, as the inset illustration in Fig. 6.8 shows. Nevertheless, some core strength variability can be anticipated due to carbon and composition variations. For many production parts, the center core strength is not too important, whereas that part of the core immediately beneath the stressed case is important. An indication of how hardness and, hence, strength vary within a section is illustrated in Fig. 6.9.

Core Yield Strength. From a designer's point of view, the core yield strength often has more significance than the tensile strength does because, in most designs, core yielding is not permitted. There are different ways of conveying yield data, for example, limit of proportionality (LoP), proof stress (offset yield), or yield strength. In

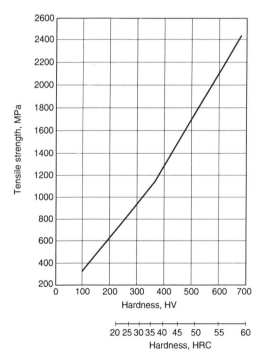

Fig. 6.6 Strength and hardness conversion. 1 psi = 0.00689476 MPa. Source: *ASM Metals Reference Book*, 1981

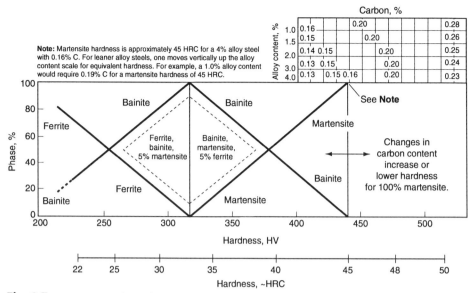

Fig. 6.7 Approximate relationship between core microstructure and hardness of a Ni-Cr-Mo carburizing steel (~4% alloy content) with ~0.16% C. The alloy content/carbon content extension (top right) permits the phase % plots to be "moved" in relation to the fixed hardness scale to approximate core strength for other steels in slide-rule fashion. Below ~250 HV represents slow-cooled (normalized) and annealed materials, and therefore, bainite could read as bainite, pearlite, or spheroidized carbides. Above ~250 HV refers to quenched materials. For the 180 °C tempered condition, there will be zero change at 360 HV and below, but there will be a 20 point HV loss at 100% martensite.

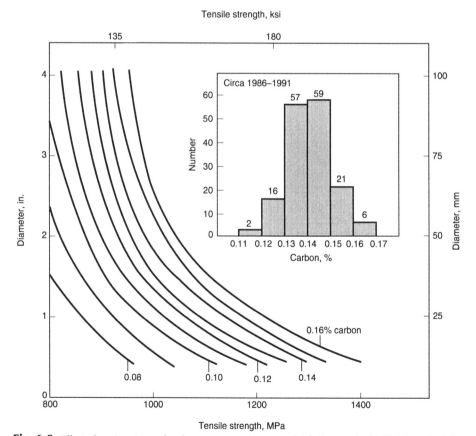

Fig. 6.8 Effect of section size and carbon content on the strength of oil-quenched 3%Ni-Cr carburizing steels 832M13 and 655M13. Source: Ref 1

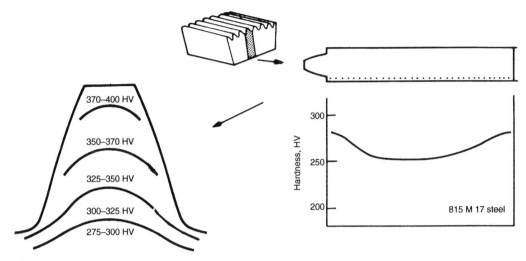

Fig. 6.9 Hardness distribution within a toothed section. Source: Ref 1

Fig. 6.10 for UK steel 822M17, the LoP is at about 50% of the UTS, the 0.1% proof stress is at ~60%, the 0.2% proof stress is at ~70%, and the AGMA yield number is about 80% of the UTS. Yield strength is influenced by microstructural constituents, grain size, or packet size or lath width, as in martensite-bainite structures. The smaller the packet size or lath width, the higher the yield strength and toughness are (Ref 4).

Another way of looking at yield strength data is with the yield ratio (YR), that is, the UTS divided by one of the yield strength indicators. In

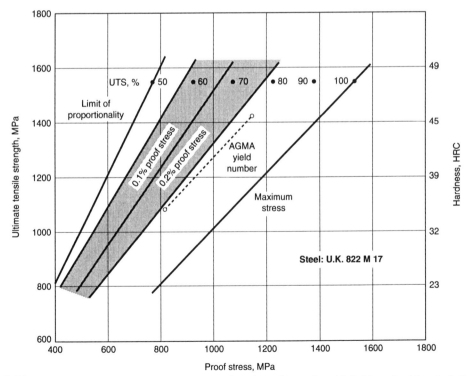

Fig. 6.10 Relationship between ultimate tensile strength and proof stress for a Ni-Cr-Mo carburizing steel. Derived from Ref 3

Fig. 6.11, the yield ratio = UTS/0.2% proof stress. This illustration represents bars of 1¼ in. diameter with carbon contents up to 0.18%, and shows that for core strengths typical of martensitic and/or bainitic microstructures, the YR is fairly low (around 1.3 to 1.5); that is, the yield strength is fairly high. With smaller section sizes and carbon contents above 0.18%, yield ratios down to 1.15 are possible. Core materials containing ferrite tend to have low yield strengths and high yield ratios (1.5 to more than 2.0); how high depends on the amount of ferrite present. (The significance of a fairly high-yield strength core material will become more apparent when fatigue resistance is considered.) Interestingly, regarding straining in tension, if the yield ratio is more than 1.4, work hardening will occur during cyclic straining. If, on the other hand, the ratio is less than 1.2, work softening will occur during cyclic straining. Between 1.2 and 1.4, the material is probably fairly stable (Ref 5).

The yielding referred to so far is macroyielding as determined during a typical tensile test. However, before that stage is reached, microplastic yielding has already commenced at a much lower stress. The onset of microplastic yielding, the true elastic limit, is when dislocation migration first occurs, and can be measured by determining the change of AC resistance while straining in uniaxial tension (Ref 6). Tests on samples of a through-hardened and low-temperature tempered case-hardening steel have indicated that the true elastic limit is only a few percent below the bending fatigue limit.

Core Ductility. The ductility indicators are the percent reduction of area and the percent elongation as derived from the tensile test; the higher the values of the two indicators, the more ductile the metal is. Ductility relates to the ability of a material to be plastically deformed without fracture; it is important to those who work metals, for example, wire manufacturers or car-body press shop operators. For case-hardened parts, the significance of core ductility is not at all clear because the applied stresses must exceed the engineering yield strength of the core before ductility becomes a consideration. Nevertheless, acceptance testing for case-hardening steels requires that the percent reduction of area, the percent elongation, and the impact resistance are determined, if only to assure the manufacturer that the steel is of acceptable quality. According to the trend, as the strength and hardenability of the steel increase, the ductility decreases (Fig. 6.12).

Core Toughness

Toughness is the ability of a metal to absorb energy. This is generally important with respect to case-hardened parts because tough case-hardened parts are more able to survive occasional

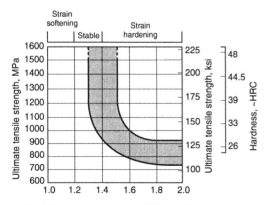

Fig. 6.11 Relationship between the ultimate tensile strength and the 0.2% proof stress (offset yield) of carburizing steels (0.08–0.18% C). Note that with carbon contents of over 0.18%, the ratio can be as low as 1.15 for strengths over about 200 ksi (44 HRC). Data from Ref 3

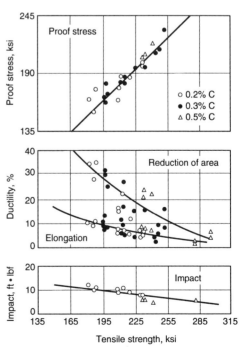

Fig. 6.12 Mechanical properties of lightly tempered plain-carbon martensites to illustrate how ductility falls as strength rises. Source: Ref 7

overloads and cyclic impact loading than are parts of relatively low toughness.

Tests for toughness include notched (severe to gentle notches) and unnotched impact bend tests, notched and unnotched slow bend tests, and precracked fracture tests. Such tests indicate the contributions of alloying elements, microstructure, and hardness to fracture resistance. The more meaningful information is obtained not from the standard impact tests (Charpy and Izod) but from those tests in which the stress raiser in the test piece is more similar to those found in actual machine components. Chesters examined the limitations of the standard impact test in an early work (Ref 8). The results of this research for a number of steels are summarized in Fig. 6.13. The standard test on its own did not supply much information on the effect of material composition on impact, whereas the simulated gear tooth did. Further, even with Izod impact test values below 10 ft · lb, the case-hardening grade of steel (En 36) still offered a good resistance to impact when the stress concentrator was similar to that for a small gear tooth. The use of the notched impact test has largely been driven by the national standards, and its value, if any, is regarding material acceptance. Even used for that purpose, a notched impact value of 20 ft · lb (27 J) is considered ample for most applications, though the appropriate material standard might call for more than 27J (Ref 9). Another limitation of the standard test is that the results do not relate to fatigue performance, nor can they be used in design formulas. The more recently adopted instrumented impact and bend tests used for research projects, while still having some limitations, employ test pieces that contain more realistic stress concentrators; therefore, these tests more usefully contribute to the understanding of the respective influences of steel composition, core microstructure, and hardness (strength).

Fett showed that steels containing little or no nickel are notch sensitive at core hardnesses of >40 HRC in slow bending, and sensitive at >30 HRC under impact loading (Ref 10). Nickel steels (>1.8% Ni) are not so sensitive at that value, which means that with the nickel grades of case-hardening steels one can develop a high-yield strength core that is adequately tough.

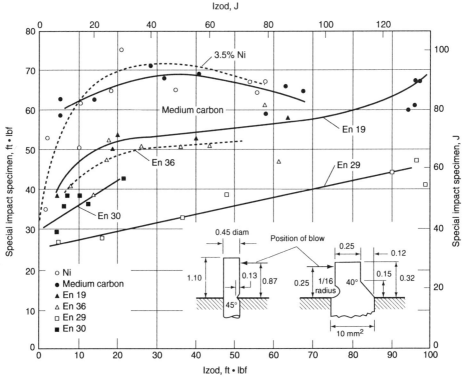

Fig. 6.13 Comparison between standard impact test results and results from a test piece designed to simulate a gear tooth. En 19, 705M40 (nominally, 0.40 C, 1.0 Cr, 0.3 Mo); En 36, 832M13 (nominally, 0.13 C, 3.2 Ni, 1.0 Cr, 0.15 Mo); En 30, 835M30 (nominally, 0.30 C, 4.0 Ni, 1.2 Cr, 0.3 Mo); En 29, 722 M24 (nominally 0.24 C, 3.0 Cr, 0.5 Mo). Source: Ref 8

The higher nickel grade (>3%) generally has superior toughness properties compared to the medium nickel grade (~1.8%), though the latter is still good. However, in practice, a steel is selected for its case and core hardenabilities, so the tendency is to use the more alloyed grade for the larger components. The effect of steel composition was determined by Cameron et al. who rated the as-carburized impact strength of Mn-Cr steels as less than that of Cr-Mo steels, which, in turn, were somewhat inferior to a Ni-Cr-Mo (PS 55) steel (Ref 11). For the Mn-Cr and the Cr-Mo steels, increasing core carbon from about 0.17 to 0.3% decreased fracture strength, but a similar variance of core carbon had no effect on the fracture strength of the PS 55 steel. The beneficial affect of molybdenum with nickel on fracture strength was demonstrated by Smith and Diesburg (Ref 12); they showed that increases of both nickel and molybdenum raise the toughness of case-hardened test pieces.

Effects of Core Properties

Effect of Core Material on Residual Stresses. The residual stress distribution within a case-hardened layer is related to the difference in volume expansion between the high-carbon martensite of the case and the low-carbon martensite, bainite, or ferrite of the core (Ref 1). The greater that difference is, the greater the likelihood is of producing high magnitudes of residual stress, provided no yielding occurs and the sequence of transformation is correct. This implies that a part with a low-carbon core should develop a more favorable residual stress distribution than a part with a high-carbon core. Sagaradze (Ref 13) carried out residual stress determinations on carburized and hardened plate samples that had been prepared from a common Cr-Mn-Ti steel base stock with carbon contents between 0.07 and 0.45%. The residual stress distribution curves obtained are shown in Fig. 6.14(a), which shows that as the carbon content of the core material increases (and, hence, the hardness increases), the value of the surface compressive stress decreases at a rate of 11 MPa per 0.01% C (Fig. 6.14b). The peak compression (here, ~75% of the case depth) follows the same trend with respect to carbon, apart from the lowest carbon level (Fig. 6.14c). The same researcher reported that similar tests with a 3.5Ni-1.5Cr steel produced the same trend, except the values of maximum compressive stress were very much higher, that is, 0.18, 0.28, and 0.38% core carbon samples gave values of 830, 860, and 730 MPa, respectively. The trend found by Sagaradze was more or less confirmed by Kern whose curve for the SAE 8600 series of steels is also shown in Fig. 6.14(c) (Ref 14).

Effect of Core Strength on Bending Fatigue Resistance. Increasing the core carbon increases the core strength for a given quench or microstructure, and increasing the quench severity increases the hardness (up to a point) for a given carbon content. Increasing the core strength can reduce the amount of compressive residual stress within the case. A reduction of surface compression can then lead to a reduction of fatigue resistance, shown in Fig. 6.15. From this, it can be concluded that there is an upper limit of desirable core strength for case-hardened parts. Low-carbon cores, while they encourage the development of high surface compressive-residual stresses, might either deform under load or locally yield enough to modify the residual stresses and increase the possibility of subcase fatigue-crack initiation. In other words, for critically

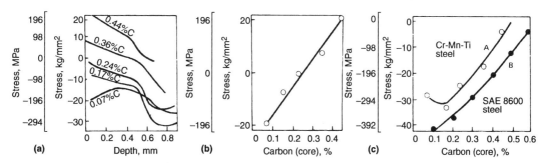

Fig. 6.14 Dependence of residual stress in carburized and hardened cases on core carbon level. (a) Residual stress distribution in samples of Cr-Mn-Ti steel of varying core carbon contents; case depth, 1.2 mm; quenched from 810 °C. (b) Relationship between surface residual stress and core carbon. (c) Relationship between peak compressive stress and core carbon. Source: Ref 13, 14

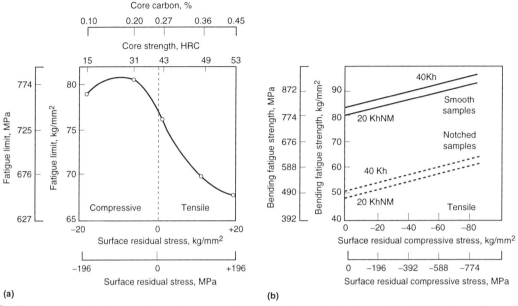

Fig. 6.15 Relationship between fatigue limit and surface residual stress for the Cr-Mn-Ti steel referred to in Fig. 6.14. Generally (a) A reduction in surface compressive stresses leads to (b) A reduction in bending fatigue resistance. Source: Ref 13, 15

loaded parts, there is a lower limit of desirable core strength.

There have been several attempts to relate the experimentally derived fatigue limit to the ultimate strength or hardness of the core material in order to arrive at the optimum value of core strength. Figure 6.16 shows that the fatigue strength rises with core strength up to a peak value, beyond which the fatigue strength falls as the core strength increases further. Four of the sources (Ref 13, 16, 17, 20) indicate an optimum core strength of ~1150 to 1200 MPa (37–39 HRC), whereas two other sources (Ref 18, 19) report 1250 to 1550 MPa (40–48 HRC). In terms of microstructures, one might expect a predominantly bainitic structure with less than 50% martensite and no ferrite for the 37 to 39 HRC range; for the 40 to 48 HRC range, the structures will contain 50 to 100%

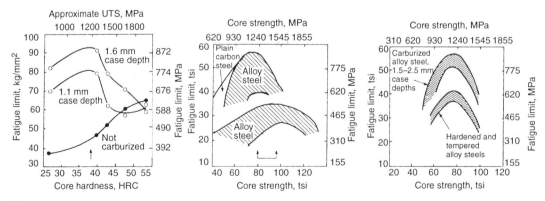

Fig. 6.16 Core properties and fatigue strength of case-hardened steels. (a) Effect of core hardness and case depth on the fatigue strength of a 1.4%Cr-3.5%Ni steel in which core carbon was varied from 0.09 to 0.42%. Arrow indicates maximum fatigue strength for Cr-Ni steels with 0.13 mm case depth. (b) Effect of core strength on the fatigue strength of gears; upper band based on MIRA tests on 7 diametrical pitch (dp) gears; lower band on 3 mm module (8.5 dp) gears. Arrows indicate the range for maximum fatigue resistance for gears. (c) Effect of core strength on the fatigue strength of carburized and hardened ~2 dp alloy steel gear-simulation test pieces (upper band) and of hardened and tempered noncarburized alloy steels (lower band). Source: Ref 13, 16, 17, 18, 19, 20

Table 6.1 Effect of quenching temperature on fatigue strength

Quenching temperature, °C	Surface hardness, HV	Core hardness, HV	Case depth to 500 HV		Fatigue limit	
			mm	0.001 in.	MPa	ksi
760	780	235	0.425	17	610	88
810	895	325	0.825	33	745	108
840	925–940	400	0.875–0.925	35–37	800–910	116–132
870	880	405	1.0	40	840	122
900	915	410	0.825	33	880	128

Tests conducted on 637 M 17 (1.07%Ni-0.88%Cr) 7dp (diametral pitch) gears pack carburized at 900 °C for 7 h at a carbon potential of 1.1% single quenched and not tempered. Source: Ref 22

martensite, again with no ferrite. Considering toughness, hardness values in excess of 40 HRC tend to be a little high, particularly for steel grades containing little or no nickel. The determination of the optimum core strength is not unreasonable; useful information is thereby provided. It is, however, unreasonable to expect the chosen optimum strength to be achieved on a commercial basis, except by chance. A wider, more achievable range is needed in practice, for instance, 1000 to 1300 MPa (32–42 HRC); in addition, ferrite must be avoided. These considerations, however, apply only to the critically loaded areas. As Kal'ner et al. related, it is undesirable for the tooth core and the main body of a gear to have the same strength because such a situation favors distortion (Ref 21). These researchers recommended that tooth cores have a hardness in the range 30 to 40 HRC, whereas the preferred hardness for the gear body is between 25 and 35 HRC. Developing a core hardness range of 30 to 40 HRC or 32 to 42 HRC immediately beneath the case ensures a more predictable local yield ratio (Fig. 6.11). That is, the yield strength will be fairly high to resist yielding and residual stress fade due to cyclic stressing.

Whereas core hardness is influenced by the chemical composition of the steel, it is also affected by the quenching temperature, as is the surface hardness. Consequently, the bending fatigue strength is also affected, as Table 6.1 shows for a small untempered 637M17 automotive gear (Ref 22). This table illustrates how quenching from above the Ac_3 of the core is important for good bending fatigue strength; in practice, however, the higher fatigue strength might not be realized if the growth and distortion are excessive due to quenching from a temperature that is too high.

Effect of Core Material on Impact-Fatigue Resistance. Undissolved ferrite in the cores of case-hardened notched test pieces significantly reduces their impact-fatigue resistance (Fig. 6.17). The difference in life between cores with ferrite (low hardness) and those without (high hardness) is entirely due to crack propagation (Fig. 6.18). In the low-cycle regime, core carbon content was significant: a core of 0.16% C required about 400 impacts to induce failure, whereas a core with 0.24% C failed in only 100 impacts (Ref 24).

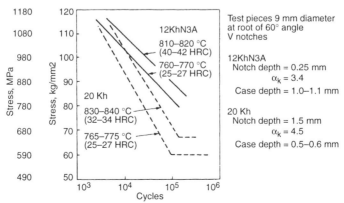

Fig. 6.17 Impact-fatigue strength of carburized test pieces. First quench was from 900–920 °C followed by tempering at 640–660 °C with final quenching from the temperatures indicated. Test pieces were tempered at 170 °C prior to testing. Core strengths are also indicated. Source: Ref 23

Impact-fatigue resistance increases with an increase of surface compressive-residual stresses. However, once a crack has propagated through the case, it is the strength of the core material that is important.

Effect of Core Material on Contact-Damage Resistance. Core properties are important to the contact-damage resistance only when the total carburized and hardened case is too shallow to adequately accommodate the contact pressure. In Fig. 6.19, the "adequate case" always has enough shear-fatigue strength to resist the applied shear stresses (even though the core strength is not particularly high). Any contact damage that occurs would be at, or close to, the surface due to surface shear stresses. The shear strengths of the "shallow case" parts shown in Fig. 6.16 more nearly coincide with the applied stresses; therefore, these parts are more susceptible to failure at or around the case-core interface. However, it is likely that the sample with the higher core strength will have a much longer life before shear fatigue damage occurs than the softer cored sample will have. The initial damage is a fatigue crack that appears to travel along the case-core interface before secondary cracks work their way to the surface. Failure due to this fatigue process, for which the time to failure can be quite short, is called deep-spalling fatigue or case crushing. Considering case-crushing, the total carburized case depth is important, not the effective case depth, and core strength only has an influence when the total case depth is inadequate.

Case Factors

Carburizing and hardening can generally provide properties that are superior to those achievable by through hardening or by the alternative surface hardening processes. This claim assumes that the correct steel is selected (for adequate case and core hardenability, and strength poten-

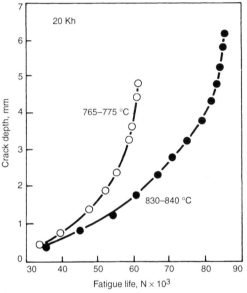

Fig. 6.18 Effect of quenching temperature and, hence, structure on the crack propagation rate during impact fatigue testing. Source: Ref 23

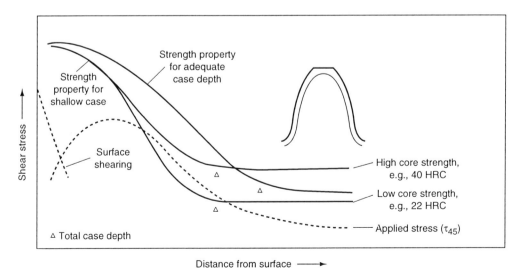

Fig. 6.19 Influence of case depth and core strength on the deep-spalling failure of gear teeth

tial), and the carburizing and hardening processes are correctly executed so that the surface carbon content and the case depth are adequate for the intended application.

Case Hardenability

The effect of carbon on the core transformation characteristics of a 3%Ni-Cr steel is depicted in Fig. 6.4(c), which shows that relatively small increases of carbon content significantly increase the hardenability. This trend continues at still higher carbon levels, as illustrated in Fig. 6.20, for a 1.3%Ni-Cr steel. In this example, the case of a carburized and quenched 50 mm (2 in.) diameter bar will be martensitic for all carbon contents above about 0.5%.

Some time ago, an attempt was made to categorize the levels of case hardenability (Ref 1):

- *Level 4:* A martensitic case occurs at all carbon levels, including the core material just beneath the case.
- *Level 3:* All carbon contents from the surface down to 0.27% C are martensitic.
- *Level 2:* All carbon contents from the surface down to 0.50% C are martensitic.
- *Level 1:* Surface carbon contents of above 0.80% are martensitic.

The respective case hardenabilities of a number of carburizing steels are compared in Fig. 6.21. This figure shows that level 4 tends to be attainable only in small sections of the more alloyed steels, whereas level 3 is more readily attained in most of the steels listed, depending on the section size. Level 2 is typical of many case-hardened parts and should be attempted as a minimum. When a steel is selected for a given component, the equivalent diameter for the critically stressed location is estimated. Then, from a chart such as Fig. 6.21, the expected level of case hardenability can be assessed.

When dealing with surface carbon contents and case hardenability, there is a carbon content (usually in the range 0.7–0.9%) for each steel above which the bainite nose time starts to decrease (i.e., the case-hardenability begins to fall) (Fig. 6.22). This means that a surface with a carbon content over about 1.0% could contain some bainite, whereas just beneath the surface where the carbon content is, for example, 0.9%, the microstructure consists of martensite and

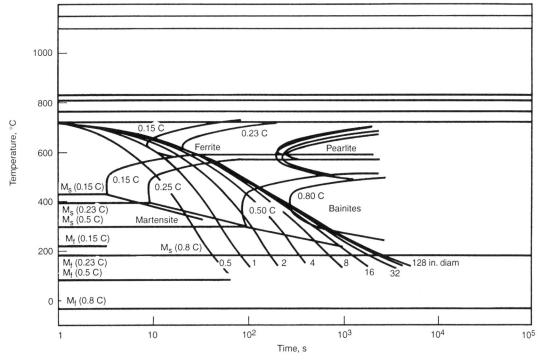

Fig. 6.20 Surface cooling rates for a number of bar diameters superimposed on the continuous-cooling diagrams for the carbon levels of the core: 0.15, 0.23, 0.5, and 0.8%. Base steel 815A16 austenitized at 830 °C for 100 minutes. Composition: 0.67 Mn, 0.90 Cr, 0.12 Mo, 1.32 Ni, 0.00 Si, 0.000 P, 0.000 S. Source: Ref 1

150 / Carburizing: Microstructures and Properties

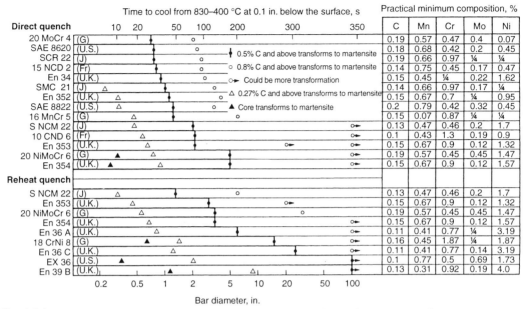

Fig. 6.21 Case hardenabilities of a number of carburizing steels with oil quenching. Source: Ref 1

austenite but no bainite. Surfaces with bainite have inferior contact fatigue properties compared to those without bainite. The bainite referred to here is not associated with decarburization or internal oxidation; it is a transformation characteristic, and is possibly related to carbide precipitation. One should also bear in mind that some bainite may be produced by austenite decomposition during low-temperature tempering.

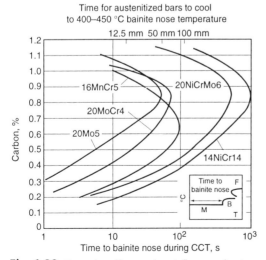

Fig. 6.22 Examples of how carbon influences the time to the bainite nose during continuous cooling. Different steels have different bainite nose temperatures within the 400–450 °C range. Derived from Ref 25

Optimizing the surface carbon content for any property requirement is only one consideration; it is also essential to have the right case depth and quenching conditions. That said, it is apparent that one surface carbon content cannot be the optimum for all loading situations; some components, such as gears, at times experience tooth bending with intermittent overloading, sliding contact and wear, and rolling contact. Therefore, if a high carbon content is necessary, the possibility of bainite formation at the higher carbon levels must be considered.

The "time to bainite nose" curves in Fig. 6.22 are additionally useful because they indicate the carbon content deep within the case at which bainite begins to appear in the microstructure. For example, for the 20MoCr4 steel depicted in the figure, a carburized part cooling at a rate equivalent to a 12.5 mm diameter bar will have some bainite in its outer case when the carbon there exceeds about 1%; the part will begin to show bainite in its lower case where the carbon content falls to about 0.47%. As the carbon content falls below 0.47%, the microstructure will contain increasing amounts of bainite.

Case Carbon Content

The total carbon penetration depth reached at a given carburizing temperature is determined by the duration of active carburizing. In single-stage carburizing at a constant gas composition, the surface carbon content is also influenced by the duration of active carburizing as Fig. 6.23 illustrates. Here the surface carbon content builds up with time so that for a given case depth, the carbon potential must be adjusted to achieve a specified surface carbon content. With two-stage (boost-diffuse) carburizing, the required surface carbon is more readily achieved, provided a reasonable minimum case-depth is exceeded. During the first stage, a high carbon potential drives carbon into the steel quickly, whereas in the second stage, when the carbon potential is lower, carbon diffuses outward and inward until equilibrium is reached with the furnace atmosphere, and the carbon gradient into the surface has the right shape (Fig. 6.24).

In developing a carbon gradient to obtain a specific case depth, an addition to the case depth must be made to account for any post-heat-treatment grinding. Furthermore, the surface carbon content may need adjustment so that after grinding, the as-ground surface has the desired carbon content. The boost-diffuse method of carburizing is perhaps the best suited for that because it has the potential to produce a high-carbon plateau can potentially be at the surface.

Commercial carburizing has produced parts with surface carbon contents between 0.6 and 1.2%, and more than 1.2% where wear resistance is the prime requirement. The most common range, however, is 0.75 to 0.95%. The choice of target surface carbon should be determined by the alloy content because, in general, an increase of alloy content reduces the eutectoid carbon content, the Ac_{cm} phase boundary is shifted to lower carbon levels, and the M_s temperature is reduced. Hence, the more alloyed a steel is, the lower the target surface carbon content should be to prevent the formation of carbides or the excessive retention of austenite. In production "jobbing" heat-treating, the frequent adjustment of carbon potential to suit alloy grade, case depth, or application of the treated parts can create problems, and the use of a standard high carbon potential might be necessary, irrespective of the other considerations.

Effect of Case Carbon on Surface Hardness. Heat treaters aim to produce parts with fine martensitic surface structures and little or no free carbide and without too much retained austenite. These parts should also have a reasonable level of product consistency as indicated by working to an as-quenched and tempered surface hardness within the 58 to 62 HRC range.

Most carburizing steels, in the fully martensitic condition, can attain 62 HRC (before tempering) with surface carbon contents as low as 0.6%, though the potential maximum hardness is achieved at still higher carbon levels, as shown in Fig. 6.25 (also see Fig. 4.11) (Ref 27). The maximum hardness value attained by any one of the steels shown probably equates to its eutectoid carbon content. At higher surface carbon levels, the hardness achieved depends on the quenching temperature; the higher the quenching temperature is, the more likely retention of austenite, and a corresponding loss of hardness, will result. There are three trends of interest shown in Fig. 6.25:

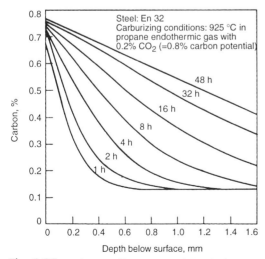

Fig. 6.23 Carbon profiles generated in single-stage carburizing in times ranging from 1–48 h. Source: Ref 26

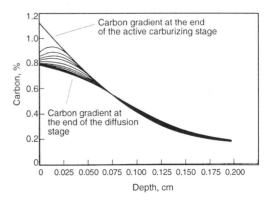

Fig. 6.24 Carbon distributions with double-stage carburizing. Carbon gradient at the end of the active carburizing stage and the carbon gradient at the end of the diffusion stage are shown along with intermediate stages of diffusion. Source: Ref 1

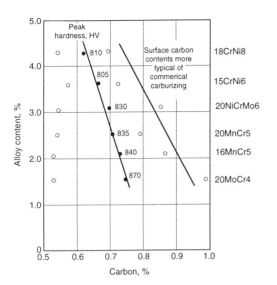

Fig. 6.25 Effect of alloy and carbon contents on peak hardness for direct quenching. ○, the limits of carbon content between which 800 HV or more can be achieved. Data from Ref 27

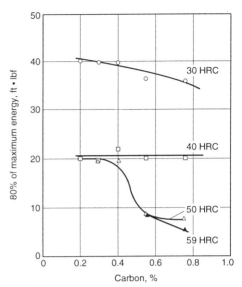

Fig. 6.26 Charpy V notch toughness tests that relate hardness, carbon content, and toughness. Source: Ref 28

- Carbon content for peak hardness value tends to fall as alloy content increases.
- The value of peak hardness itself tends to rise as the alloy content falls.
- As the alloy content increases, the carbon range for high hardness becomes smaller.

In practice in commercial carburizing, surface carbon contents are typically 0.1 to 0.2% higher than those for peak hardness (Fig. 6.25).

Effect of Case Carbon on Case Toughness. Tests used to measure or grade the steel toughness usually involve either slow bending or impact bending. In each of these tests, the initial crack forms at the surface, and therefore, the condition of the surface is important (but only in terms of crack initiation). Once a crack has started to propagate, the condition of the material and the residual stresses ahead of the crack provide any resistance to its development.

For virtually all applications involving case-hardened parts, surface hardnesses in excess of 58 HRC are essential. These values are achieved with high surface carbon contents. Based upon notched impact tests, high-carbon materials lose toughness at hardness levels above 40 HRC (Fig. 6.26); therefore, typical case-hardened surfaces have limited toughness (Ref 28). For high-carbon materials typical of a case-hardened surface (0.7–1.0% C), the toughness tends to decrease as the carbon content increases (Table 6.2) (Ref 29). Although lowering the surface carbon content to ~0.6% will improve the apparent toughness, the initial crack strength will be increased by increasing the core strength (Ref 30). Using precracked impact tests, Smith and Diesburg (Ref 12) obtained fracture toughness values of about 20 MPa$\sqrt{m}$ for high-carbon material close to the surface. As the carbon content decreased, these values increased to about 50 to 90 MPa$\sqrt{m}$ for material at the case-core interface. The spread in toughness at the case-core interface is influenced by the alloy content and the residual stresses. On their own, the main alloying elements manganese, chromium, and nickel have a negative

Table 6.2 Effect of carbon content on K_{Ic}

Steel	Carbon, %	Hardness, HRC	RA, %	K_{Ic}, MPa$\sqrt{m}$
PS-15	0.99	60.0	39	16.6
PS-15	0.86	60.5	23	22.4
PS-15	0.72	60.5	16	21.7
4895	0.95	55.5	40	24.5
4870	0.70	57.0	21	34.5

RA, retained austenite. Source: Ref 29

effect on fracture toughness, whereas molybdenum has a slightly positive effect. In combination, the effects are different: nickel with molybdenum enhances toughness appreciably, especially when the nickel content exceeds about 2%. The impact and bend strengths of Mn-Cr steel can also be improved by the addition of 2% Ni (Ref 31). Most investigators find that the best toughness properties, regardless of test method used, were obtained with the 3%Ni-Cr-Mo steels.

Of the several variations of bending and impact testing, only the static bend strength appears to relate to bending fatigue limit (Fig. 6.27) (Ref 32). Here, the impact strength and fracture strength fall as the case depth increases. The static bend strength, on the other hand, peaks at a certain case depth, which coincides with a peak in the endurance limit. Other studies show the impact fracture stress to correlate with the number of constant-load impacts to failure (low cycle) (Ref 12).

Effect of Case Carbon on Impact Fatigue. Impact-fatigue resistance benefits from a lower surface carbon content. Brugger for example, found that 20NiCrMo6 steel test pieces with 0.6% surface carbon had an approximately 10% better fatigue limit under impact-fatigue testing conditions than did test pieces of the same steel with 0.8% C surfaces (Fig. 6.28) (Ref 33).

Regardless of surface carbon, it is important to ensure that the quenching conditions are right. The impact-fatigue life can be optimized by employing a quenching temperature that will produce a case microstructure of fine martensite with a small amount of well-dispersed retained austenite (Fig. 6.29) (Ref 34). Hot-oil quenching can also improve toughness, as can tempering (Ref 12, 35).

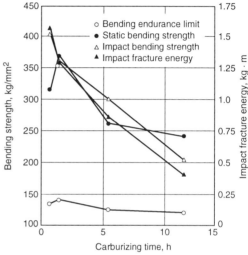

Fig. 6.27 Relation between carburizing time and bending strength. Source: Ref 32

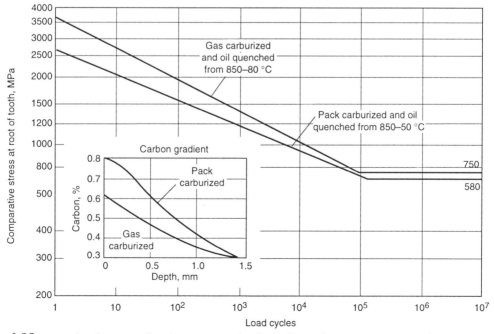

Fig. 6.28 Material performance of steel 20NiMoCr6 related to surface carbon content. Source: Ref 33

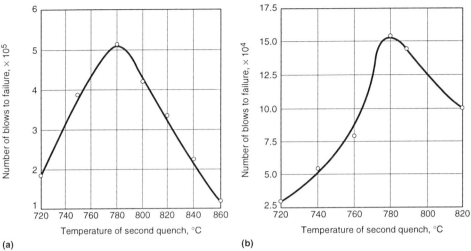

Fig. 6.29 Relation between repeated impact resistance on temperature of second hardening. (a) Steel 20N3MA. (b) Steel 20KhN3A. Source: Ref 34

The low-cycle impact-fatigue tests carried out by DePaul (Ref 36) confirmed the benefit of alloying with nickel and molybdenum, and the test results suggest that to extract the best from a 2% nickel content, the case-depth would need to be optimized (Fig. 6.30). Smith and Diesburg (Ref 12), however, warn that a high nickel content does not guarantee good low-cycle impact-fatigue properties.

Aida et al. (Ref 32) investigated occasional impacts to gear teeth that otherwise experience only bending fatigue. A single blow impact, whether administered before or during the fatigue test, had no effect on the fatigue life or fatigue limit, provided the impact did not exceed 0.63 of the impact fracture stress, P, and 0.4 of the impact energy, E. Similarly, 5 to 500 impacts at 0.32 P and 0.1 E had no adverse effect on the bending fatigue resistance. However, 5 impacts at 0.39 P and 0.15 E slightly increased the fatigue life and limit, whereas 10 impacts at that stress increased the life but not the limit. Note that the impact load 0.39 P was equal to a normal tooth load of 1430 kg, which is slightly higher than the bending-fatigue limit. Between 10 and 40 impacts delivered at a load of 0.45 P and 0.2 E brought about a 7% reduction of both bending fatigue life and limit.

Effect of Case Carbon on Residual Stresses. The austenite to martensite transformation resulting from quenching involves a volume expansion, and the amount of expansion increases with carbon content. Therefore, a high-carbon surface layer that completely transforms to martensite will expand appreciably more than the core does; this difference causes compressive-residual stresses in the outer case and balancing tensile-residual stresses in the core. If the carbon content of the quenched outer case is high enough to retain austenite, the volume expansion in the outer case will be less by an amount based on the amount of austenite present; consequently, the residual stress distribution will be adversely affected. The quantity of austenite retained relates to the martensite transformation range (M_s–M_f): when the M_f temperature is below the quenchant temperature, some austenite will be retained. The theoretical limiting carbon contents for essentially zero retained austenite for a number of steels are shown in Table 6.3. However, this does not imply that one should carburize to low surface-carbon levels to avoid austenite retention

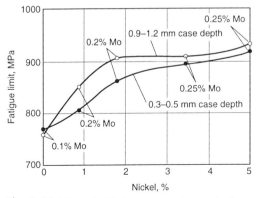

Fig. 6.30 Effect of nickel content and case depth on the bending fatigue strength of case-hardened steels. Source: Ref 36

because the volume expansion of a 0.9% C surface with 20% retained austenite will still be much greater than that of a 0.5% C surface with zero retained austenite.

The Effect of Case Carbon on Bending Fatigue. Fatigue strength is greatly influenced by the case microstructure and surface residual stresses; these, in turn, are affected by the surface carbon content and the quenching method. A direct-quenched carburized surface contains more retained austenite and less carbide than a comparable reheat-quenched surface. Consequently, the maximum surface carbon content for direct quenching should be about 0.95% for the lean alloy grades and 0.75% for grades with ~4% total alloy content. That said, in general, the lean-alloy grades are direct quenched, and the more alloyed grades are reheat quenched. Reheat-quenched surfaces are usually more tolerant to higher surface-carbon contents, though the quantity and distribution of any carbides surviving the quench need to be considered.

Diesburg (Ref 37), working with a number of SAE and Ex steels, observed that the bending-fatigue resistance of test pieces with 0.8% surface carbon contents was superior to those with 1.0% surface carbon. It is possible that good bending-fatigue results would have been obtained with carbon contents even lower than 0.8% if the carbon content remained above the eutectoid for each steel, and the microstructure was martensitic. Unfortunately, the lower carbon levels do not favor contact-fatigue resistance or wear resistance, so a compromise is necessary. For many applications, 0.85 to 0.95% should give acceptable general properties, assuming that the final surface microstructure does not contain excessive austenite and/or excessive carbides.

Most case-hardened parts contain stress raisers of one shape or another, and as the stress concentration factor increases, it is prudent to choose a steel with good toughness characteristics and work to the highest surface carbon content consistent with an acceptable microstructure.

The Effect of Case Carbon on Contact Damage. Until a certain case depth is attained for a part, contact-fatigue life increases as the case depth increases (Ref 38). Once adequate case depth is achieved, other metallurgical variables, such as carbon content and microstructure, can be considered.

The carbon content of a case-hardened surface must ensure high hardness both at the surface to resist wear, adhesion, and surface shearing, and deeper to resist pitting and shallow spalling. Generally, an essentially martensitic microstructure is preferred, where most, if not all, of the carbon is in solution, where the martensite is fine, and where any austenite is fine and evenly dispersed. The steel should have an alloy content adequate to deter the formation of high-temperature transformation products (HTTP) such as bainite, a few percent of which is deleterious (Ref 14). Where roll-slide contact fatigue is involved, the surface carbon must be moderately high: approximately 0.9 to 0.95%, as Vinokur reported in Ref 39 (Fig. 6.31). However, in high-speed gearing where the scoring and scuffing potential is high, a dispersion of carbides in martensite will likely be more resistant to wear and adhesive-wear processes than martensite alone will be. This is due to the high hardness of the carbide and its low weldability characteristics under sliding contact conditions.

Case Depth

The effective case depth of a case-hardened part is taken as the perpendicular distance from the surface to a depth where a specified hardness value is attained (e.g., 50 HRC). The total case depth is the perpendicular distance from the surface at which the case merges with the core. Methods for assessing case depth are to be found in SAE standard J423a.

It is important that with any organization the meaning of the term *case depth* is fully understood by all concerned. For example, a designer specifies a case depth for a gear tooth (using

Table 6.3 The estimated carbon content for zero retained austenite

Steel	Direct quench in oil		Reheat quench in oil	
	60 °C	25 °C	60 °C	25 °C
835M 15 (U.K.)	0.35	0.42	0.37	0.46
9310 (U.S.)	...	0.46	...	0.50
PS.55 (U.S.)	...	0.50	...	0.58
17CrNiMo6 (G)	0.44	0.52	0.48	0.61
8617 (U.S.)	0.50	0.59	0.52	0.63
665 M 17 (U.K.)	...	0.60	...	0.66
20MoCr4 (G)	...	0.63	...	0.70
1017 (U.S.)	0.58	0.67	0.60	0.73

AGMA 2001) and, therefore, expects the specified case depth to be attained in the dedendum-pitch line area of the finished tooth. The heat treater must understand what is needed, and must adjust the carburizing time to account for any grinding allowance. The inspector, who tests for case depth, must also have the same knowledge. Discrepancies can arise if, for the same designer requirement, the heat treater considers only the basic specification or the inspector does not allow for the differences between test pieces and parts.

If an actual part is sectioned for case depth (and quality) assessment, the case depth measurements must be at the locations specified by the designer. If no locations are specified, the primary test location must be at a critical area (for gear teeth this is near the lowest point of single tooth contact). If test pieces are employed for case depth determination, the nearer the test piece is to the part in terms of material and cooling rate, the more reliable are the test results, although some adjustment for surface curvature may be needed (Fig. 6.32). Alternatively, if a standard test piece (e.g., 1 in. diameter, SAE 8620) is used for all occasions, sufficient correlation work must exist to confidently derive a case depth for the part from the measured case depth of the test piece; case hardenability and cooling rate differences can be significant in this respect.

The assessment of case depth is made just before, or just after, the parts are removed from the carburizing furnace. The first assessment may involve fracturing the as-quenched test piece (which can be done quickly); this assessment is moderately accurate (to the experienced eye). Such a test, along with a surface hardness test, indicates whether the carburizing cycle has been typical, and that the parts treated do not need any further carburizing. A more accurate case depth assessment, although more time consuming, can then be made by grinding, polishing, and nital etching one of the fracture faces. An experienced operator can judge the effective and total case depths from the etching response of different zones within the case. A more accurate, but more time consuming, assessment is achieved by conducting a hardness traverse through the case. Case depth assessments using test pieces should be correlated with one another, with any carbon gradient data, and with test data from actual components.

Dependence on Shape and Size. The depth to which carbon atoms penetrate during carburizing is primarily determined by the temperature, the carbon potential, and the duration of carburizing. Also influential are the gas flow rate, and whether a surface is in or out of the direct stream of the furnace gas. The depth to which the carbon profile will harden is determined by the steel composition (case hardenability) and the rate of cooling during the postcarburizing quench. The rate of cooling for a given quench relates to the quenching temperature, the quenchant temperature, and the mass of metal being quenched.

However, product shape also significantly influences case depth. In a gear tooth, for example, the depth of carbon penetration into the flat end of the tooth will be more than the penetration at the tooth fillets, less than at the tooth dedenda, and appreciably less than at the top edge of the tooth. Figure 6.32 indicates how the curvature of a surface affects the depth of carbon penetration for case depths of 1 to 2 mm. Therefore, if a tooth has a fillet radius of 1.5 mm, a dedendum radius of 5 mm, and an addendum radius of 13 mm, the respective depths of carbon penetration when the target depth is 1.0 mm on a flat surface will be 0.79, 1.125, and 1.04 mm. Such differences in carbon penetration result because different surface curvatures and edges have different surface area-to-volume ratios.

In addition to product shape, the case depth (in terms of the depth to a given hardness, typically 50 HRC) will also depend upon size. If the steel hardenability is high for the part being considered, resulting in the transformation to martensite of both the case and the adjacent core materials at all locations during quenching, then 50 HRC will be attained at the depth within the

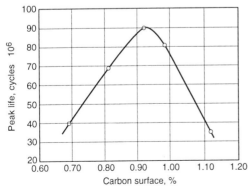

Fig. 6.31 Effect of surface carbon content on the resistance to pitting fatigue of a case-hardened alloy steel. The samples were fully austenitized prior to quenching.

case where the carbon content is, for example, 0.25%. Actually, this varies depending upon the alloy content, as Fig. 6.7 indicates. The differences in effective case depth between the tooth dedendum and the tooth fillet are similar to the differences of carbon penetration depth as outlined above. If, on the other hand, the hardenability is not excessive or the size of the component is large so that the inner case consists of martensite and bainite, the carbon content necessary to develop 50 HRC will be more than 0.25%. How much more depends on the cooling rate that determines the proportions of martensite and bainite. A gear tooth fillet cools more slowly during a quench than a gear flank does, and even more if the teeth are integral with a large mass, for example, as with a pinion. Therefore, at the tooth flank, the carbon content to give 50 HRC at quenching might be 0.29%, whereas at the tooth fillet it might be 0.32%. On this basis alone, the

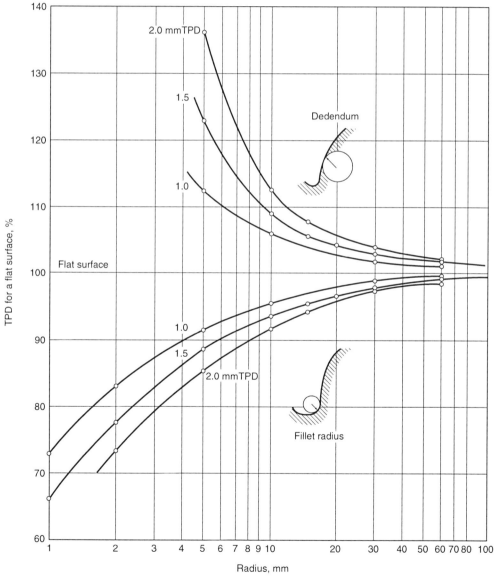

Fig. 6.32 Effect of surface curvature on the total depth of carbon penetration (TPD). 1.0, 1.5, and 2.0 TPD refer to the total carbon penetration depth at a flat surface. The curves themselves represent departures from the flat. If a gear tooth is carburized to give a nominal total penetration depth of 1.5 mm (0.06 in.), and the tooth has a fillet radius of 2 mm and a dedendum radius of 10 mm, then the TPD at the fillet will be ~78% of 1.5 mm, or 1.17 mm (~0.047 in.) and the TPD at the dedendum will be 109% of 1.5 mm, or 1.635 mm (~0.065 in.).

effective case depth at the fillet could be 20% less than at the flank, and, taking into account the difference in carbon penetration due to shape, the difference of effective case depth (50 HRC) between fillet and flank could be around 30%. This aspect is important with respect to gear teeth, and it also applies to keyway corners and to changes of diameter on shafts.

At the top and side edges of gear teeth, or the edges of keyways, there is a large area of surface feeding carbon into a small volume of metal; this ratio results in a buildup of carbon at the edge or corner and carbon penetration depth appreciably greater than at the tooth flank or in a flat surface. This can become a problem, particularly when the tops of gear teeth are slender, the steel is of a relatively high hardenability, and subcase cracking can ensue as a result of quenching.

Case Depth and Properties. Crude representations of how case depth and core strength can be manipulated to counter applied bending stresses are provided in Fig. 6.33(a) to (c). These figures indicate that there are two likely sites at which failure can initiate: at the surface and near the case-core interface. Therefore, attention must be paid to the metallurgical quality of the case and the adjacent core at these two locations in particular. Figure 6.33, however, does not take into account the influence of residual stresses within the case and core regions.

Residual stresses and applied stresses are additive; therefore, if the residual stresses at the surface are highly compressive (−ve), they will detract from the applied stresses (+ve) and thereby offer some protection against crack propagation from the surface. If, on the other hand, the surface residual stresses are tensile due to HTTP (associated with internal oxidation, for example), then the residual stresses will augment the applied stresses and failure will be more likely to initiate at, and propagate easily from, the surface. Metallurgical variability can be a problem, and the use of shot peening more or less ensures that the critical surface areas are in compression and are, therefore, reinforced against failure.

Often, the compressive residual stresses peak at some distance beneath the surface (about midcase), and whereas they are unlikely to influence the crack initiation process at the surface, they retard cracks propagating from the surface.

Just beyond the case-core interface, the residual stresses are tensile and balance the compressive residual stresses within the case. A survey of data suggested that the tensile stress peaks beneath carburized and hardened cases fell within the range 40 to 150 MPa (Ref 1). These tensile stresses add to the applied stresses acting just beneath the case, thereby encouraging either yielding or high-cycle fatigue damage. The deeper the carburized case is, the further from the surface the potentially damaging tensile residual stress peak is. This relationship, along with the fact that applied stresses diminish with distance from the surface, reduces the chances of a fatigue failure initiating in the core. Of course, raising the core strength has a similar effect. However, note that excessive increases of case depth and/or core strength can have an adverse influence on surface compressive-residual stresses and the residual stress distribution through the case.

Case Depth and Residual Stresses. Increasing the case depth is more likely to favor an increased depth of internal oxidation and increased amounts of retained austenite and free carbides, each of which can adversely affect the residual stress distribution. For example, in Fig. 6.34 retained austenite and HTTP associated with internal oxidation have each caused the surface residual stresses to be tensile. Presumably, these three test pieces were carburized at one carbon potential; therefore, because approximately 15 h at 925 °C are required for the surface carbon to more or less reach equilibrium with the carburizing atmosphere, the surface carbon contents of

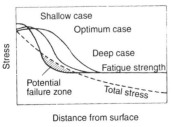
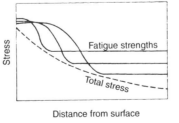
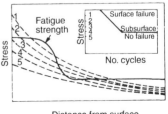

Fig. 6.33 Schematic diagrams representing the relationship between the total stress (applied and residual) and fatigue strength of carburized and hardened steel. (a) Effect of case depth with constant core strength. (b) How case depth can be decreased by increasing core strength. (c) Effect of stress on the location of fatigue failures

the test pieces differed. When the process is controlled so that, although the case thickness might be varied, the surface carbon content is essentially maintained, a different picture emerges (Fig. 6.35). Here, the peak compressive-residual stresses are very similar, and whereas changing the surface carbon content influences the magnitude of the compressive stresses, it does not alter the pattern. The lower compressive-residual stresses at the surface in those samples containing 0.9% C at the surface are attributed to the higher retained-austenite content there.

Therefore, when carburizing to produce deep cases, this aspect of carbon potential control should be observed.

Residual stresses, which act as a mean stress onto which cyclic applied stresses are superimposed, are not altogether stable and can fade during aging (Ref 42) or be modified during service (Ref 40). With all case-hardened components, loading will cause a small readjustment of residual stresses, which then become, in many instances, more or less stable. Therefore, in any calculations involving residual stresses, the readjusted residual stresses will be more relevant than the initial residual stresses. Residual stress modification during bending-fatigue loading tends to affect the part of the case where the hardness falls below 500 HV (see Fig. 6.36) and where ferrite is present in the lower case and core (Ref 40, 43). In such cases, a significant residual stress modification will occur in the zone in which the sum of the applied stress and residual stress has exceeded the microplastic yield strength or the fatigue strength of the weaker structural constituents.

For surface residual stresses to undergo change, some deformation of the austenite and martensite is required. Such deformation could result from, for example, shot peening or deformation rolling, which increase compression. Without deformation, rolling can cause a reduction of the surface compression (Ref 44).

Bending Fatigue. Although increasing the case depth above a certain amount improves resistance to deep contact spalling (case crushing) and pitting fatigue, it does not necessarily ensure that the bending-fatigue resistance of the component will be improved, even if the metallurgical quality of the case is good.

Bending-fatigue strength is influenced by section size. For example, Dawes and Cooksey (Ref 45), using Ni-Cr-Mo steels, showed that for 7.6 mm (0.3 in.) diameter test pieces, the maximum bending-fatigue strength was obtained with case depths of about 0.6 mm (0.024 in.), that is, when the case depth-to-section thickness ratio, CD/t, was approximately 0.07 to 0.08 (Fig. 6.37). Aida et al. (Ref 32) obtained an optimum CD/t ratio of 0.076 working with case-

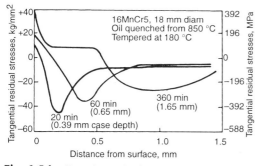

Fig. 6.34 Effect of case depth on residual stress. Influence of internal oxidation at the surface of the deep-case test piece is also indicated. Source: Ref 40

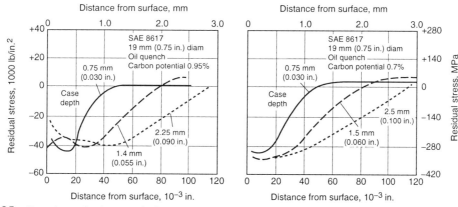

Fig. 6.35 Effect of case depth on residual stress. Effect of carbon potential is also indicated. Source: Ref 41

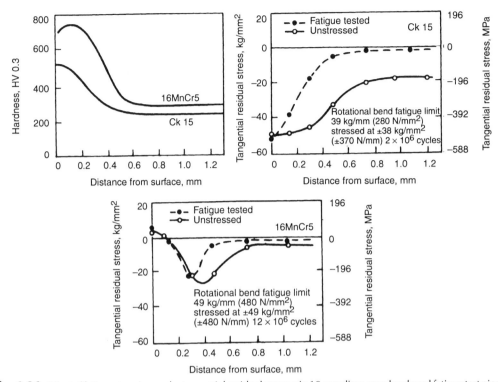

Fig. 6.36 Effect of fatigue stressing on the tangential residual stresses in 18 mm diam case-hardened fatigue test pieces. Source: Ref 40

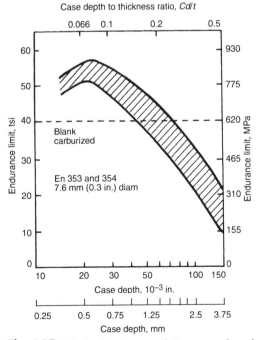

Fig. 6.37 Relationship between fatigue strength and case depth for two carburized lean-alloy case-hardening steels. Source: Ref 45

hardened gear teeth of 7.25 mm root thickness. Contrary to this, Tauscher (Ref 46) noted, after reviewing several sets of published data, that the optimum values for the CD/t ratio ranged from 0.014 to 0.21. This scatter was attributed to differences in the residual stress distributions from one set of tests to another. DePaul's results, as plotted in Fig. 6.30, suggest that alloy content or hardenability might contribute because the more alloyed a steel is, the less important the CD/t ratio is (Ref 36). Weigand and Tolasch (Ref 40) showed that component size and geometry and the method of hardening all determined the optimum value of the CD/t ratio (Fig. 6.38a, b). In those tests, which involved 6 mm diameter test pieces, the CD/t ratio for maximum fatigue strength of alloy steels was 0.07 to 0.075. For notched specimens ($\alpha_k = 2$) the ratio was about half this value. With 12 mm diameter test pieces, the relationships were more difficult to establish. Significantly, in Fig. 6.38(a) and (b) the maximum value of fatigue strength almost coincides with the changeover from subsurface to surface initiated failure. In other words, the ideal case depth (in terms of bending fatigue) appears to be reached at the value where the failure initiation point is transferred from the core to the surface. However, in some bending-fatigue investigations, when subsurface failures have occurred, they have occurred at stresses just above the fatigue limit, whereas at higher stresses, the fracture initiation points have been at the surface. This suggests that for high-cycle applications, the CD/t ratio is more relevant than it is for high-stress, low-cycle fatigue applications, where aspects related to case toughness and core yield strength might be even more significant.

In general, a fatigue fracture will initiate either at an engineering or a metallurgical stress raiser at, or close to, the surface of the component.

The fatigue life of a part is composed of three stages:

1. The load cycles that initiate a fatigue crack
2. The cycles that expand the initial crack to the critical size
3. The cycles that propagate the critical crack through the section to total failure

The number of cycles in any of the three stages depends on the applied stress. At high stresses, most of the life is taken up by the crack initiation stage, and toughness is an important requirement to counter failure. At stresses close to the fatigue limit, the third stage predominates, and strength rather than toughness is the main requirement.

For initial microcrack development, one of four mechanisms can be involved (Ref 47):

- *Intrusion/extrusion.* Slip along persistent slip bands; featureless fracture surface
- *Transgranular.* Microcrack perpendicular to the applied stress; flat surface with no local deformation
- *Shear deformation.* Initially parallel to the maximum shear stress, then extends to be normal to the applied stress. Possibly dimpled fracture surface near nucleation site
- *Intergranular.* Microcrack nucleation at a prior austenite grain boundary or an interface (with a nonmetallic inclusion, for example). Initial microcracks may be smaller than a grain-boundary facet, but grow rapidly to the size of the facet (Ref 48).

Whereas the initial cracks in low- and medium-carbon alloy steels tend to develop by the intrusion/extrusion mechanism or by the transgranular mechanism at low applied stresses, shear deformation is the main mechanism at high applied stresses (Ref 47). In such materials, the martensite is lath, providing a combination of high yield strength and toughness (Ref 4).

The initial crack to develop in direct-quenched high-carbon surfaces tends to form by the intergranular mechanism. The microstructure is predominantly plate martensite, so the material has high strength and low toughness; consequently, the first stage of crack initiation is short at high applied stresses. Reheat-quenched surfaces rarely exhibit intergranular crack initiation; however, high nickel contents tend to reduce the susceptibility to intergranular cracking (Ref 49), as does fine grain size.

The crack initiation process (stage 1) and its extension to critical crack size (stage 2) are thought not to be dependent on residual stresses (Ref 50). The transition from stage 2 to stage 3, and stage 3 itself (crack propagation), are dependent on the residual stress state; compression tends toward keeping the crack closed and countering tension ahead of it.

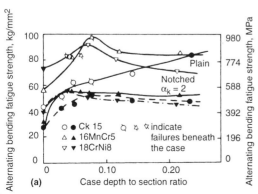
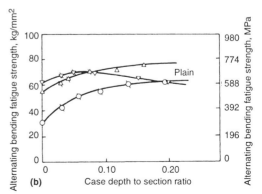

Fig. 6.38 Alternating bending fatigue strength of carburized test pieces in relation to case depth and section ratio. (a) 6 mm diam. (b) 12 mm diam. Source: Ref 40

Contact Damage. Case depth requirements for case-hardened parts vary according to the application. For gear teeth, the case depth specification is determined by the needs of the tooth dedenda/pitch-line region with respect to deep-spalling (sometimes referred to as case-crushing) resistance. Sufficient case depth for deep-spalling resistance is regarded as more than enough for bending-fatigue resistance. However, remember that at the gear-tooth root fillet, the case depth is typically shallower than at the dedendum. For bending-fatigue resistance, the quality of the immediate surface and the residual stresses at the surface are important. In relation to wear and surface pitting, surface quality and lubrication are important. For deep-spalling resistance, the total case depth is most influential; where the case depth is shallow or marginal, the strength of the material near the case-core interface then becomes important. Considering the large numbers or gear units that are, and have been, in service, deep spalling remains a fairly rare occurrence (except, perhaps, in special applications such as the final reduction train of heavy tractors and off-highway equipment for which there has been a history of such failures) (Ref 51). In general, therefore, case depth specifications must have been correct, or even a little generous.

Surface pitting, shallow spalling, and deep spalling are fatigue failures in which the applied shear stresses have overcome the shear strength of the material in the respective failure zones. There are, however, distinct differences between the different types of damage, as listed in Table 6.4, as there are differences in the shear stresses that cause the damage. In Fig. 6.39, a composite illustration compiled by Sharma et al. (Ref 53), the plots of surface shear stresses due to sliding and contact spikes are added to the τ_{45} and orthogonal shear stress (τ_{yz}) range curves (due to rolling) to give a profile of the greatest stress types. Thus, any damage occurring due to the shear stresses in zone I will be shallow, for example, gear tooth pitting and surface flaking. Shallow spalling damage, which is common in bearings and bearing tracks, will occur due to the τ_{yz} shear stresses of zone II. Deep spalling failures are thought to occur due to the τ_{45} shear stresses, which are shown here as predominant in zone III.

In their analysis of deep-spalling failures, Pederson and Rice (Ref 54) considered the τ_{45} shear stresses to be the most relevant because these have the greater magnitude at any depth (though not the greatest stress range). Therefore, these researchers calculated the τ_{45} shear stresses to a depth greater than that of the case, then they took the hardness values of the case and converted them to shear yield values. This allowed them to compare the applied stress with the appropriate material strength property. After running tests on gears for which case depth was the primary variable, Pederson and Rice concluded that if the maximum shear stress-to-shear yield strength ratio exceeded 0.55, deep-spalling failures would eventually occur. They also believed that the maximum bending strength was obtained by using the thinnest case in order to resist case crushing. Because this opinion stemmed from their tests on gears, the case depth required to resist crushing would have been approximately 25 to 33% more than at the tooth fillet. It is implied that if there is enough case on the dedendum to resist crushing, there should be sufficient in the fillet to resist bending fatigue, provided the root fillet radius is reasonable

Table 6.4 Summary of contact failures descrpitive and visual for through-hardened and surface-hardened gears

Property	Surface pitting	Subsurface pitting/spalling	Case crushing
Location of origin	Surface, often at micropits	Short distance below surface may be at nonmetallics	Probably at case-core interface
Appearance	Shallow	Shallow	Deep ridged
Initial size	Small	Small	Large
Initial area-depth ratio	Small	Small	Large
Initial shape	Arrowhead then irregular	Irregular	Gouged and ridged (longitudinal gouging)
Crack angle with respect to surface	Acute	Roughly parallel at bottom perpendicular sides	Roughly parallel at bottom perpendicular sides
Distribution	Many teeth	Maybe many teeth	One or two teeth
Apparent occurrence	Gradual	Sudden	Sudden

Source: Ref 52

(see Fig. 6.32), and the steel has adequate case hardenability.

Sharma et al. similarly determined the limiting total case depth required to avoid deep spalling in gear teeth. In their study, the τ_{45} shear stresses were used to represent the applied stresses; however, it was reasoned that the shear-fatigue endurance was the correct material property to employ (Ref 53). The shear-fatigue endurance was calculated from the hardness data and found to be, for a gear tooth, equal to 155 HB or 0.31 UTS. Then, by adopting an appropriate safety margin, Sharma et al. were able to determine the case depth and core strength requirements to avoid deep spalling failures.

Fujita et al. (Ref 55) concluded that contact failures occur where the ratio of the maximum amplitude of orthogonal shear stress to the Vickers hardness is at a maximum [$A(\tau_{yz}/HV)$]. An example of how this ratio varies with depth is shown in Fig. 6.40; the ratio increases with loading to give failure in a shorter time, as shown in Fig. 6.41, so it is comparable to a standard fatigue plot. In many past studies by Fujita of contact failures in surface-hardened gears (not referenced here), this ratio has been found to be valid for both shallow and deep spalling failures, irrespective of case depth. Therefore, with respect to deep spalling failures, this approach (like those previously referenced) allows the case depth and core strength to be manipulated to ensure that the ratio maximum is not deep in the case, but much nearer to the surface. The ratio of τ_{yz}/HV does not account for residual stresses im-parted by heat treatment; hence, if failure initiation always occurs at a point where the ratio is at a maximum, it is implied that residual stresses are not especially involved in the contact fatigue process.

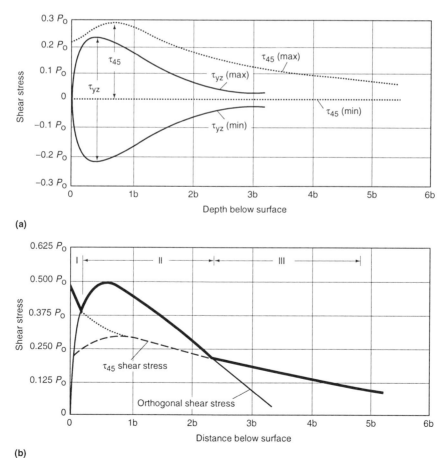

Fig. 6.39 Composite shear stress range gradient. Fatigue-crack initiation in carburized and hardened gears controlled by the 45 shear stress in zones I and III and by the orthogonal shear stress in zone II. P_o, maximum pressure at the surface; b, half the contact width. Source: Ref 53

Quenching Methods

The hardening of a carburized part is achieved by cooling it at a rate that produces the desired metallurgical condition at, or within, the most critically loaded areas of the part. The most used techniques are direct quenching and reheat quenching. Double reheat quenching was common before the introduction of grain-size control, but is now less frequently employed, though some regard it as capable of producing high-durability components. The downside is the cost of the extra quenching treatment and the risk of increased distortion with excessive corrective grinding.

When high production runs are involved, direct quenching from the carburizing temperature or single quenching from approximately the Ac_{cm} temperature are common. These methods are suitable for those small, lean alloy steel items that are not required to be die or plug quenched.

With high production runs, the chosen steels have just enough hardenability to consistently produce good quality parts; consequently, the cost per item is kept to a minimum.

Reheat quenching requires that parts be cooled, rather than quenched, after carburizing, followed by quenching at a later stage. This method is preferred for the more alloyed case-hardening grades of steel, and with heat-treatment situations with one-off or small numbers of parts ("jobbing"). There are a few reasons why this method of quenching might be used:

- To ensure metallurgical quality, for example, grain-size and retained austenite control
- When an intermediate subcritical heat treatment is required, either to condition the carbides within the case or to facilitate additional machining
- When parts are to be plug or die quenched to control distortion
- When it is not possible to direct quench (as in pack carburizing)

With low production work, matching the hardenability of the steel to the workpiece may not be precise, and there is a tendency to err on the side of safety by selecting steels with more than the minimum required hardenability.

Distortion

In general, case hardening significantly improves the load-carrying capacity and wear resistance of parts. Unfortunately, these benefits can be undermined if the distortions that accompany the carburizing and hardening processes prevent the parts from complying with design tolerances, or require an unreasonable amount of grinding to restore the size and shape to within acceptable limits. Excessive corrective grinding could lead to unacceptable thinning of the case and, possibly, the step formation at a critical location, such as the root fillet of a gear tooth.

There are two types of distortion: size distortion, which refers to growth or shrinkage, and shape distortion, which is essentially warpage. Growth and shrinkage relate to the volume changes that accompany microstructural phase transformations, while warpage relates more to asymmetrical thermal effects. If uniform growth or shrinkage were the only concerns, then, with

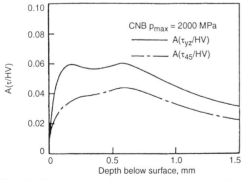

Fig. 6.40 (a) Curves showing the distribution of the τ_{45} and τ_{ortho} stresses during one cycle of pure rolling contact. Note that τ_{45} stress has the highest magnitude, but τ_{ortho} has the greatest range. (b) Distributions of amplitude of ratio of shear stress to Vickers hardness $A(\tau/HV)$ below tooth surface at working pitch point. Derived from (a). Source: Ref 55

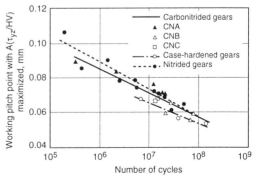

Fig. 6.41 Curves of the ratio of the maximum amplitude of orthogonal shear stress to Vickers hardness $[A(\tau_{yz}/HV)]$-N2. Source: Ref 55

experience and planning, a process such as grinding would only be needed to optimize the surface finish. As it is, growth and shrinkage are not always uniform, and warpage adds to the problem. Warpage includes loss of roundness (ovality) and loss of parallelism (warping, bowing, tapering). Thus, one could have a gear wheel for which the side faces are "dished," the bore and outer surface are tapered, the outside diameter is oval, and the tooth pitch is variable.

Growth and shrinkage are influenced by:

- *Chemical composition:* hardenability and the relative proportions of the different microconstituents
- *Steelmaking:* grain size, hardenability
- *Hot working:* hot reduction, length and direction of "fiber"
- *Prior heat treatment:* grain size and microstructure uniformity for in-batch size consistency
- *Geometry:* cheese blanks, shafts, rims
- *Heat-treatment aspects:* heating rates, cooling rates, jigs and fixtures, plug quenching

Warpage is affected by:

- *Uneven residual stresses in the original blank:* like prior heat treatment above to remove stresses
- *Lack of uniformity of heating or cooling:* furnace shape, part shape, heat control
- *Time in the furnace:* undersoaking can be detrimental
- *Creep:* hanging versus standing

When a part is heated in a furnace, thermal gradients are created that give rise to thermal stresses. The hotter the furnace, the steeper the early thermal gradients. If the thermal stresses at any stage of the heating process exceed the "hot" yield stress of the steel, then some yielding will take place to relieve the thermal stresses. Once the part has reached the furnace temperature, there will be no thermal gradients and, therefore, no thermal stresses; however, any distortions that have taken place during heating will remain. If the heating causes austenitizing, the ferrite to austenite transformation, which progresses from the surface to the interior as dictated by the thermal gradient, will be accompanied by transformation stresses. Again, if these stresses exceed the high-temperature yield strength of the steel, some yield deformation takes place. Similarly, austenite transformation and thermal gradients occur during cooling (steep gradients if the cooling involves a quench); this results in thermal stresses, transformation stresses, yielding, and, hence, distortion.

Parts that are pack carburized (both heated and cooled in compound) distort very little because the thermal gradients involved are quite shallow. The same parts, if gas carburized and quenched, generally grow or shrink more because the thermal gradients are steeper both in heating and in cooling.

The shape and size distortions of small components are affected by the transformation behaviors of both the case and the core, whereas for larger parts, the presence of the case has less of an influence. That is, what happens in the core, in terms of thermal gradients and transformation, determines how the part will change its size or shape as a result of carburizing and hardening. The thermal gradients are determined by type of quench and the mass and geometry of the component, whereas the transformation behavior is related to the hardenability and the mass and geometry of the component. The trends are illustrated in Fig. 6.42; note the effect of edges.

The composition range for an individual steel specification is wide enough for appreciable differences of hardenability to occur on a batch to batch basis; the effect of such variability on distortion is shown in Fig. 6.43 for 1 meter diameter discs, and in Fig. 6.44 for 132 and 76 mm diameter washer-type test pieces. Such variability is perhaps a little extreme, but, nevertheless, these examples illustrate how distortion trends can be influenced by hardenability.

If a selected lean grade of steel has borderline hardenability for a given design, the variability regarding distortion can be significant, as the previous examples have shown; therefore, to some extent, the use of an H grade should help keep the distortion within acceptable limits. When the steel has adequate hardenability, the use of the H grade might not be as vital, but a narrow carbon range could be useful for attaining distortion consistency. A more detailed discussion of the subject of shape and size distortions is provided in (Ref 1).

Summary

Core Properties

Core strength and associated properties (e.g., toughness) are regarded as important to the overall strength of a case-hardened part. Core strength is controllable within limits, for which a tensile strength range (minimum to maximum) of about 155 MPa (22 ksi) is normal. Impact toughness must be better than 27 J (20 ft lbf).

- *Preprocess considerations:* Steels are selected according to part size and the eventual duty requirement. Small sections in the leaner grades of steel are best made from the H grades to avoid excessive variability of core properties.
- *In-process considerations:* Core properties are affected by grain size; therefore, some thought is needed if high-temperature carburizing is selected. More importantly, core properties are affected by the makeup of the as-quenched microstructure, that is, the relative proportions of martensite, bainite, and ferrite. Lean-steel grades tend to produce ferritic cores, and quenching from below the Ac_3 temperature results in some undissolved core ferrite.
- *Postprocess corrections:* There are no corrective treatments when the core strength is out of specification following faultless quenching. Tempering or refrigeration have no significant effect.
- *Core properties:* For heavy-duty parts, core hardness over 30 HRC is recommended because it implies an absence of ferrite in the microstructure and provides a useful yield ratio. Hardness above 40 HRC is, in general, considered too hard. Having just enough hardenability to provide the required strength (hardness) at the critical locations is preferable to having excessive hardenability, which could lead to distortion problems.
- *Standards:* ANSI/AGMA: for contact loading (S_{ac}), there is no specification for grade 1, and a 21 HRC minimum for grades 2 and 3. For bending (S_{at}), grade 1 has a 21 HRC minimum; grade 2, a 25 HRC minimum; grade 3, a 30 HRC minimum. The hardness values quoted for both contact and bending strength relate to the center of the tooth at the root diameter. ISO 6336-5.2: grade ML has a 21 HRC minimum; grade MQ, a 25 HRC minimum; grade ME, a 30 HRC minimum.

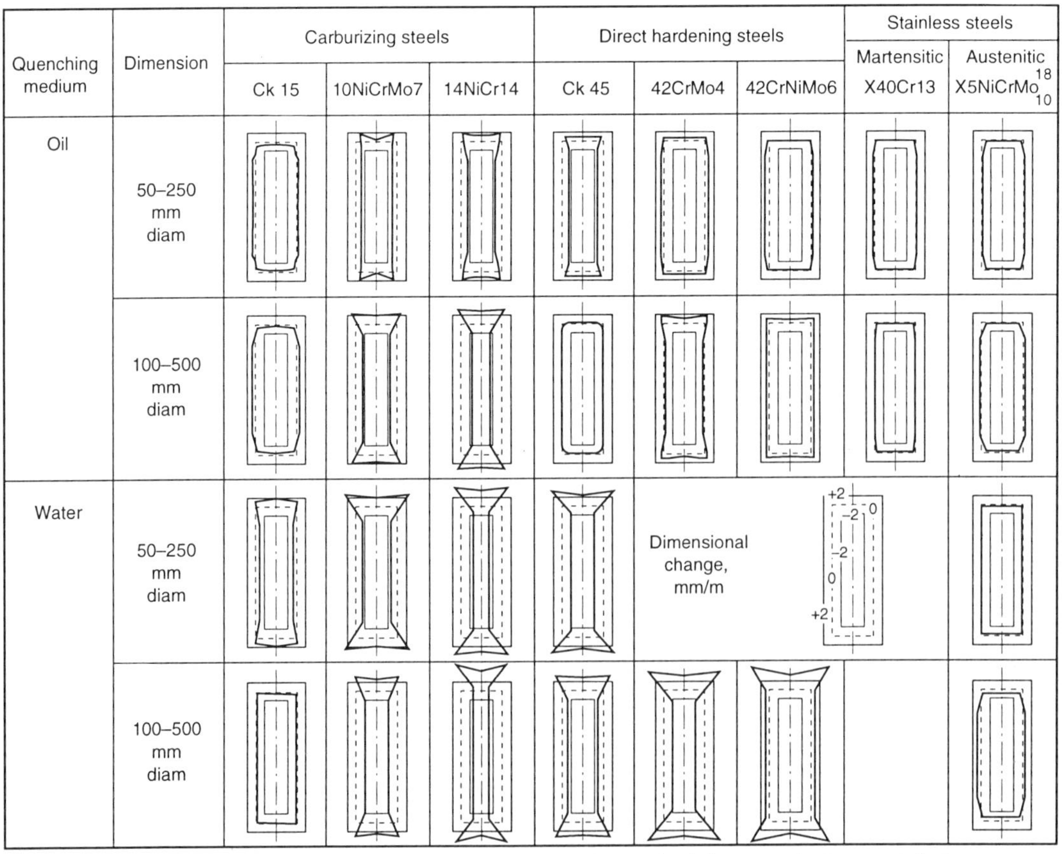

Fig. 6.42 Heat-treatment deformations after quenching in oil and water for different steels. Source: Ref 56

Case Depth

Effective case depth refers to the distance from the surface to a point within the case where the hardness is 50 HRC (or a comparable value on a Vickers or microindentation hardness scale). This distance depends on the carbon gradient and the case hardenability. For well-alloyed case-hardening steels, the carbon content for 50 HRC is likely in the range 0.25 to 0.30% C, for intermediate alloy grades it is approximately 0.30 to 0.35% C, and for the leaner grades, approximately 0.35 to 0.40% C. Total case depth refers to the total depth of carbon penetration.

- *Preprocess considerations:* The case depth for a given part is determined by the service requirements. For a gear, the specified case depth relates to the dedendum/pitch-line region of the tooth, where case depth is specified to resist deep spalling. In general, the case depth increases as the tooth size and the contact stresses increase. An allowance should be made for grinding after case hardening.
- *In-process considerations:* At a given carburizing temperature, the depth of carbon penetration is controlled by the duration of active carburizing. Good atmosphere circulation and good parts distribution within the furnace are essential to minimize the expected case-depth differences between gear tooth flanks and root fillets (due to shape and size differences).
- *Postprocess considerations:* When case depth is outside specification limits, acceptance of a deviation from specification must be determined on a case-by-case basis involving the designer. If the case depth is too shallow, reclamation is possible. There are no corrective measures when the case depth is too deep.
- *Effect on properties:* Bending-fatigue strength increases with increasing compressive-residual stresses. Unfortunately, compressive-residual stresses within a case can be adversely affected when cases are deep. For contact-fatigue situations, a shallow case can result in deep spalling failure. Deep cases can lead to subcase cracking of the as-quenched part.
- *Standards:* AGMA effective case depths for finished items are specified in terms of 50 HRC, whereas ISO considers 550 HV (52 HRC). For AGMA grade 2 specified gear teeth, the root-fillet case depths (effective) should not be less than 50% of the case depth at the midtooth height. For grade 3, the

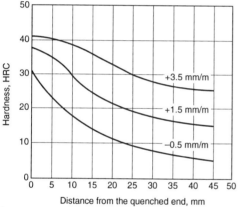

Fig. 6.43 Jominy diagram for steel 14NiCr14 showing the change of diameter by oil quenching forged disks of case-hardening steels. High, medium, and low hardenability due to differences in chemical analysis. Forged disks, 1000 mm outside diam, 200 mm thick. Source: Ref 56

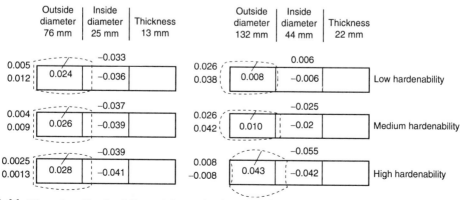

Fig. 6.44 Effect of steel hardenability and size on the distortion of case-hardened washer-like test pieces made of En 353 steel. Dimensional ratio for both test pieces is ~3:2:1. Source: Ref 57

root-fillet case depth should be not less than 66% of that at the midtooth height.

Case Carbon

The content target for surface carbon generally falls in the range between 0.75 and 0.95wt% C, where the actual target depends on the alloy content of the steel. Lean grades (<2% total alloy) are carburized to carbon contents at the top of that range, whereas the more alloyed grades (~4% total alloy) are carburized to the lower end of the range.

- *Preprocess considerations:* Consider the steel grade: for a given carbon potential, the alloy content of the steel can influence the target surface carbon content. Also consider that with carburizing, obtaining a surface carbon content within, for example, 0.05wt% C of the target value can be difficult in practice, especially when aiming for a shallow case depth. The quenching method (direct or reheat) affects the types of surface microstructures (e.g., high austenite, grain refinement, carbides) that are produced from a given carbon content. Also consider surface carbon content after grinding.
- *In-process considerations:* Ensure that the atmosphere generator catalyst is in good order; ensure good carbon potential control. Consider experience with previous work and use a test piece for carbon analysis if needed.
- *Postprocess considerations:* When surface carbon content is too high, parts can be conditioned and reheated in an atmosphere with a lower carbon potential. However, this increases the case depth, possibly adding to distortion.
- *Effect on properties:* For a given set of process parameters, the surface carbon content determines the as-quenched microstructure, which has a significant effect on properties. See Chapters 2, 3, and 4 and the section "Microcracking" in Chapter 5.
- *Standards*: ANSI/AGMA grades 1 and 2 set a carbon range of 0.6 to 1.1wt% C; grade 3 requires, 0.6 to 1.0wt% C. ISO 6336 has no limits for its grade ML, but requires the surface carbon for the MQ and ME grades to be within +0.2 to −0.1wt% C of the eutectoid.

REFERENCES

1. G. Parrish and G.S. Harper, *Production Gas Carburising*, Pergamon Press, 1985
2. Properties and Selection: Irons, Steels, and High-Performance Alloys, Vol 1, *ASM Handbook,* ASM International, 1990, p 549, 567
3. J. Woolman and R.A. Mottram, Vol 2 and 3, *The Mechanical and Physical Properties of the British Standard En Steels,* Pergamon Press, 1964, p 540–559
4. J.P. Naylor, The Influence of the Lath Morphology on the Yield Stress and Transition Temperature of Martenistic-Bainitic Steels, *Metall. Trans. A,* Vol 10, July 1979, p 861–873
5. How to Pick the Right Alloy to Resist Low Cycle Fatigue, *Materials Engineering,* July 1969
6. A. Esin and W.J.D. Jones, The Effect of Strain on the A.C. Resistance of a Metal: A Method of Studying Microplasticity, *J. Appl. Phys.,* Vol 18, 1967, p 1251
7. K.J. Irvine, F.B. Pickering, and J. Garstone, The Effect of Composition on the Structure and Properties of Martensite, *J. Iron Steel Inst. London, U.K.,* Sept 1960, p 66–81
8. W.T. Chesters, "The Metallurgical Interpretation," paper presented at *Why Carbon Case Harden?* (Coventry), Metals Society, 1975
9. G.V. Cleare, *Int. Conf. on Gearing* (London), Institution of Mechanical Engineers, 1958, p 490
10. G. Fett, Bending Properties of Carburising Steels, *Adv. Mater. and Process. Inc. Met. Prog.,* April 1988, p 4345
11. T.B. Cameron, D.E. Diesburg, and C. Kim, Fatigue and Overload Fracture of a Carburised Case, *J. Met.,* May 1983
12. Y.E. Smith and D.E. Diesburg, Fracture Resistance in Carburising Steels, Part 1, *Met. Prog.,* May 1979, p 6873; Part 2, *Met. Prog.,* June 1979, p 3538; Part 3, *Met. Prog.,* July 1979, p 6771
13. V.S. Sagaradze, Effect of Carbon Content on the Strength of Carburised Steel, *Met. Sci. Heat Treat. (USSR),* March 1970, p 198–200
14. R.F. Kern, Selecting Steels for Heat-Treated Parts, Part 2: Case Hardenable Grades, *Met. Prog.,* Vol 94 (No. 6), p 71–81
15. D.A. Sveshnikov, I.V. Kudryavtsev, N.A. Gulyaeva, and L.D. Golubovskaya, Chemico-thermal Treatment of Gears, *Met. Sci. Heat Treat. (USSR),* July 1966, p 527–532
16. M.A. Balter and I.S. Dukarevich, The Relationship Between the Properties of Steels Subjected to Chemicothermal Treatment and the Fatigue Limit, *Met. Sci. Heat Treat (USSR),* Vol 13 (No. 9), p 729–732

17. "Bending Strength of Gear Teeth—A Comparison of Some Carburising Steels," MIRA Report 1952/5
18. H. Sigwart (quoting Ulrich and Glaubitz), Influence of Residual Stress on the Fatigue Limit, *Int. Conf. on Fatigue of Metals,* Institution of Mechanical Engineers/American Society of Mechanical Engineers, 1956, p 272–281
19. I. Y. Arkhipov and M.S. Polotskii (quoting Semencha), Tooth Bending Life of Case-Hardened Gears with Cores of Different Hardness, *Russian Engineering Journal,* 1972, Vol LII (No. 10), p 28–30
20. W.T. Chesters, Contribution to Session IV: Materials Selection Criteria, *Low Alloy Steels,* ISI Publication 114, The Iron and Steel Institute, 1968, p 246–250
21. V.D. Kal'ner, V.F. Nikonov, and S.A. Yarasov, Modern Carburizing and Carbonitriding Techniques, *Met. Sci. Heat Treat. (USSR),* Vol 15 (No. 9), p 752–755
22. R.J. Love, H.C. Allsopp, and A.T. Weare, "The Influence of Carburising Conditions and Heat Treatment on the Bending Fatigue Strength and Impact Strength of Gears Made from En 352 Steel," MIRA Report No. 195–9/7
23. G.V. Kozyrev, G.V. Toporov, and R.A. Kozyreva, Effect of Structurally-Free Ferrite in the Core on the Impact-Fatigue Strength of Carburized Steels, *Met. Sci. Heat. Treat. (USSR),* Vol 14 (No. 4), p 337–339
24. R.A. DePaul, "Impact Fatigue Resistance of Commonly Use Gear Steels," SAE 710277, 1971
25. A. Rose and H. Hougardy, *Atlas Zur Wärmbehandlung Der Stähle,* Vol 2, *Verlag Stahleisen mb, (Atlas of the Heat Treatment of Steels),* Düsseldorf, 1972
26. F.A. Stills and H.C. Child, Predicting Carburising Data, *Heat Treat. Met.,* Vol 3, 1978, p 67–72
27. K. Bungardt, E. Kunze, and H. Brandis, Betrachtungen Zur Direkthärtung Von Einsatzstählen (Consideration of the Direct Hardening of Carburized Steels), *DEW-Technische Berichte,* Vol 5 (No. 1), 1965, p 112
28. H. Schwartzbart and J. Sheehan (Requoted from Ref 29), "Impact Properties of Quenched and Tempered Alloy Steel," Final Report, ONR Contract N60NR244T.0.1, I.T.T. Chicago, Sept 1955
29. G.H. Sharma, V.K. Walter and D.H. Breen, The Effect of Alloying Elements on Case Toughness of Automotive Gear Steels, *Heat Treatment '87,* London Institute of Metals, 1987
30. C. Razim, "Survey of Basic Facts of Designing Case-Hardened Components," unpublished private print, 1980
31. B. Thodin and J. Grosch, Crack Resistance of Carburised Steels Under Bend Stress, *Carburising: Processing and Performance Conf. Proc.* (Lakewood), G. Krauss, Ed., ASM International, July 1989, p 303–310
32. T. Aida, H. Fujio, M. Nishikawa, and R. Higashi, Influence of Impact Load on Fatigue Bending Strength of Case-Hardened Gears, *Bull. Jpn. Soc. Mech. Eng.,* Vol 15 (No. 85), 1972, p 877–883
33. H. Brugger, Effect of Material and Heat Treatment on the Load Bearing Capacity of the Root of Gear Teeth, *VDI Berichte,* Vol 195, 1973, p 135–144
34. S.S. Ermakov and V.I. Kochev, Effect of Temperature of a Second Hardening on Resistance to Repeated Impact of Carburised Steels, *Metalloved Term. Obrab. Met.,* Aug 1959, p 49–50
35. G.A. Fett, Tempering of Carburised Parts, *Met. Prog.,* Sept 1982, p 53–55
36. R.A. DePaul, High Cycle and Impact Fatigue Behaviour of Some Carburised Steels, *Met. Eng. Quart.,* Nov 1970, p 25–29
37. D.E. Diesburg, "High Cycle and Impact Fatigue Behavior of Carburised Steels," SAE Technical Paper 780–771, 1978
38. D. Zhu, Z. Li, H. Zhai, The Application of Mathematical Statistics to the Analysis of Contact Fatigue Tests of Carburised Specimens, *5th Int. Congress of Heat Treatment of Materials* (Budapest), 1986, p 1200–1204
39. B Vinokur, The Composition of the Solid Solution, Structure and Contact Fatigue of the Case Hardened Layer, *Met. Trans. A,* May 1993, p 1163–1168
40. H. Weigand and G. Tolasch, Fatigue Behaviour of Case-Hardened Samples, *Härt.-Tech. Mitt.,* Vol 22 (No. 4), p 330–338
41. W.S. Coleman and M. Simpson, Residual Stresses in Carburized Steels, *Fatigue Durability of Carburized Steel,* American Society for Metals, 1957, p 47–67
42. L. Salonen, The Residual Stresses in Carburised Layers in the Case of an Unalloyed and a Mo-Cr Alloyed Case Hardened Steel After

Various Heat Treatments, *Acta Polytech. Scand.*, Series 109, 1972, p 7–26
43. Y. Udegawa, An Experimental Investigation of the Effect of Induction Hardening of Fatigue Strength: Behaviour of Residual Stress During Application of Cyclic Stress, *Nihon Kikai Gakkai Rombunshu,* 1969, Vol 35 (No. 272), p 693–700
44. M. Weck, A. Kruse, and A. Gohritz, "Determination of Surface Fatigue on Gear Materials by Roller Tests," Paper 77, Det. 49, ASME, Sept 1977
45. C. Dawes and R.J. Cooksey, Surface Treatment of Engineering Components, *Heat Treat. Met. Special Report 95,* The Iron and Steel Institute, 1966, p 77–92
46. H. Tauscher, "Relationship Between Carburised Case Depth, Stock Thickness and Fatigue Strength in Carburised Steel," paper presented at the *Symposium on Fatigue Damage in Machine Parts,* 1960 (Prague)
47. X. Jin and B. Lou, The Effect of the Heat Treatment and Carbon Content on Fatigue Crack Initiation in Cr-Ni-Mo Steels, *5th Int. Congress on Heat Treatment of Materials* (Budapest), 1986, p 427–434
48. M. Ericsson, *Heat Treatment '87, London Institute of Metals,* 1987
49. G. Krauss, Bending Fatigue of Carburized Steels, *Fatigue and Fracture,* Vol 19, *ASM Handbook,* ASM International, 1996, p 680–690
50. M. Hasegawa, S. Kodama, and Y. Kawada, *A Study of the Effect of Residual Stress on the Fatigue Crack Propagation,* 1972, p 227–235
51. M. Jacobson, Gear Design, *Automot. Des. Eng.,* Aug 1969, Sept 1969, Oct 1969, Nov 1969, Dec 1969
52. D.J. Wulpi, *How Components Fail,* in *Metal Progress Bookshelf Series,* ASM International, 1966
53. G.H. Sharma, V.K. Walter, and D.H. Breen, An Analytical Approach for Establishing Case Depth Requirements in Carburised Gears, *J. Heat Treat.,* Vol 1 (No. 1), 1980, p 48–57
54. R. Pederson and S.L. Rice, Case Crushing of Carburized and Hardened Gears, *SAE Trans.,* 1961, Vol 69, p 370–380
55. A. Yoshida, K. Miyanishi, Y. Ohue, K. Yamamoto, N. Satoh, and K. Fujita, Effect of Hardened Depth on Fatigue Strength of Carbonitrided Gears, *JSME Int. J. C,* Vol 38 (No. 1), 1995
56. P. Bloch, "Heat Treatment Distortions," *Eleventh Round Table Conf. On Marine Gearing,* Oct 1977 (Chatham, MA)
57. D.T. Llewellyn and W.T. Cook, Heat Treatment Distortion in Case Carburising Steels, *Met. Technol.,* May 1977, p 565–578

SELECTED REFERENCES

- M. Dubois and M. Fiset, Evaluation of Case Depth on Steels by Barkhausen Noise Measurement, *Mater. Sci. Technol.,* Vol 11 (No. 3), 1995, p 264–267
- J. Killey and T. Guler, Control of the Interaction between Case Depth and Hardenability of Carburising Steels Using Modern Instrumentation, *IMMA Conference "The Heat Is On!"* 24–25 May 1995 (Melbourne, Victoria, Australia), Institute of Metals and Materials Australasia Ltd., 1995, p 35–41
- C.M. Klaren and J. Nelson, Methods of Measuring Case Depth, *Heat Treating,* Vol 4, *ASM Handbook,* ASM International, 1991, p 454–461
- T. Mihara and M. Obata, Carburized Case Depth Estimation by Rayleigh-Wave Backscattering, *Mater. Eval.,* Vol 49 (No. 6), June 1991, p 696–700
- S. Singh, R. Mitra, D. Leeper, and R. Fuquen, Ultrasonic Evaluation of Case Depth in Case-Carburized Steel Components, *Conf.: 22nd Symposium on Quantitative Nondestructive Evaluation, Review of Progress in Quantitative Nondestructive Evaluation 15B,* 30 July to 4 Aug 1995 (Seattle), Plenum Publishing Corp., p 1589–1596
- B. Vandewiele, Influence of the Base Material Hardenability on Effective Case Depth and Core Hardness, *Mater. Sci. Forum,* Vol 102-104 (No. 1), p 169–181

Chapter 7

Postcarburizing Thermal Treatments

Thermal treatments, such as tempering and refrigeration (subzero cooling), performed on case-hardened parts subsequent to quenching are considered necessary by some to optimize material properties. Others argue that thermal treatments are merely corrective measures, and that if the carburizing and quenching processes are executed properly neither tempering nor refrigeration treatments are necessary. This chapter does not favor either viewpoint; rather, it discusses what these processes do to carburized and quenched parts and how the properties of those parts are improved or impaired by these treatments.

Tempering

Generally, the tempering of steels can be carried out at any temperature up to about 700 °C (1290 °F). This range is divided into two more specific ranges: low-temperature tempering (up to ~300 °C), which modifies the characteristics of the quenched structure, and high-temperature tempering (~550 to 700 °C), which removes many of the characteristics of the quenched structure. With respect to carburized steels, high-temperature tempering is only important if adequate softening is to be induced to facilitate an intermediate machining operation, or as a preparation for a reheat quench. Low-temperature tempering, on the other hand, is of much greater interest because it directly affects the properties of the finished part.

Following the carburizing and quenching operations, components are usually heated to between 140 and 250 °C (285 and 480 °F) (more specifically, in the range 150 to 200 °C, or 300 to 390 °F), and held at temperature for between 2 and 10 h. This tempering operation generally renders components more amenable to subsequent manufacturing operations, more structurally and dimensionally stable, and for some applications, more durable in service than they would have been had they remained in the quenched condition.

Tempering Reactions

The As-Quenched Microstructure. The carbon content of a carburized layer is high at the surface, and decreases with depth until it reaches that of the original steel. Therefore, the range of carbon contents is typically close to 1.0% at the surface and decreases to the core carbon content, for instance, 0.2% C. In the quenched condition, the high-carbon surface region will consist of finely twinned plate martensite and retained austenite. Each plate of plate martensite is confined to the austenite grain in which it grows, and, therefore, the largest plate, which is usually the first plate to form, equates to the grain diameter. Thereafter, smaller plates subdivide the remaining grain volume. A martensite plate grows as an individual and has an orientation different from adjacent plates. Any retained austenite associated with the plate martensite exists as irregular volumes between martensite plates in the high-carbon regions of the case.

As the carbon content of the case decreases with distance from the surface, the amount of plate martensite and retained austenite will also decrease and be replaced by lath martensite, as illustrated in Fig. 7.1 (Ref 1). Therefore, in the mid-carbon range of the carburized layer, say at approximately 0.6% C, the microstructure will contain a high proportion of lath martensite. In lath martensite the laths grow more or less parallel to one another to form bundles or packets, though there may be several packets of differing orientation formed within an austenite grain. Retained austenite associated with this type of martensite exists as films that separate the laths or surround the bundles of laths in the low-carbon regions. These films have thicknesses of 30 to 400 Å (Ref 2). Below about 0.3% C, some of the lath martensite can appear as individual needles, and in the low-carbon core material, the martensite will often be of the needle type. This assumes that the cooling rate during the quench and the hardenability of the steel are adequate to promote the martensite reaction into the core.

The low-carbon needle and lath martensite of the core material beneath a carburized and hardened layer have likely experienced some carbide precipitation during the quench due to carbon diffusion to low energy sites (autotempering). The degree of autotempering and the shape and size of the precipitated carbides are, to a large extent, determined by the M_s temperature (when M_s is greater than 300 °C, autotempering is unavoidable), the alloying elements present (which can inhibit the nucleation or the growth of precipitates), and the cooling rate (the slower the cooling rate, up to the critical cooling rate, the greater the amount of autotempering) (Ref 2). These precipitates may be either highly dispersed granular Fe_3C carbides or rod-like carbides. The precipitation of η-carbide is unlikely during autotempering unless the carbon content is over 0.25% and the steel is more complicated in terms of alloying elements. At about 0.3% C, little autotempering is expected (Ref 3).

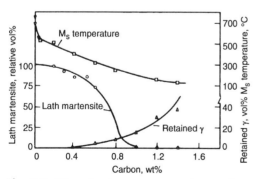

Fig. 7.1 Effect of carbon content on relative volume percent of lath and plate martensite, M_s temperature, volume percent of retained austenite in Fe-C alloys. Source: Ref 1

During tempering, the three structural features—lath martensite, plate martensite, and retained austenite—will respond differently, or at different temperatures, to one another.

Generally, during carburizing and hardening, only the smallest sections or most drastically quenched parts will have a case and a core that are both martensitic. Often cores contain bainite, or even ferrite when the hardenability of the steel is low for the section concerned or when quenching begins below the Ac_3 temperature of the core material. Also, for the same reasons it is not uncommon for the lower reaches of the case itself to contain bainite. Because bainite is composed of a dispersion of precipitated carbides in ferrite, it is more or less unaffected by low-temperature tempering. However, in the following discussion, only the tempering of martensitic microstructures is considered.

Influence of Temperature. Tempering can be divided into three stages:

- *Stage I:* Temperature range of 80 to 200 °C in which transitional carbides form
- *Stage II:* Temperature range of 150 to 300 °C in which much of the retained austenite transforms
- *Stage III:* Temperatures above 200 °C in which the transitional carbides give way to more stable carbides, and matrix recovery and recrystallization take place

These ranges overlap and may shift somewhat depending on the amount of added alloying elements; however, they are regarded as applicable for typical case-hardening steels. Some researchers have suggested that secondary hardening is stage IV of tempering, which is not unreasonable, but it will not be considered further here.

The term *stage I* is a little misleading because there is an extremely important conditioning stage that precedes it. In this preprecipitation stage, which takes place at temperatures below about 80 °C, carbon atoms segregate to dislocations, and some preprecipitation clustering of carbon atoms occurs in the as-quenched microstructure.

Stages I and II are the most important in carburizing and hardening where low-tempering temperatures of 150 to 200 °C are the most common. Nevertheless, there are occasions when an intermediate high-temperature tempering (or subcritical annealing) operation is required, and therefore stage III is also of interest. The entire tempering process is summarized in Table 7.1

and in Fig. 7.2. The composition of the transitional carbide and the temperature at which transitional carbides form depend on the composition of the steel. The amount of precipitation in a given time depends on the temperature, as Fig. 7.3 illustrates for an unalloyed steel. This figure shows how the first carbides formed are sacrificed to form other carbides. It may be noted that some of the carbides are brittle as they develop, as would be a steel with a predominance of such a carbide. However, such carbides are likely to develop in the temperature range 300 to 400 °C, and are of little interest in relation to case hardening.

Table 7.1 shows that the low-carbon core of a carburized part is little changed by tempering at temperatures below 200 °C because much of the carbon has already been precipitated during the quench (autotempering). The high-carbon case is different. There has been little or no autotempering, and the plate martensite and retained austenite, because they are saturated with carbon atoms, are somewhat unstable. Therefore, the application of some energy, whether thermal or even mechanical, causes microstructural changes. Here the application of thermal energy is considered. Tempering at the temperatures normally used for carburized parts (up to 200 °C) causes the coherent precipitation of η-carbides. This leads to a darkening of the martensite plates when examined metallographically. Retained austenite begins to transform to bainite at about 150 °C with typical or short-duration tempering, although the reaction can occur at temperatures below 150 °C, depending on the time. At such a temperature, and even up to 180 °C, only a small amount of the austen-

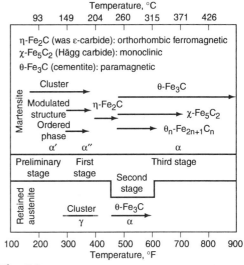

Fig. 7.2 Structural changes in martensitic steel resulting from tempering. Source: Ref 4

Table 7.1 Structural changes during the tempering of martensite

Temperature, °C	Core material	Case material
Room temperature (as quenched)	Needle and lath martensite structure. When the carbon content of the steel is <0.2%, the martensite is usually body-centered cubic. With >0.2% C, it is body centered tetragonal. When the M_s temperature is >300 °C, carbon diffusion will occur during the quench so that the martensite is autotempered	Generally the case structure contains plate martensite (body-centered tetragonal) and retained austenite. Lath martensite is present in the lower carbon sections of the case (below ~0.6%). The plate martensite contains fine internal twins. Alloying elements present in the steel inhibit autotempering, but if it does occur it is in the first formed plates.
<80	Little, if any reaction. Much of the carbon segregation will have already taken place during the quench	Carbon segregation and/or preprecipitation clustering takes place at pre-existing grain boundaries (originally termed unidentified carbides).
80–200	When carbon is <0.2%, precipitation is sluggish up to ~150 °C. With higher carbon contents, precipitation is rapid at ~150 °C; the carbides are Fe_3C.	Coherent precipitation of η-carbide by nucleation at martensite twin interfaces or from existing clusters. Precipitation heavy at 200 °C.
150–300		Retained austenite transforms to lower bainite. Any austenite surviving at medium to high temperatures will transform to upper bainite.
>200	Noticeable softening of martensitic core from ~200 °C. (Bainitic cores soften noticeably above ~300 °C).	η-carbides disintegrate to form intermediate χ-carbides, Fe_5C_2 of Fe_9C_4, which give way to θ-carbide, Fe_3C, with a corresponding loss of coherency. Loss of tetragonality of the martensite begins early in tempering, although some may persist up to 300 °C
500–700	Development of Fe_3C with coalescence at higher temperatures (spheroidization)	Smaller Fe_3C develops into Fe_3C cementite. Coalescence and growth take place as the temperature and time increase (spheroidization). Ferrite in the matrix recrystallizes.

ite is affected. As a rule, most of the austenite retained during quenching survives a typical low-temperature tempering, and though austenite transformation proceeds more easily and more rapidly above about 180 °C, some austenite might survive to quite high tempering temperatures. Austenite transformation during low-temperature tempering produces lower bainite, whereas medium- and high-temperature tempering will cause an austenite to upper bainite reaction.

Over the years there have been numerous studies of the tempering process; of these, Ref 5 to 9 have been used for this review, though not as any specific statement.

Influence of Time. The time dependence of carbide precipitation during tempering is illustrated in Fig. 7.4, and an example of the time dependence of retained austenite transformation is shown in Fig. 7.5. These figures show that for a given amount of reaction, temperature and time can be traded against one another. Aston, working with medium-carbon through-hardening steels, favored using higher tempering temperatures for shorter durations (Ref 12). Unfortunately, this is not typically a good idea for case-hardened parts in view of the need to expel hydrogen absorbed during the carburizing operation. Also note that carbon migration can continue at room temperature following the tempering operation (Ref 13, 14). This is aging. Aging is mentioned later in connection with hydrogen effusion. Consider also the effect of tempering time on the state of the martensite matrix, that is, the loss of tetragonality that takes place at temperatures below 300 °C.

Volume Changes during Tempering. In quenched steels, the austenite-to-martensite transformation is accompanied by an increase of volume; the higher the carbon content is, the greater the volume increase. Thus, the high-carbon martensite in the surface of a carburized layer will

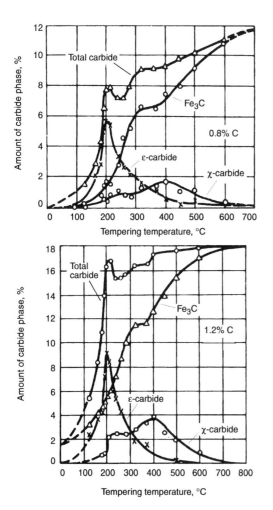

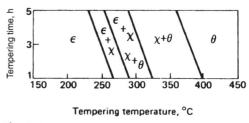

Fig. 7.4 Precipitation of ε-, χ-, and θ- (Fe₃C) carbides related to tempering time and temperature. (1.34% C steel). Source: Ref 10

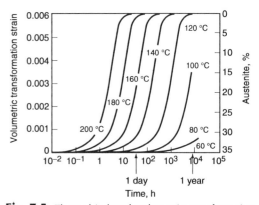

Fig. 7.3 Variations in the amounts of carbide phases with tempering temperature. Tempering time, 5 h. Source: Ref 5

Fig. 7.5 Thermal-induced volumetric transformation strain for carburized 4320 steel (35% retained austenite) as a function of temperature and time. Source: Ref 11

expand more than a low-carbon martensite in the core. In practice, the full expansion potentials of both the outer case and the core are not realized due to retained austenite in the outer case and autotempering of the core. Only the intermediate carbon levels within a case (between 0.3% and 0.6%) approach their potential volume expansions, and this could affect why some residual stress distributions peak at about the middle of the case. Many carburized and quenched parts do not transform to martensite throughout their sections; often their cores and lower cases are bainitic. Bainite has only about half the volume expansion of martensite at any carbon level.

When tempering at above approximately 200 °C, low-carbon martensite of the core material "gives way" to ferrite and precipitated carbides, which are accompanied by a decrease in volume. With tempering at temperatures below approximately 200 °C, the volume of the low-carbon material likely decreases by only a very small amount by additional precipitation of carbides because much of the core carbon is already tied up as precipitates due to autotempering. Below approximately 200 °C, any bainite in the core is relatively stable.

In the high-carbon case, tempering in the temperature range 80 to 200 °C (stage I) causes the precipitation of transitional carbides within the martensite, which is accompanied by a decrease of volume. Tempering in the range 150 to 300 °C (stage II) causes retained austenite to decompose to bainite (by interstitial carbon diffusion), which is accompanied by a volume increase (Ref 11). However, carburized and hardened parts are typically tempered at 180 °C, which is high in the temperature range of stage I and low in the temperature range of stage II. Therefore, the contraction due to the stage I reactions far outweighs the expansion due to any austenite transformation likely to occur at that temperature. Further, the volume increase due to stage II is likely insignificant for initial retained austenite contents of less than about 25%.

As a point of interest, Zabil'skii et al. argued that the overall volume change that accompanies tempering cannot be fully accounted for by transformation and precipitation processes (Ref 15). The difference, they suggest is due to the healing of defects in the martesite structure.

Volume changes that take place during both quenching and tempering are significant because they influence the residual stress distribution in the case region (see the following section) and the growth or shrinkage of the part as a whole (see the section "Distortion" in Chapter 6).

As-quenched, carburized parts are not quite dimensionally stable, in terms of either shape or size. Tempering can induce a certain measure of microstructural and dimensional stabilization for those components where a high degree of precision and stability are vital. Furthermore, low-temperature tempered parts remain essentially stable at service operating temperatures approaching those used for tempering.

Effects of Tempering

Influence on Hardness. Tempering temperatures to about 200 °C have little effect on the martensitic core hardnesses of carburized and quenched lean-alloy steels because much carbide precipitation has taken place by autotempering (Fig. 7.6). Often the cores are essentially bainitic and are hardly affected by tempering at temperatures up to about 300 °C; this, more or less, applies to the lower regions of the case up to about 50 HRC (Fig. 7.7). The hardness of the outer case, on the other hand, becomes noticeably affected as the tempering temperature exceeds 100 °C when η-carbides precipitate, and even more so at 150 °C when precipitation is more advanced and some of the retained austenite might transform.

The hardness reduction in the outer case due to low-temperature tempering depends on the austenitizing temperature; the higher the quenching temperature, the greater the fall in hardness during low-temperature tempering. Reductions of 50 to 150 HV are typical.

The influence of tempering time at a given temperature is illustrated in Fig. 7.8 (Ref 17), which

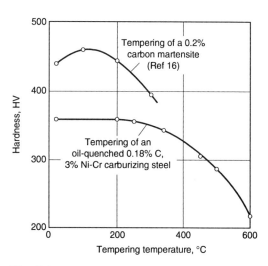

Fig. 7.6 Effect of tempering on core hardness

shows that above approximately 120 minutes, little further change of hardness is likely to occur.

Influence on Tensile Properties. Just as the hardness of the core material is essentially unchanged by tempering at temperatures up to 250 to 300 °C, so too the tensile strength remains unchanged. The local yield strength for lean-alloy and alloy carburizing steels tends to rise during tempering at temperatures between 100 and approximately 250 °C, whereas plain-carbon carburizing steel are hardly, if at all, affected (Fig. 7.9).

Within the quenched carburized case, an increase of both tensile and yield strengths results from tempering at about 100 °C, with the yield strength approaching the ultimate tensile strength. At higher tempering temperatures, the

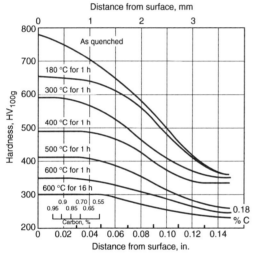

Fig. 7.7 Effect of tempering on case hardness of a 3%Ni-Cr carburizing steel with 0.18% C

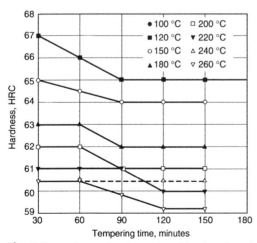

Fig. 7.8 Hardness of carburized and hardened steel 30KhGT as a function of tempering temperature and time. Source: Ref 17

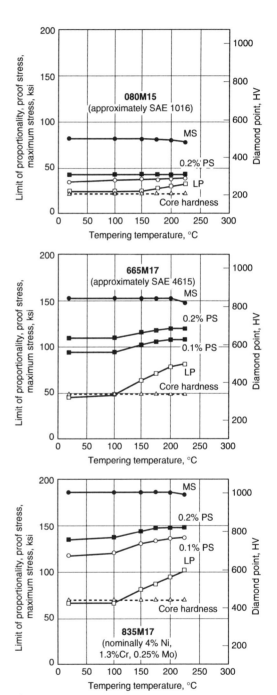

Fig. 7.9 Effect of tempering on the core tensile properties of three steels. Blank carburized core steel: 920 °C oil quench, reheated 780–830 °C, oil quenched, cooled to −78 °C, and tempered. LP, limit of proportionality; PS, proof stress; MS, maximum stress. Source: Ref 18

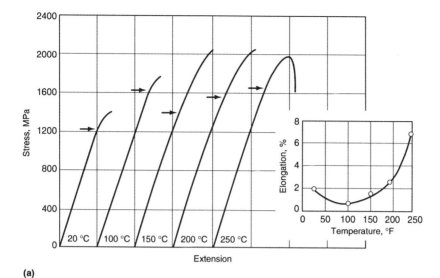

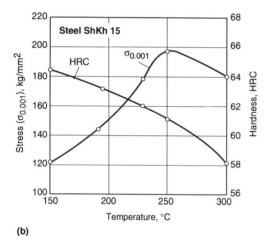

Fig. 7.10 Effect of tempering temperature on the tensile yield strength of two steels. (a) Composite stress-strain curve for a Ni-Cr steel (0.57% C, 3.07% Ni, 0.9% Cr) where arrows denote limit of proportionality. Source: Ref 19. (b) Stress for 0.001 plastic deformation ($\sigma_{0.001}$) for a high-chromium bearing steel (ShKh15) after quenching and cold treatment (soak time 3 h). Source: Ref 20. 1 kg/mm² = 9.8 MPa

fall in hardness that accompanies tempering is reflected by a fall in tensile strength. However, the yield strength of the high-carbon case, in keeping with that of the core material, tends to rise as the tempering temperature rises to 250 °C (Fig. 7.10a, b). This particular feature can cause problems during any post-case-hardening shaft-straightening operation. Straightening to correct heat-treatment distortion requires that the shaft be plastically strained in a direction that opposes the heat-treatment distortion. If the yield stress is closer to the fracture stress due to tempering, then straightening without cracking can be difficult to achieve. This aspect of tempering has been examined by Vogel, who obtained fewer reject shafts by straightening in the untempered condition (Ref 21).

Residual Stresses. Compressive-residual stresses within the carburized layer contribute greatly to the useful properties of case-hardened parts. As-quenched, the residual stress distribution through a carburized case varies largely according to the relative proportions of retained austenite and martensite (discounting surface anomalies). The magnitude of compressive-residual stresses at or near the surface (where the austenite may be present in significant quantities) is generally less than at some distance below the surface, at locations where the microstructure is wholly martensitic and the carbon content is approximately 0.5 to 0.6%.

Tempering reduces the magnitudes of both the compressive stresses within the case and the balancing tensile stresses within the adjacent core (Fig. 7.11). Moreover, tempering tends to shift the location of the peak compression nearer to the case-core interface. Several factors contribute to the as-quenched residual stress distribution within a case and core, and these influence how much the stress magnitudes fall due to tempering at some appropriate temperature. Figure 7.12 shows examples of peak stress reduction due to tempering.

A comparison of Table 7.1 and Fig. 7.12 reveals that carbide segregation and precipitation clustering coincide with the initial minor decrease of compressive residual stresses, and that

178 / Carburizing: Microstructures and Properties

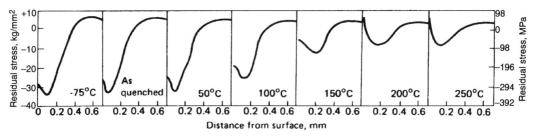

Fig. 7.11 Residual stresses (tangential) in cyanide-hardened 40Kh rings before and after tempering (for 1.5 h). Ring dimensions: 80 mm outside diam × 66 mm inside diam × 15 mm high. Case depth 0.22 mm (on outside diam only). Source: Ref 22

precipitation of η-carbides and loss of martensite tetragonality coincide with the major reduction of surface compressive residual stresses. A decrease of volume accompanies the formation of these carbides. The change of slope that takes place above approximately 160 °C is the sum of the contractions due to precipitation and the expansions accompanying the onset of retained austenite transformation (Fig. 7.12). However, when the η-carbide disintegrates to form other transitional carbides or Fe_3C above about 200 °C, a volume contraction occurs, and the peak compressive stresses again fall and continue to fall as carbide coalescence advances, as cohesion between carbide and matrix is lost, and as the ferrite matrix attempts to recrystallize with rising temperature, up to approximately 700 °C.

Influence on Bending Fatigue. Tempering temperatures below 100 °C do not greatly influence the bending-fatigue strength of small case-hardened gears. With higher temperature treatments (to 200 °C), the fatigue limit progressively reduces by up to approximately 20% (Table 7.2). A similar trend was found when small beam samples were fatigue tested (Table 7.3). At tempering temperatures between 200 and approximately 250 °C (the range in which retained austenite transforms to bainite), the fatigue limit

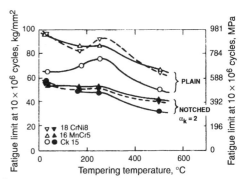

Fig. 7.13 Effect of tempering temperature on the alternating bending fatigue strength of 6 mm diam case-hardened test pieces. Carburized at 930 °C for 1 h, water quenched, reheated to 850 °C for 10 minutes, and oil quenched. Note: Ck15 steel was water quenched from 850 °C. Source: Ref 25

Fig. 7.12 Change in peak compression due to tempering after carburizing and (a) oil quenching or (b) water quenching. Source: Ref 22–25

Table 7.2 The effect of tempering on the fatigue and impact resistance of En352 pack carburized at 900 °C and quenched from 870 °C

| Temperature, °C | Fatigue limit | | Tooth impact | | Bar Impact | | | |
| | | | | | Uncarburized | | Carburized | |
	MPa	ksi	J	ft · lbf	J	ft · lbf	J	ft · lbf
Untempered								
NA	800	116	32.5	24	12	9	21.5	16
NA	910	132	35	26	13.5	10	8	6
NA	910	132	36.5	27	12	9	6.5	5
NA	880	128	23	17	9.5	7	4.5	3.5
NA	880	128	34	25	11.5	8.5	12	9
NA	827	120	36.5	27	11	8	4.5	3.5
Tempered								
100	854	124	36.5	27	13.5	10	4	3
100	840	122	31	23	15	11	4.5	3.5
150	780	113	47	35	17.5	13	4.5	3.5
150	705	102	27	20	15	11	7.5	5.5
200	760	110	40.5	30	16	12	6.5	5
200	637	92	28.5	21	17.5	13	4	3

NA is not applicable. Source: Ref 26

may increase a little before decreasing again at still higher tempering temperatures when most of the retained austenite has transformed (Fig. 7.13).

Gu et al., showed that under low-cycle fatigue conditions, the fatigue-crack initiation life increases as the residual compression within the case increases (Fig. 7.14) (Ref 27). Tempering, however, reduces compression (Fig. 7.11) (Ref 22), and consequently one might expect it to reduce the number of stress cycles for crack initiation. Fett stated that tempering at 180 °C increased the low-cycle fatigue life (without affecting the high-cycle fatigue life) (Ref 28). Therefore, one conclusion that can be derived from these observations is that for low-cycle fatigue, at least, tempering must induce some significant resistance to crack propagation.

The retained austenite content appears to affect how a steel responds to tempering. For example, Razim examined the extremes of austenite content and found, using notched rotating beam test pieces, that tempering at 180 °C reduced the fatigue limit of a fully martensitic (<2% retained austenite) case structure by only approximately 3%, but that it also appeared to reduce the low-cycle fatigue strength considerably (Ref 29). When the case contained large quantities of retained austenite (~80%), tempering at 180 °C caused an 18% increase in fatigue limit, and pushed the knee of the S-N curve to a longer time (Fig. 7.15). Note, however, that the test pieces with the high austenite content were

Table 7.3 Effect of tempering on fatigue

| Condition | Fatigue limit | | Mean surface hardness, HV |
	tsi	MPa	
As carburized and quenched	±38	±570	860
Tempered at 60 °C	±38	±570	866
Tempered at 100 °C	±34	±510	858
Tempered at 125 °C	...	...	847
Tempered at 150 °C	±35	±525	823
Tempered at 185 °C	±33	±495	767

Test pieces 2.38 mm thick × 12.7 mm wide in SAE 8620H. Case depth 0.375–0.45 mm. Source: Ref 24

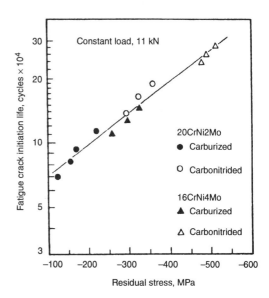

Fig. 7.14 Variation of fatigue-crack initiation lives with residual stress at the notch of tested steels. Source: Ref 27

Table 7.4 Influences on the toughness of case-hardened structures

Property	Influence		
	Lowering hydrogen content	Tempering	Storage after tempering at 200 °C for 2 h
Crack toughness	No effect	Positive effect	Positive effect
Fracture toughness	Positive effect	Positive effect	Positive effect
Time to delayed failure	Positive effect	Positive effect	Positive effect
Pulsating fatigue toughness	No effect	Positive effect	Positive effect
Fatigue strength under reversed stress	No effect	Negative effect	No effect

Source: Ref 13

always inferior to those essentially free of austenite. Most case-hardened components have retained austenite contents of 10 to 30%.

It is apparent that many variables (e.g., steel composition, microstructure, test-piece design, and type of loading) have a bearing on the fatigue strength and how it is affected by tempering. Figure 7.16 shows how tempering to 180 °C reduces the unnotched bending-fatigue strength of a case-hardened alloy steel, yet raises the same property of a carburized plain-carbon steel. In the notched and carburized condition, the fatigue limits of both the alloy steel and the plain-carbon counterpart are reduced by tempering. Because many actual components have stress concentrators, for example, gear tooth fillets, the notched test-piece results perhaps indicates the right trend. Streng et al. determined that tempering has a favorable effect on pulsating fatigue toughness and a deleterious effect on the fatigue strength under reversed stress (Table 7.4) (Ref 13).

Most laboratory fatigue tests are operated under constant load conditions and, as such, do not take into account the likelihood of occasional high-loading events. Real-life components can experience occasional overloads in addition to their normal operating load, whether that is high-cycle low load, low-cycle high load, or somewhere in between. Considering the advantages and disadvantages of tempering, designers and operators require guidance about whether or not to temper. Rosenblatt attempts to provide such guidance with respect to gears in Fig. 7.17 (Ref 30). His work implies that tempering is essential for the high-load low-cycle requirement

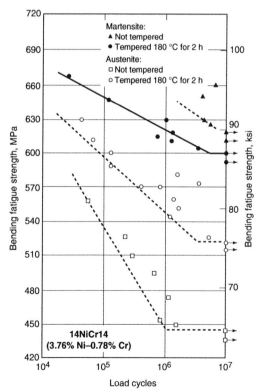

Fig. 7.15 Bending-fatigue strength of notched test pieces with and without retained austenite. Source: Ref 29

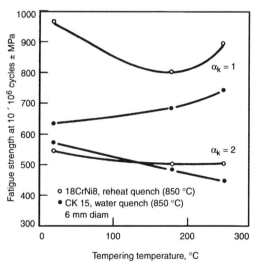

Fig. 7.16 Effect of tempering on the alternating bending-fatigue strength of two case-hardened steels. Source: Ref 25

and where impact loading is involved. For the low-load high-cycle situation, tempering might have an adverse effect. However, other considerations, such as dimensional stability and grindability, might determine whether or not a part should be tempered.

Influence on Contact Fatigue. Under pure rolling-contact fatigue conditions, that is, those comparable to the contact at the pitch line of a gear tooth, it was found that the fatigue limit of reheat-quenched and tempered test rollers increases as the tempering temperature increases, up to 250 °C. However, in a low-cycle high-load regime, the contact-fatigue life diminishes as the tempering temperature increases (Fig. 7.18). It was noted that the highly loaded test tracks of the untempered disks increased in hardness by up to 85 HV and those of the disks tempered at 100 °C increased by up to 120 HV, whereas those tempered at 150 °C or more hardened by only 35 HV. This effect is considered to be related to the increase of yield strength due to tempering (Fig. 7.10).

It should be remembered that tempering temperatures above about 150 °C (300 °F) may transform some of the austenite to bainite; the higher the temperature is, the greater the amount of austenite transformed. According to the standard ANSI/AGMA 2001-C95, bainite has a detrimental effect on the resistance to both contact and bending fatigue.

Influence on Bending and Impact-Fracture Strength. A study to determine the impact-fracture stress of case-hardened steels with differing chemical compositions revealed that the impact-fracture stress increased as the maximum compressive-residual stress of the case increased (Fig. 7.19) (Ref 31). All samples had been tempered at 170 °C. Therefore, it might be reasoned that if tempering reduces the compressive-residual stresses, then the absence of tempering should lead to higher fracture stress. This does not seem to be so. Shea observed that tempering at 165 °C increases the energy required to initiate a crack in a conventionally carburized and quenched test piece under impact loading, but it had little, if any, influence on the crack-propagation energy (Ref 32). Because the

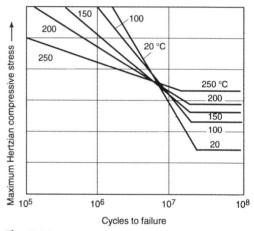

Fig. 7.18 Influence of tempering temperature on the rolling-contact fatigue limit of a carburized and hardened alloy steel.

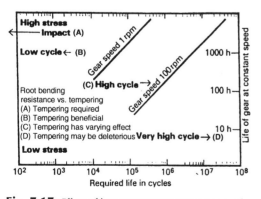

Fig. 7.17 Effect of low-temperature tempering on the service life of carburized and hardened gears. Root bending resistance vs. tempering: (A) impact, tempering required; (B) low cycle, tempering beneficial; (C) high cycle, tempering has varying effect; (D) very high cycle, tempering may be deleterious. Source: Ref 30

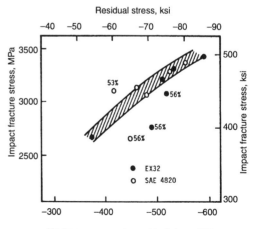

Fig. 7.19 Relationship between impact-fracture stress and compressive-residual stress (percent values indicate maximum amount of retained austenite content in the carburized case). Source: Ref 31

crack-initiation energy increased, and the crack-propagation energy was essentially unaffected as a result of tempering, which had probably reduced the surface compression, one can reason that residual stresses have no effect when the loading is by impact.

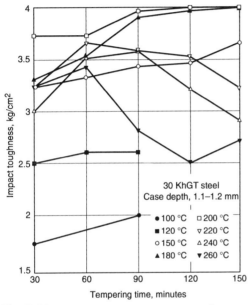

Fig. 7.20 Effect of tempering temperature and time on impact toughness of unnotched test pieces. Note: untempered impact toughness ~2 kg/cm². Source: Ref 17

Impact resistance is increased by the correct tempering temperature and time, as Fig. 7.20 illustrates. In this example, a tempering temperature of 180 to 200 °C for a minimum of 90 min produced the best results. Using steels of varying nickel content (≤4.6%), Thoden and Grosch showed that the bending-crack stress and the impact strength were each improved by tempering at 180 °C for 1 h, as was the case for both direct-quenched and double-quenched samples (Ref 33). Another study using unnotched test pieces (Table 7.5), indicated that impact toughness of 8620 steel was little affected by tempering in the wider range applicable for case-hardened parts (150 to 250 °C). However, the bending strength was improved (Fig. 7.21) (Ref 34). Subsequent gear-set life tests with impact loading showed that tempering had been beneficial.

In fracture-toughness tests on a 0.85% C lean-alloy steel, as-tempered samples exhibited fractures with larger areas of transgranular fracture surface than did the fractures of untempered samples (Ref 35), implying that tempering can benefit fracture toughness. Other factors, such as austenitizing temperature, cooling rate, and phosphorus content, were found to affect the appearance of the fracture and the K_{Ic} value. A slow quench rate and a high phosphorus content together contributed to grain-boundary carbon enrichment or iron-carbide precipitation, causing more intergranular fracture surface and a lower K_{Ic}.

Table 7.5 Data on as-quenched and tempered unnotched Charpy bars following gas carburizing

Sample No.	AISI grade	Tempering temperature, °C (°F)	Hardness, HRC Surface	Hardness, HRC Core	Case depth, mm (in.) Effective	Case depth, mm (in.) Visual	Charpy Impact energy, J (ft·lbf)	Yield, kN (lb)	Slow bend test results Ultimate, kN (lb)	Deflection, mm (in.)
1	8615	As-quenched	66	36	0.89 (0.0035)	1.02 (0.040)	16–20 (12–15)	19.6 (4400)	30.2 (6780)	0.086 (0.034)
2	8615	150 (300)	63–64	37	0.97 (0.038)	1.02 (0.040)	24–26 (18–19)	27.6 (6200)	33.2 (7460)	1.02 (0.040)
3	8615	205 (400)	59–61	35–36	0.91 (0.036)	1.02 (0.040)	26–30 (19–22)	27.6 (6210)	35.1 (7900)	1.07 (0.042)
4	8615	260 (500)	58–59	35–36	0.91 (0.036)	1.02 (0.040)	19–31 (14–23)	34.3 (7700)	39.2 (8820)	1.42 (0.056)
5	8615	315 (600)	55–56	36	0.084 (0.033)	1.02 (0.040)	43–56 (32–41)	32.0 (7200)	42.9 (9640)	1.45 (0.057)
6	8615	370 (700)	51–53	34	0.58 (0.023)	1.02 (0.040)	53–144 (39–106)	28.0 (6300)	42.2 (9480)	2.39 (0.094)
7	8615	425 (800)	48–49	32	0.036 (0.013)	1.02 (0.040)	175–231 (129–170)	...	...	...
8	8615	480 (900)	45–46	29–30	...	1.02 (0.040)	264–302 (195–223)	23.6 (5300)	22.2 (5000)	5.08 (0.200)
9	8620	As-quenched	64–66	45	1.17 (0.046)	1.14 (0.045)	24–30 (18–22)	22.2 (5000)	34.6 (7780)	1.09 (0.043)
10	8620	150 (300)	62–65	45–46	0.91 (0.036)	1.14 (0.045)	34–39 (25–29)	32.9 (7400)	37.4 (8400)	1.09 (0.043)
11	8620	205 (400)	59–60	45–46	1.09 (0.043)	1.14 (0.045)	33–60 (24–44)	29.8 (6700)	38.7 (8700)	1.12 (0.044)
12	4320	As-quenched	64	46	1.40 (0.055)	1.52 (0.060)	26–28 (19–21)	26.7 (6000)	34.3 (7700)	1.17 (0.046)
13	4320	150 (300)	61–63	46	2.65 (0.065)	1.52 (0.060)	38–41 (28–30)	27.1 (6100)	36.9 (8290)	1.14 (0.045)
14	4320	205 (400)	58–59	46–47	1.40 (0.055)	1.52 (0.060)	43–47 (32–35)	30.2 (6800)	38.4 (8640)	1.17 (0.046)
15	8617	150 (300)	60–61	38	0.99 (0.039)	0.91 (0.036)	22–45 (16–33)	28.9 (6500)	36.1 (8100)	1.12 (0.044)
16	4815	150 (300)	58	42–43	1.22 (0.048)	0.91 (0.036)	53–79 (39–58)	...	...	...
17	4820	150 (300)	58	40–41	0.89 (0.035)	0.86 (0.034)	58–68 (43–50)	28.0 (6300)	37.0 (8320)	1.40 (0.055)

Bars 1–14 carburized in one group. Bars 15–17 carburized in one group. All bars cold-oil quenched. Spectrographic analysis of Charpy bars by weight percent: 8615M: 0.18 C, 0.64 Mn, 0.36 Ni, 0.48 Cr, 0.12 Mo; 8620: 0.19 C, 0.74 Mn, 0.59 Ni, 0.43 Cr, 0.17 Mo; 4320: 0.21 C, 0.49 Mn, 1.89 Ni, 0.38 Cr, 0.28 Mo; 8617: 0.20 C, 0.91 Mn, 0.45 Ni, 0.51 Cr, 0.17 Mo; 4815: 0.16 C, 0.57 Mn, 3.42 Ni, 0.24 Mo; 4820: 0.18 C, 0.75 Mn, 3.13 Ni, 0.24 Mo

Postcarburizing Thermal Treatments / 183

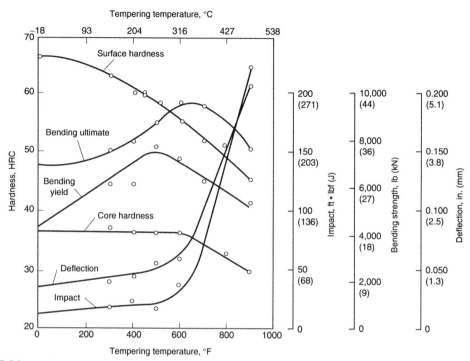

Fig. 7.21 Mechanical properties of unnotched AISI 8615 Charpy bars at various tempering temperatures. Bars were gas carburized to 1.02 mm (0.040 in.) visual depth, cold oil quenched to 50 °C. Source: Ref 34

Influence on Wear. The abrasive wear resistance of high-carbon surfaces is related to hardness and the distribution of dispersed carbides (Ref 36). As the tempering temperature increases, the wear resistance decreases (Fig. 7.22). The main fall in resistance results from tempering in the temperature range 125 to 225 °C (~400 to 500 K), the temperature range generally used for carburized and hardened parts. The shape of the plots in Fig. 7.22 appears to relate to the changes that take place during tempering, that is, the carbon clustering, carbide precipitation, concurrent austenite transformation, and matrix change from martensite (α') to ferrite (α).

Figure 7.22 applies to dry abrasive wear and, to some extent, is applicable to adhesive wear. However, adhesive wear is influenced by the chemistry of the mating surfaces, where, for example, nickel tends to favor adhesion, and, in case-hardened surfaces, retained austenite might also favor adhesive wear (e.g., scoring and seizure).

Machines with parts that move relative to one another generally require a lubricant of adequate viscosity and flash temperature to inhibit or control both abrasive and adhesive wear of either of the contacting surfaces. Therefore, in the event of excessive wear, the condition of the lubricant and the effectiveness of the lubricating system, as well as aspects of design, should be examined as much, if not more than, the metallurgical quality of the wearing surfaces.

Additional Process Factors

Hydrogen Content. The outcome of bending and impact tests on case-hardened steels can be complicated by the presence of hydrogen in the steel. Hydrogenation results directly from high-temperature heating in an atmosphere containing molecular hydrogen and a hydrogen

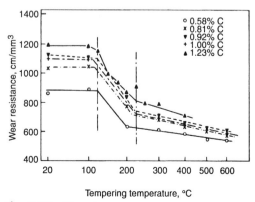

Fig. 7.22 Effect of tempering temperature on wear resistance. Source: Ref 36

Table 7.6 Tensile properties of 20Cr and 18CrNiW steels

Steel	Heat treatment	Vacuum hardening		Blank carburized and quenched	
		Strength, MPa	Reduction of area, %	Strength, MPa	Reduction of area, %
20Cr	As-quenched	1520	38	1442	10
	Quench hardening plus 190 °C × 2 h tempering	1500	38	1471	15
18CrNiW	As-quenched	1471	56	1412	13
	Quench hardening plus 190 °C × 2 h tempering	1422	61	1392	43

Source: Ref 37

Table 7.7 The effect of hydrogen on the toughness of the core material of 0.2%C, 3.5Ni-Cr carburizing steel

Treatment	Reduction of area, %	Crack initiation, kgf · cm	Deflection, mm	Bending load, kgf
Heated in air	58 (53)	78 (80)	0.8 (1.1)	840 (850)
Heated in endo-gas	27 (58)	10 (58)	0.1 (0.8)	640 (840)

Numbers in parentheses show the test values after 14 days aging. The original values were obtained soon after heat treating. The bending load and the deflection were those at the time of crack initiation. Source: Ref 39

compound, that is, water vapor in an endothermic gas.

The hydrogen content of steel surfaces after gas carburizing and hardening has been measured as approximately 2 ppm (Ref 37, 38). Subsequent tensile tests on as-quenched samples and on quenched and tempered samples (190 °C for 2 h) showed that tempering raises the percent reduction of area value (Table 7.6) and is also beneficial regarding delayed fracture under load (Ref 37). The influence of tempering on the time to delayed failure was confirmed where tempering time was considered more influential than tempering temperature (Ref 38). Dukarevich and Balter observed that hydrogen has an adverse influence on toughness and ductility (Table 7.7) and that tempering improves bending strength depending on tempering time and temperature (Table 7.8) (Ref 39). Streng et al. established that the amount of stress required to induce a crack is not affected by the hydrogen content, whereas the stress required for crack propagation is affected to a degree dependent on the hydrogen content (Fig. 7.23) (Ref 38). Further, it was observed that tempering, according to its duration, raises the crack-initiation stress; at least four hours at the tempering temperature is necessary to reach the maximum stress value. Adequate tempering is considered effective for reducing the hydrogen content and removing the adverse effects of hydrogen; vacuum tempering is more effective than air tempering.

Aging. Two aspects of the aging of carburized and hardened steels must be considered: its effect, if any, on any residual hydrogen in the steel, and the possible additional migration of carbon atoms.

Carburized bend or toughness test pieces heated in endo-gas will have relatively low toughness and ductility following a quench. During testing to fracture, the initial crack is more likely to be influenced by the surface carbon content than the hydrogen present. Crack propa-

Table 7.8 Effect of tempering on the bend test results of a 3.5%Ni-Cr steel heated in endothermic gas and quenched

Tempering temperature, °C	Tempering time, h	Approximate maximum load, kgf	Approximate maximum deflection, mm
150	0	880	0.02
	5	1210	0.1
	10	1380	0.22
180	5	1500	0.4
	10	1580	0.48

Source: Ref 39

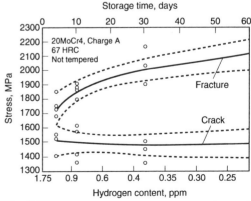

Fig. 7.23 Influence of hydrogen content on the crack stress and fracture stress. Source: Ref 38

Table 7.9 Carbon content of martensite and the changes of residual stress during a test period of 16 months for a 20MoCr4 steel

Quenching conditions	Carbon, %		Compressive residual stress, MPa	
	Start	End	Start	End
Oil at 50 °C	0.57	0.49	510	440
Oil at 50 °C plus −196 °C subzero cool	0.62	0.46	550	500
Oil at 200 °C	0.24	0.21	450	420
Oil at 10 °C plus 175 °C temper	0.40	0.40	250	250

Carburize 900 °C for 1 h 20 minutes for effective case depth (0.4 C) of 0.35 mm and a surface carbon content of 0.8–0.85%. 0.25 mm layer removed to reach zone of maximum residual stress. Peak residual stress coincides with the point where the carbon content equals 0.60%. Source: Ref 14

gation, on the other hand, is affected by the toughness of the core material, which is influenced by the hydrogen content. Aging permits outgassing of hydrogen with a corresponding improvement of toughness and ductility (Table 7.7).

Tests carried out after a 16 month room-temperature hold revealed that in as-carburized and quenched parts, the carbon content of the martensite and the compressive-residual stresses fell (Table 7.9) (Ref 14). Samples tempered at 175 °C exhibited no carbon change in the martensite or in residual stress values. Therefore, tempering is important both for hydrogen effusion and for carbon stabilization within the carburized and hardened case.

The fatigue strength of case-hardened and tempered (220 °C) pieces tested under reversed loading was unaffected by subsequent storage, whereas the pulsating-fatigue strength increased during holding for approximately 100 days (Fig. 7.24).

Grinding. Ground surfaces are the product of gouging, rubbing, and rolling by hard abrasive particles bonded into the surface of the grinding wheel. A ground surface is plastically deformed and heat affected, and as a result, its surface properties are modified. Even in a burn-free ground surface, the local yield strength and the residual stresses are altered with grinding.

Tempering, typically at 180 °C, reduces the tetragonality of the martensite and induces precipitation of transitional carbides within the martensitic structure of the entire case. Thus, the surface is made more structurally uniform, to some extent, and is preconditioned against surface temperatures of up to 180 °C that might be generated during grinding or in service. An un-

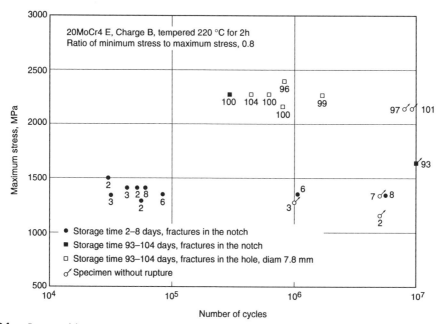

Fig. 7.24 Influence of the storage time at room temperature on the pulsating-fatigue strength. Source: Ref 38

tempered part, on the other hand, is sensitive to any temperature rise above ambient.

Tempering after grinding should be considered. It could reduce favorable surface compressive stresses generated during grinding; likewise, it could take the peak off tensile stresses similarly developed. Further, tempering is believed to raise the yield strength of the ground surface material and, consequently, would be expected to improve the resistance of the surface to fatigue-crack initiation (Table 7.10) (Ref 20).

Refrigeration

It is not uncommon for a refrigeration (subzero temperature) treatment to be included in the carburizing and hardening program, either as a standard procedure or as an optional operation. It transforms to martensite excess austenite retained after the hardening quench, thereby (a) increasing the surface hardness, (b) reducing the propensity to produce burns and cracks during surface grinding, and (c) at least partially ensuring the dimensional stability of critical precision parts.

The latter quality (c) is not usually regarded as important for general engineering applications, and refrigeration is used without concern about its effect on dimensions. However, there are special applications for which dimensional stability is vital, and refrigeration is therefore justified. Where grinding is involved (b), even moderate amounts of retained austenite can be tolerated, especially if the austenite is fine and well distributed and good grinding techniques are employed. It is primarily the surface hardness requirement (a) that dictates the need to treat at subzero temperatures. Whatever the reason is for refrigeration, it is prudent to low-temperature temper after the treatment.

Influence on Hardness. The specified minimum surface hardness of case-hardened components does not normally fall below 650 HV (58 HRC). However, with alloy grades of carburizing steel, particularly those containing over 2% Ni, it can be difficult to realize the specified minimum surface hardness when the surface carbon content is high. In such cases, the retained austenite content at or near the surface exceeds 30% (assuming that softening has not resulted from decarburization, high-temperature transformation products associated with internal oxidation, or poor quenching); therefore, a refrigeration treatment is likely necessary.

Cesarone examined the effects of both tempering and refrigeration on the retained austenite and hardness of carburized and hardened SAE 9310 steel samples (Ref 40). In the as-quenched conditions (of which there were twelve) the retained austenite contents varied between 30 and 80%. Refrigeration of the as-quenched samples reduced the retained austenite by amounts related to the original retained austenite content and the refrigeration temperature (Fig. 7.25a). Whereas tempering the as-quenched material led to only a 10% reduction of retained austenite, tempering after refrigeration was far more effective for reducing the austenite, particularly for austenite contents over approximately 50%. This suggests that refrigeration had destabilized the remaining austenite. The as-quenched hardness fell in the 55 to 65 HRC range. Refrigeration narrowed that range to 62 to 66 HRC (averaging 63.9 HRC) (Fig. 7.25b), and subsequent tempering modified it to 60.5 to 65 HRC (averaging 62.5 HRC) (Fig. 7.25c). Subzero treating at −196 °C in liquefied natural gas (LNG) reduced the retained austenite of the as-quenched material by about 10%γR more than did freezing at −80 °C, and it raised the hardness by approximately 1 HRC.

Whether or not an "in-line" refrigeration treatment should be employed for all processed components is somewhat debatable because so much depends on the steel grade and the processing in general. It is prudent to consider the effects of the process variables and use them to control austenite retention before resorting to refrigeration as a programmed treatment. The process variables that might be manipulated are: quenching temperature (Table 7.11), surface carbon content (Table 7.12), the chemical composition of the steel (Table 7.13), and the use of reheat quenching (Table 7.14). Based on the work of Koistinen and Marburger, the quenchant temperature should

Table 7.10 Microyield stress ($\sigma_{0.001}$) of ground surfaces

Process	$\sigma_{0.001}$, MPa
Quench, subzero treat, 150 °C temper for 3 h, grind	114
Quench, subzero treat, 150 °C temper for 3 h, grind, 140 °C temper for 3 h	133
Quench, subzero treat, 150 °C temper for 3 h, grind, 140 °C temper for 24 h	153

Source: Ref 20

Postcarburizing Thermal Treatments / 187

theoretically also have an influence on the quantity of austenite retained (Table 7.15) (Ref 42); experimental work using quenchant temperatures of 60 to 140 °C essentially agreed (Ref 44). Shea, on the other hand, determined that there was no significant variation of retained austenite when quenchant temperatures were 50 to 270 °C for a carburized SAE 4130 steel or when quenchant temperatures were 50 to 235 °C for an SAE 1526 steel (Ref 32).

The effectiveness of subzero treating at either −120 or −196 °C is indicated in Table 7.14 where appreciable quantities of retained austenite are transformed, and significant hardness increases are achieved. These macrohardness values represent the hardnesses of aggregates of martensite and austenite (and carbides, if present). What they do not show is that the austenite surviving refrigeration is strengthened. Microhardness tests on retained austenite in a 1.2 to 1.3% C surface, performed before and after refrigeration at −120 °C, indicated an increase of hardness from 520 to 650 HV (Ref 39).

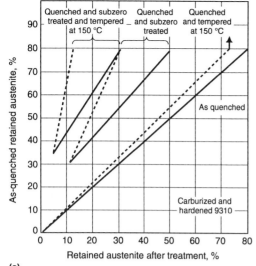
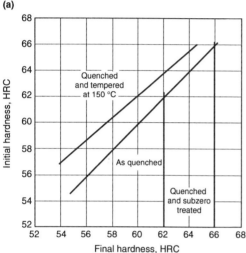
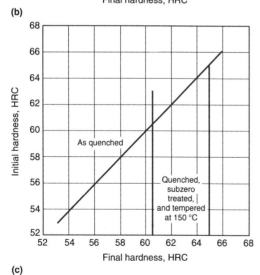

Fig. 7.25 Changes of retained austenite content and hardness due to refrigeration (−196 and −80 °C) and tempering (150 °C). (a) Effect of refrigeration and tempering on retained austenite content. (b) Individual effects of tempering and refrigeration on surface hardness. (c) Effect of refrigeration and tempering on surface hardness. Source: Ref 40

Table 7.11 Effect of quenching temperature on retained austenite

Steel	Quenching temperature (into oil), °C	Tempering temperature, °C	Surrface hardness, HRC	Probable austenite content, %
18Kh2N4VA	850	140	54	>30
18Kh2N4VA	800	140	59	~20

Source: Ref 41

Table 7.12 Effect of surface carbon content on retained austenite (estimated for a 3.5%Ni steel)

Surface carbon, %	Heat treatment	Retained austenite, %	Surface hardness, HRC
0.9	Oil quenched from 820 °C. Tempered at 150 °C	32	54
0.8	Oil quenched from 820 °C. Tempered at 150 °C	28	56
0.7	Oil quenched from 820 °C. Tempered at 150 °C	20	58

Table 7.13 Effect of alloy content on retained austenite

Steel	M_s, °C	$M_s - T_q$, °C	Retained austenite, %	Estimated hardness, HRC
0.8%-4.5%Ni-1.4%Cr	50	30	60	48
0.8%-2.0%Ni-0.3%Cr	110	90	35	56

Steels quenched from 850 °C. T_q, quenchant temperature, taken to be 20 °C

Table 7.14 Effect of subzero cooling after quenching

Heat treatment after carburizing	Condition	Retained austenite, %	Hardness, HRC	Bending strength MPa	Bending strength kg/mm²	Impact MPa	Impact kg/mm²
Oil quenched from 800 °C, low-temperature tempered	As-quenched	62	54	1530	156	25.5	2.6
	Subzero treated	20	62	1442	147	19.5	2.0
Tempered at 650 °C, oil quenched from 800 °C, low-temperature tempered	As-quenched	34	60	1697	173	40	4.1
	Subzero treated	10	62	1608	164	...	...
Air cooled from 900–750 °C, oil quenched, low-temperature tempered	As-quenched	90	47	1618	165	59	6.0
	Subzero treated	20	60	1353	138	19.5	2.0

Steel 18Kh2N4VA. Subzero treatments carried out at –120 °C. For fatigue data, see Fig. 7.27. Source: Ref 43

Influence on Tensile Properties. The ultimate tensile strength of a steel in the quenched condition is little affected by refrigeration, even to –196 °C. The yield strength, on the other hand, rises as the refrigeration temperature decreases, so much that the yield strength (0.2% proof stress) of a high-carbon steel can be marginally lower than the ultimate tensile strength (Table 7.16) (Ref 45). This increase of yield strength is likely a consequence of reduced austenite particle size brought about by the subdivision of austenite volumes by the martensite produced during refrigeration. The ductility indicators, elongation and area reduction, decrease as the temperature of the refrigeration treatment decreases.

Influence on Fatigue Resistance. The current opinion is that moderate to high quantities of retained austenite are beneficial to the low-cycle fatigue resistance of carburized and hardened parts, and are detrimental to the high-cycle fatigue resistance (Ref 46). Refrigeration to reduce the austenite content, therefore, is expected to adversely affect the low-cycle fatigue strength and favor the high-cycle fatigue strength. This, however, is not the case: refrigeration has often been found to have an adverse effect on both high- and low-cycle fatigue resistance.

Roberts and Mattson showed that a subzero treatment at -75 °C reduces the low-cycle bending fatigue strength of case-hardened SAE 8620 test pieces (Fig. 7.26) (Ref 47). Using notched test pieces, Sveshnikov et al. found that the high-cycle fatigue limit of a 20KhNM steel is improved by refrigeration at -75 °C, whereas the same treatment on the leaner Kh40 steel reduces the fatigue limit (Table 7.17) (Ref 22). In another study, refrigeration at -120 °C marginally increased the torsional fatigue strength of a case-hardened 3.5%NiCr steel (Fig. 7.27, curves

Table 7.15 Effect of quenchant temperature on retained austenite

M_s, °C	T_q, °C	$M_s - T_q$, °C	Approximate retained austenite(a), %	Estimated hardness, HRC
150	80	70	45	52
150	60	90	35	56
150	40	110	29	57
150	20	130	25	58

(a) M_s, martensite start temperature; T_q, quenchant temperature. This refers to the austenite content of T_q. With further cooling, more transformation will take place, but a significant amount of stabilization will have occurred. Source: Ref 42

Table 7.16 Effect of subzero treating on mechanical properties

Material(a)	Treatment temperature, °C	Ultimate tensile strength, MPa	Yield strength, kg/mm²	RA, %	Elongation, %	Fatigue, kg/mm²	K_{Ic}, kg/mm$^{3/2}$	Retained austenite, %	$(\Delta a)/a \times 10^{-3}$(b)
50Kh tempering 150 °C	...	235	190	16.5	6.6	96	82	6.1	4.45
	–50	237	194	14.5	6.1	102	88	5.9	2.3
	–90	238	196	13.5	4.3	104	90	5.4	2.57
	–196	237	207	10.0	3.2	97	83	2.2	2.98
50KhN tempering 150 °C	...	230	188	10.5	7.2	92	76	5.7	4.08
	–70	238	200	13.0	5.7	99	83	...	...
	–196	239	206	12.0	4.2	90	73	...	...
ShKh15 tempering 200 °C	...	241	198	...	3.08	88	65	15.8	4.24
	–50	243	206	...	2.97	100	75	14.9	3.44
	–90	246	224	...	1.7	96	70	10.5	3.93
	–196	239	229	...	1.04	90	66	8.2	4.01

(a) Through-hardened materials. Nominal compositions are 50Kh: 0.5% C, 0.6% Mn, 1.0% Cr, <0.3% Ni, <0.3% Cu; 50KhNi: 0.5% C, 0.6% Mn, 0.6% Cr, 1.2% Ni, <0.3% Cr; ShKh15: 1.0% C, 1.5% Cr, <0.3% Ni, <0.25% Cu. (b) Distortions in crystal lattice in α solid solution

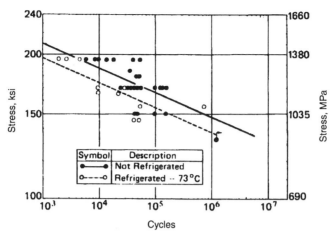

Fig. 7.26 Bending-fatigue strength of a carburized SAE 8620 steel (6.35 mm diam). Source: Ref 47

3 and 4). In this instance 20% austenite survived the cold treatment (Table 7.13). Further freezing to −196 °C reduced the fatigue strength appreciably (Fig. 7.27, curve 5). Using an AISI E9310 steel, Panhans and Fournelle observed that after carburizing and quenching, the austenite content was 56% and the surface compression was −595 MPa (Ref 48). Subzero treatment for one hour at −196 °C reduced the austenite to 31% and increased the surface compressive-residual stresses to −760 MPa. In this instance, refrigeration reduced both the very low- and the high-cycle fatigue lives, but not the intermediate life (Table 7.18). However, Razim found that with carburized 14NiCr14 steel test pieces, the better high-cycle fatigue results came from the test pieces for which subzero treatment had been used to achieve virtual austenite freedom (Fig. 7.15) (Ref 29).

It seems, therefore, that the refrigeration temperature, or perhaps the amount of retained austenite surviving refrigeration, might influence fatigue strength. Working with through-hardened unnotched specimens of 0.5 and 1.0% C lean-alloy steels, Romaniv et al. showed that subzero treating at −196 °C had only a small effect on the fatigue limit (negative or positive depended on steel grade) (Ref 45). However, 10 to 30 minute refrigeration treatments at −50 to −70 °C decidedly benefited the fatigue limit and fracture toughness (K_{Ic}) (Fig. 7.28, 7.29). These improvements from a shallow refrigeration treat-

Table 7.17 Effect of refrigeration on hardness and fatigue

Steel and treatment	Hardness, HRC	Fatigue limit MPa	Fatigue limit kg/mm²
20KhNM (Cr-Ni-Mo)			
Carburized and quenched	63.1	524	53.5
Carburized, quenched, and refrigerated	67.5	564	57.5
40Kh (Cr)			
Carburized and quenched	61	524	53.5
Carburized, quenched, and refrigerated	63.8	466	47.5

Source: Ref 22

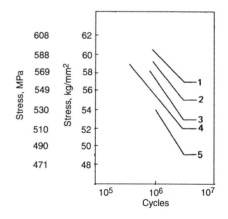

		Treatment	
Curve	Temper	Oil quench	Subzero
1	650 °C	800 °C	...
2	650 °C	800 °C	−120 °C
3	...	800 °C	−120 °C
4	...	800 °C	...
5	...	800 °C	−196 °C

Fig. 7.27 Torsional fatigue curves for carburized 18Kh2N4VA steel. Case depth, 1.5 mm. See also Table 7.14. Source: Ref 43

Table 7.18 Fatigue of 9310 steel at two retained austenite levels

Life cycles	Stress for failure	
	Retained austenite, 56%	Retained austenite, 31%
10^7	880	806
10^6	~880	~880
10^5	~900	~970
10^4	1060	1040
10^3	1400(a)	1120
Knee of S/N curve	4×10^4 to 10^5	4×10^6 to 7×10^6

(a) Extrapolated. Source: Ref 48

ment were attributed to transformation of the least stable austenite of the austenite regions; a deeper treatment (say at –196 °C) would also transform some austenite that was stable at –70 °C. This additional transformation would develop local high-magnitude tensile microstresses and areas of excessive microdistortions, which in turn would reduce the duration for crack initiation (Fig. 7.30). Note that there was not a great amount of retained austenite in any of the samples, and it is unclear how effective the –50 °C treatment would have been had the retained austenite been, say, 40%. Nevertheless, in this instance refrigeration had a significant effect on the high-cycle fatigue life, and incidentally, it marginally improved the low-cycle fatigue life.

A recent innovation in subzero treating omits thermal tempering and replaces it with subzero treating in the presence of a cyclic magnetic field (Ref 49). Rotating beam test results indicate that with this treatment, fatigue lives comparable to those of conventionally tempered parts could be achieved (Fig. 7.31).

Influence on Residual Stresses. The resistance of retained austenite to fatigue cracking is determined by the amount of applied energy it absorbs and uses in the formation of martensite because energy used for the martensite reaction and for heating is not available for crack initiation or propagation (Ref 45). If refrigeration raises the level of tensile microstresses within the austenite

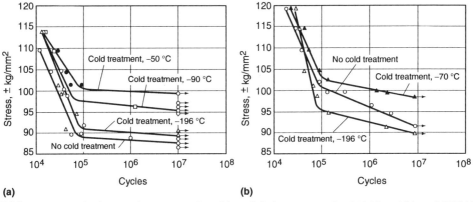

Fig. 7.28 Fatigue graphs for smooth specimens from (a) steel ShKh15 tempered at 200 °C and (b) steel 50KhN tempered at 150 °C. Source: Ref 45

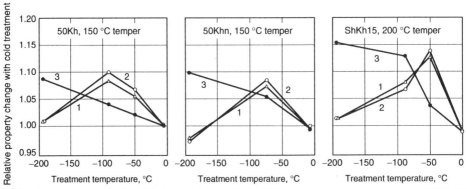

Fig. 7.29 Relative variations in the (1) fatigue endurance limit, (2) fracture toughness parameter, and (3) 0.2% yield strength as a result of cold treatment at various temperatures. The relative change is a ratio of properties after treatment and properties prior to treatment. Source: Ref 45

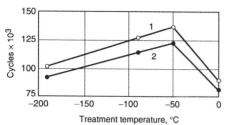

Fig. 7.30 Effect of cold treatment temperature on the total fatigue life (1) and the fatigue life to crack initiation (2) in notched specimens of steel ShKh15 (tempered at 200 °C) at a nominal stress of 605 MPa (88 ksi). Source: Ref 45

without triggering the martensite reaction, then the ability of the austenite to absorb energy is reduced.

Kim et al. determined the residual stresses within the austenite and martensite phases of a carbonitrided surface layer and showed that refrigeration at −85 °C for four hours resulted in high magnitudes of tensile-residual stress in the austenite (Fig. 7.32) (Ref 50). Consequently, the fatigue strength was greatly impaired. In a way, these findings contradict the fatigue results of Romaniv et al. (Fig. 7.29) (Ref 45); hence, one

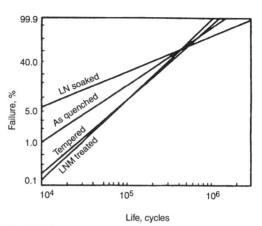

Fig. 7.31 Weibull plot of data obtained at 1061.79 MPa (154 ksi) test stress on carburized SAE 8620 steel fatigue specimens. Source: Ref 49

can propose that although the temperature of a subzero treatment is important, the duration of the treatment is crucial.

Macroresidual stresses reportedly are not greatly affected by subzero treatments, even though the amount of austenite transformed is appreciable (Table 7.14) (Ref 22, 39). Other re-

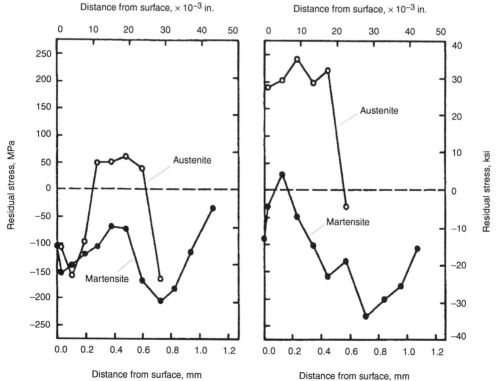

Fig. 7.32 Residual stresses in the carbonitrided case of EX55 (a) without subzero treatment and (b) with subzero treatment. Source: Ref 50

searchers observed significant changes of residual stress distribution due to refrigeration with liquid nitrogen (Fig. 7.33) (Ref 51).

The effect of double subzero cooling is shown in Fig. 7.34. The change in residual stress distribution as a result of increasing the duration of the −75 °C treatment indicates that more austenite transformation has occurred accompanied by, presumably, an increase in the magnitude of local micro-tensile-residual stresses in the surviving austenite volumes.

Influence on Contact Fatigue. Under contact conditions (either roll or roll-slide), at least one of the mating surfaces should be able to be deformed slightly, thereby allowing a uniform distribution of the applied load. Fully martensitic carburized and hardened surfaces (especially those that have been refrigerated and in which any traces of residual austenite are highly strained) will resist this deformation. Microstructures containing some retained austenite, which has more ability to deform than martensite alone, should more readily accommodate the applied loads. Further, surfaces containing austenite, compared with martensitic surfaces, will prefer to deform rather than crack at critically stressed locations, and perhaps transform to martensite or reject interstitial carbon as precipitates, in order to reduce microstresses. Hence, the initiation and propagation of prepitting cracks should be slower. It is reasoned, therefore, that under contact-fatigue conditions, subzero treatments that produce essentially martensitic structures might not improve the contact-fatigue resistance, despite the trend of contact-fatigue strength increasing with surface hardness.

Roller tests have shown that with pure rolling or extreme sliding, the loss of metal in surfaces containing austenite (54 HRC) is less than for a martensitic surface (59 HRC) or a subzero-treated surface (63 HRC) (Ref 41). Other roller tests determined that as-quenched carburized surfaces survive longer than surfaces that have been frozen in LNG (Table 7.19) (Ref 50). Testing gears, Razim concluded that retained austenite is beneficial to contact fatigue, as illustrated

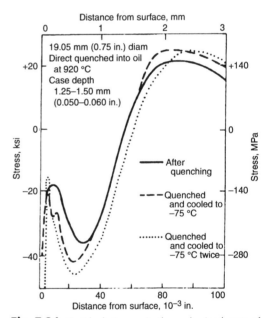

Fig. 7.34 Residual stresses in the carburized case of SAE 9310 before and after subzero treatments. Source: Ref 52

Table 7.19 Contact fatigue life of case-hardened 16CD4 steel

Additional treatment	Hardness, HV Core	Hardness, HV Case	Retained austenite at surface, %	Life, h
None	445	830	20–25	51
LNG treated	445	810	10–12	40.5
LNG plus 150 °C temper	443	865	No data	48

Rolling contact (balls on plane surface) = 6.2×10^5 loadings per hour. Source: Ref 51

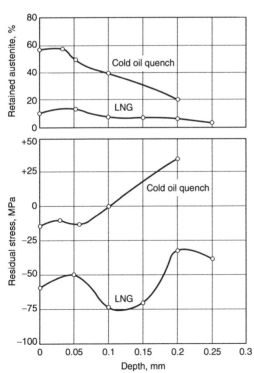

Fig. 7.33 Increase of compressive-residual stresses due to subzero treatment. Source: Ref 51

in Fig. 7.35 (Ref 53). In this study, refrigeration was used to arrive at essentially zero retained austenite within the cases of some of the test gears. From such a test it is not clear how much of the difference between the results for refrigerated gears (containing small amounts of austenite) and the results for as-quenched gears (containing, say, 30% austenite) is due to the discrepancy in austenite contents and how much of that difference is due to other effects of subzero treating. If refrigeration produces local tensile microstresses in the resultant microstructure, might not these have an adverse effect on the crack-initiation and growth times? And if the carbides observed to have precipitated under the contact tracks differ (η-carbides or monoclinic Hagg carbides) depending on whether or not refrigeration has been used, might not such differences have some significance?

Influence on Bending and Impact Toughness. Table 7.14 and 7.16 indicate how some toughness properties and the ductility are adversely influenced by refrigeration, though Fig. 7.29 indicates that the K_{Ic} fracture toughness may be improved by a shallow refrigeration treatment.

Influence on Wear Resistance. In general, subzero treating a case-hardened surface to reduce the austenite content, and thereby raise the hardness, would be expected to have a positive influence on the resistance of that surface to abrasive wear. However, roller tests indicated the opposite, and showed that as-quenched or as-quenched and tempered surfaces had approximately three times the wear resistance of quenched and –50 °C refrigerated surfaces (Table 7.20) (Ref 54). The changes in hardness and matrix carbon content due to wear are far more dramatic in the refrigerated surfaces than in the others. This potentially confirms that austenite surviving a refrigeration treatment is destabilized by the treatment and thereby responds more readily to subsequent thermal and mechanical actions (Fig. 7.25).

The adhesive wear situation, however, involves many factors and cannot adequately be covered by generalizations. The alloy content of the steel is important. Some added elements (e.g., nickel)

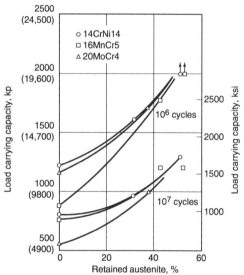

Fig. 7.35 Gear test results show that within the conditions of the tests, the load carrying capacity increased with retained austenite content. Source: Ref 53

Table 7.20 Results of slide/roll wear tests on 12KhN3A steel

Loading cycles	Hardness, HV	Carbon in martensite, %	Wear, mg	
			Small roller	Large roller
Carburized at 1000 °C to a depth of 1.8 mm; oil quenched from 900 °C and 800 °C				
0 (before test)	792	0.60	...	...
1 × 10⁶	760	0.55	60	68
2 × 10⁶	758	0.41	68	75
3 × 10⁶	738	0.49	74	90
Carburized at 1000 °C to a depth of 1.8 mm; oil quenched from 900 °C and 800 °C; tempered at 230 °C for 5 h				
0 (before test)	679	0.50	...	...
1 × 10⁶	697	0.31	18	52
2 × 10⁶	758	0.32	39	85
3 × 10⁶	729	0.39	45	126
Carburized at 1000 °C to a depth of 1.8 mm; oil quenched from 900 °C and 800 °C; refrigerated at –50 °C for 1 h				
0 (before test)	763	0.69	...	...
1 × 10⁶	517	0.25	120	262
2 × 10⁶	597	0.35	140	287
3 × 10⁶	619	0.32	170	353

Tests conducted at 12,400 kg/cm² contact pressure. Test pieces wer tempered at 650 °C for 4 h between carburizing and hardening. Source: Ref 54

tend to increase the susceptibility of a surface to adhesive wear (scoring, scuffing) more so, it is claimed, than a high austenite content does (Ref 55). Even so, retained austenite does have an effect. Friction tests on a carburized 4%NiCr steel with either 5 or 25% retained austenite showed that the coefficient of friction (μ) of the 5% austenite surface stayed fairly constant over the temperature range of the tests, whereas the μ value for the sample with 25% austenite rose appreciably at 160 °C (Ref 56). This implies that under certain conditions (e.g., pressure, temperature, speed, and lubrication), a higher austenite content can favor adhesive wear; therefore refrigeration might have a beneficial effect.

Unfortunately, laboratory tests seem to produce conflicting results when it comes to assessing the effects of retained austenite. Kozlovskii et al. found that a surface containing approximately 50% retained austenite has a superior resistance to seizure when compared to a surface containing less than 20% retained austenite (Fig 4.28) (Ref 57). Seizure tests by Manevskii and Sokolov showed that the scoring resistance was, to some extent, hardness related; for the hardest surface tested (~700 H_{100}) the scoring resistance was less than it was for a softer surface (~600 H_{100}). Terauchi and Takehara observed the opposite trend (Fig. 4.29) (Ref 59).

For gears, adhesive wear relates to the conditions prevailing during rolling with sliding contact under pressure; generally a designer can estimate if a design has a tendency to score in service. It is not certain if laboratory tests can adequately simulate real-life conditions, although Naruse and Haizuka claim that the limiting load for scoring by means of the FZG spur gear test, the four-ball test, and disk tests could be compared (Ref 60). Without resorting to the metallurgy of the test pieces, they concluded that the limiting load for scoring is a function of sliding velocity, specific sliding, and the type of lubricant used.

When scoring occurs on gear tests, it is usually high on the tooth addenda at the point of disengagement where the amount of sliding is high. Therefore, if a design is considered prone to scoring, lubricant choice is important, and "running-in" to precondition the surface is advisable to remove asperities that might otherwise penetrate the lubricating film thickness at higher pressures, and to induce work hardening of the mating surfaces. Thus, the ability of a surface to work harden is significant, but it must be given the opportunity to work harden without an excessive surface temperature rise that could lead to scoring. If refrigeration reduces the capacity of a surface to work harden or to shed its asperities, then in certain situations it could be regarded as detrimental.

Retained Austenite Standards. It has been stated before, and is restated here, that it is better to manage other process variables to control retained austenite before resorting to refrigeration. Some laboratory tests have indicated that high austenite contents are beneficial, whereas others conflict with that view. For the time being, the gear industry, through tests and experience, has chosen to restrict the acceptable amount of retained austenite to 25% (ISO 6336-5) or 30% (AGMA 2001-C95) for all but the lowest grade of commercially carburized gear. That said, if the basic allowable stresses used by designers can be regarded as generous enough to include parts in the refrigerated condition, then it is the decision of the manufacturer whether or not to use the subzero process (unless specified), provided the recommended maximum for retained austenite is not exceeded and the surface hardness is adequate. Note that aerospace and marine gears (AGMA 246-02[1983] and AGMA 6033) permit up to 20% retained austenite on the highest rated gears.

Summary

Tempering

Most case-hardened parts are tempered at a temperature above 130 °C (265 °F) but rarely exceeding 250 °C (480 °F). The most common tempering temperature is about 180 °C (350 °F). At this temperature, any austenite in the case is affected only a little. At higher tempering temperatures, some austenite transforms to bainite; this is not necessarily beneficial. Tempering times are usually from 2 to 10 h, depending on the size of the component and the structural stability required. Such tempering, in additon to precipitating coherent carbides and reducing the tetragonality of the martensite, also drives off hydrogen taken up during the case-hardening process.

- *Preprocess considerations:* None
- *In-process considerations:* If intermediate machining is to be carried out to locally re-

move case, then a high-temperature temper (or anneal) is used to impart adequate machinability. The temperatures for this are in the range of 600 to 650 °C (1110 to 1200 °F).
- *Post-process considerations:* Low-temperature tempering is done after hardening; however, it should be repeated after refrigeration (if used). A repeat low-temperature temper should be considered if the surface is still exceptionally hard after the initial temper. Some manufacturers temper after grinding.
- *Effect on properties:* Low-temperature tempering reduces the hardness into the normally accepted range of 58 to 62 HRC. It reduces the surface compressive-residual stresses and, therefore, lowers the high-cycle bending-fatigue strength, and it is thought to improve the low-cycle bending strength. In terms of rolling-contact fatigue, tempering is believed to raise the high-cycle fatigue life, but may have an adverse effect on the low-cycle endurance. The hardness-strength properties of the core are affected only a little by tempering, although the yield strength is raised.
- *Standards:* The ANSI/AGMA standard has no tempering specification for grade 1, but recommends tempering for grade 2 and requires it for grade 3.

Refrigeration

Case-hardened parts are refrigerated to transform retained austenite in the outer case. This raises the surface hardness and induces structural stability. Refrigerants available for the subzero treatment of heat-treated parts are dry ice (solid CO_2) and liquid nitrogen gas (LNG). Temperatures down to about −80 °C (112 °F) are achievable with CO_2, whereas temperatures of −196 °C (−320 °F) are possible with LNG refrigeration. It is not normal to place heat-treated parts into LNG; rather LNG is used to chill a compartment to whatever temperature is considered reasonable for the job, say −80 °C or even −120 °C.

- *Preprocess considerations:* Refrigeration is carried out after the quenching operation; it generally follows the initial tempering operation. It is prudent to allow parts to cool to room temperature before subjecting them to subzero temperatures. Hardness test before treatment.
- *In-process considerations:* Record the minimum temperature and the duration of the process.
- *Postprocess cosiderations:* It is advisable to retemper soon after the part has attained room temperature. Perform a hardness test after refrigeration, and if the part is still soft after refrigeration, then other possible reasons for the softness must be considered.
- *Effect on properties:* Refrigeration transforms case austenite to martensite, which raises the hardness. It also raises the surface macro-compressive-residual stresses, but induces microresidual tensile stresses in any remaining austenite volumes. Tests to determine the effect of refrigeration on bending fatigue have produced mixed results; the steel grade, the subzero temperature, and the duration of refrigeration can all influence the results. Contact fatigue and case ductility each appear to be adversely affected by the process. Refrigeration, coupled with tempering, is nonetheless a valuable process where dimensional and microstructural stability are important.
- *Standards:* ANSI/AGMA permit refrigeration following the tempering operation, and it is followed by a retemper. The purpose is to obtain a 1 to 2 HRC increase of hardness. Refrigeration to transform high amounts of retained austenite (say 50%) should not be considered, as this might cause microcracking.

REFERENCES

1. G.R. Speich and W.C. Leslie, Tempering of Steels, Met. Trans., Vol 13, May 1972, p 1043–1054
2. H.-J. Zhou and Y.J. Li, Low Carbon Martensites and Their Application, *Proc. 6th Int. Congress on Heat Treatment of Metals* (Chicago), ASM International, 1988
3. P.M. Kelly and J. Nutting, The martensite transformation in carbon steels, *Proc. R. Soc. (London) A,* Vol 259, 1960, p 45–58
4. S. Nagakura, Y. Hirotsu, M. Kusonoki, T. Suzuki, and Y. Nakamura, Crystallographic Study of the Tempering of Martensitic Carbon Steel by Electron Microscope and Diffraction, Metall. Trans. A, Vol 14, June 1983, p 1025–1031
5. I.Y. Kagan, S.V. Bronin, and I.Y. Sidorenko, Tempering of Quenched Carbon Steels, *Met. Sci. Heat Treat. (USSR),* No. 1, Jan/Feb 1964, p 87–91
6. O. Yasuya and T. Imao, Epsilon Carbide Precipitation During Tempering Plain Carbon

Martensite, *Metall. Trans. A,* Vol 23, Oct 1992, p 2737–2751

7. E.J. Mittemeyer and F.C. Van Doorn, Heat Effects of Pre-Precipitation Stages on Tempering of Carbon Martensites, *Metall. Trans. A,* Vol 14, May 1983, p 976–977

8. Y. Nakamura, T. Mikami, and S. Nagakura, In Situ High Temperature Electron Microscope Study of the Formation and Growth of Cementite Particles of the Third Stage of Tempering of Martensitic High Carbon Steel, *Trans. Jpn. Inst. Met.,* Vol 26 (No. 12), Dec 1985, p 876–885

9. C.-B. Ma, T. Ando, D.L. Williamson, and G. Krauss, Chi-Carbide in Tempered High Carbon Martensite, *Metall. Trans. A,* Vol 14, June 1983, p 1033–1044

10. Y. Imai, T. Oguro, and A. Inoue, Formation of χ-Carbide in Carbon Steel, *Tetsu-to-Hagané (J. Iron Steel Inst. Jpn.),* Vol 57 (No. 4), 1971, S113–S114

11. R.W. Neu and H. Sehitoglu, Thermal Induced Transformation of Retained Austenite in the Simulated Case of a Carburised Steel, *J. Eng. Mater. Technol. (Trans. ASME),* Vol 115, Jan 1993, p 83–88

12. J.L. Aston, The Influence of Tempering Time on Some of the Mechanical and Physical Properties of Steel, *Bull. Jpn. Inst. Met.,* Vol 192 (No. 4), Aug 1959, p 377–382

13. H. Streng, C. Razim, and J. Grosch, Influence of Hydrogen and Tempering on the Toughness of Case Hardened Structures, *Carburising: Processing and Performance* (Lakewood, CO), G. Krauss, Ed., ASM International Conference, 1989

14. L. Salonen, The Residual Stress in Carburised Layers, *Acta Polytech. Scand.,* Vol 109, 1972, p 7–26

15. V.V. Zabil'ski, V.I. Sarrak, and S.O. Suvorova, The Role of Relaxation Processes on the Volume Change Occurring in Steels During Tempering, *Fiz. Met. Metalloved.,* Vol 48 (No. 2), Aug 1979, p 323–331

16. K.J. Irvine, F.B. Pickering, and J. Garstone, The Effect of Composition on the Structure and Properties of Martensite, *Bull. Jpn. Inst. Met.,* Vol 196, Part 1, Sept 1960, p 66–81

17. V.V. Babayan, Low-Temperature Tempering of Carburized and Hardened Steel 30KhGT, *Metalloved. Term. Obrab. Met.,* Aug 1959, p 41–43

18. G. Meyer, The Influence of Heat Treatment on the Structure and Mechanical Properties of Carburised Steels, *Conf. on Heat Treatment Practice,* July 1960 (Harrogate), BISRA, p 13–23

19. G. Parrish and G.S. Harper, *Production Gas Carburising,* Pergamon, 1985

20. A.G. Ran'kova, V.S. Kortov, M.L. Khenkin, A.I. Saprindashvili, and G.M. Guseva, Stability of the Surface Layer of Bearing Steel ShKh15, *Met. Sci. Heat Treat. (USSR),* (No. 6), June 1975, p 71–73

21. P.M. Vogel, Alignment of Case-Hardened Parts, *Carburising: Processing and Performance* (Lakewood, CO), G. Krauss, Ed., ASM International Conference, 1989

22. D.A. Sveshnikov, I.V. Kudryavstev, N.A. Gulyaeva, and L.D. Golubovskaya, Chemicothermal Treatment of Gears, *Met. Sci. Heat Treat. (USSR),* (No. 7), July 1966, p 527–532

23. M. Motoyama and S. Yonetani, On the Influence of Tempering and Hydrostatic Pressing on the Residual Stress of Carburised and Quenched Steel, *J. Jpn. Inst. Met.,* Vol 33 (No. 1), 1969, p 109

24. D. Kirk, P.R. Nelms, and B. Arnold, Residual Stresses and Fatigue Life of Case-Carburised Gears, *Metallurgia,* Vol 74 (No. 446), Dec 1966, p 255–257

25. H. Weigand and G. Tolasch, Fatigue Behaviour of Case-Hardened Samples, *Härt.-Tech. Mitt.,* Vol 22 (No. 4), Dec 1967, p 330–338

26. R.J. Love, H.C. Allsopp, and A.T. Weare, "The Influence of Carburising Conditions and Heat Treatment on the Bending Fatigue Strength and Impact Strength of Gears made from En352 Steel," MIRA Report 1959/7, 1959

27. C.Q. Gu, B.Z. Lou, X.T. Jing, and F.S. Shen, Microstructures and Mechanical Properties of the Carburised CrNiMo Steels with Added Case Nitrogen, *Proc. 6th Int. Congress on Heat Treatment of Metals* (Chicago), ASM International, 1988

28. G. Fett, Bending Properties of Carburising Steels, Adv. Mater. Proc., Vol 4, 1988, p 43–45

29. C. Razim, Influence of Residual Austenite on the Strength Properties of Case-Hardened Test Pieces During Fatiguing, *Härt.-Tech. Mitt.,* Vol 23, April 1968, p 1–8

30. D. Rosenblatt, Controlling Variables Which Affect the Tempering of Carburised Gears, *4th Int. Congress on Heat Treatment of Materials,* IFHT, 1985

31. D.E. Diesburg and C. Kim, Microstructural and Residual Stress Effects on Impact Fracture Resistance of Carburised Cases, *5th ASM Heat Treating Conf. and 18th Int. Conf. On Heat Treating of Materials*, ASM, 1980
32. M.M. Shea, Influence of Quenchant Temperature on the Surface Residual Stresses and Impact Fracture of Carburised Steel, *J. Heat Treat.*, Vol 3 (No. 1), June 1983, p 38–47
33. B. Thoden and J. Grosh, Crack Resistance of Carburised Steel Under Bending Stress, *Carburising: Processing and Performance* (Lakewood, CO), ASM International Conference, 1989, p 303–310
34. G. Fett, Tempering of Carburised Parts, *Met. Prog.*, Sept 1982, p 53–55
35. H.K. Obermeyer and G. Krauss, Toughness and Intergranular Fracture of a Simulated Carburised Case in Ex 24 Type Steel, *J. Heat Treat.*, Vol 1 (No. 3), June 1980, p 31–39
36. J. Larsen-Badse, The Abrasion Resistance of Some Hardened and Tempered Carbon Steels, *Trans. Metall. Soc. AIME*, Vol 236 (No. 10), Oct 1966, p 1461–1466
37. X. Yaomin and F. Dongli, Hydrogen Embrittlement of Steels Heat Treated in Hydrogen Rich Atmosphere, *Proc. 6th Int. Conf. On Heat Treatment of Metals* (Chicago), ASM International, 1988
38. H. Streng, C. Razim, and J. Grosch, Diffusion of Hydrogen During Carburising and Tempering, Proc. *6th Int. Conf. On Heat Treatment of Metals* (Chicago), ASM International, 1988
39. I.S. Dukarevich and M.A. Balter, Thermomechanical Treatment in Hydrogen-Containing Atmospheres Improves Carburised-Steel Qualities, *Russ. Eng. J.*, Vol 53 (No. 8), 1973, p 62–65
40. J. Cesarone, Increasing Hardness Through Cryogenics, *Gear Technol.*, March 1997, p 22–27
41. M.A. Balter and M.L. Turovskii, Resistance of Case-Hardened Steel to Contact Fatigue, *Met. Sci. Heat Treat. (USSR)*, (No. 3), March 1966, p 177–180
42. D.P. Koistinen and R.E. Marburger, A General Equation Prescribing the Extent of the Austenite-Martensite Transformation in Pure Iron-Carbon Alloys and Plain Carbon Steels, *Acta Metall.*, Vol 7, 1959, p 59–60
43. M.A. Balter and I.S. Dukarevich, The Relationship Between the Properties of Steels Subjected to Chemicothermal Treatment and the Fatigue Limit, *Met. Sci. Heat Treat. (USSR)*, Vol 13 (No. 9), Sept 1971, p 729–732
44. E. Szpunar and J. Birelanik, Influence of Retained Austenite on Properties of Fatigue Cracks in Carburised Cases of Toothed Elements, *Proc. Heat Treating 1984*, The Metals Society, London
45. O.N. Romaniv, Y.N. Gladkii, and N.A. Deev, Some Special Features of the Effect of Retained Austenite on the Fatigue and Cracking Resistance of Low-Temperature Tempered Steels, *Fiz. Khim. Mekh. Mater.*, Vol 11 (No. 4), 1975, p 63–70
46. M.A. Zaccone and G. Krauss, Fatigue and Strain Hardening of Simulated Case Microstructure in Carburised Steel, *Proc. 6th Int. Congress on Heat Treatment of Metals* (Chicago), ASM International, 1988
47. J.G. Roberts and R.L. Mattson, *Fatigue Durability of Carburized Steel*, American Society for Metals, 1957, p 68–105
48. M.A. Panhans and R.A. Fournelle, High Cycle Fatigue Resistance of A.I.S.I., E. 9310 Carburised Steel with Two Different Levels of Retained Austenite and Surface Residual Stress, *J. Heat Treat.*, Vol 2 (No. 1), 1981, p 56–61
49. J.D. Collins, The Effects of Low Temper Magnetic Cycling on the Properties of Ferromagnetic Materials, *J. of Eng. Mater. Technol. (Trans. ASME)*, Vol 102, Jan 1980, p 73–76
50. C. Kim, D.E. Diesburg, and R.M. Buck, Influence of Sub-Zero and Shot-Peening Treatments on Impact and Fatigue Fracture Properties of Case-Hardened Steels, *J. Heat Treat.*, Vol 2 (No. 1), June 1981, p 43–53
51. A. Diament, R. El Haik, R. Lafont, and R. Wyss, "Surface Fatigue Behaviour of the Carbo-Nitrided and Case-Hardened Layers in Relation to the Distribution of the Residual Stresses and the Modifications of the Crystal Lattice Occurring During Fatigue," *15th Colloque Int.*, May 1974 (Caen, France), International Federation for the Heat Treatment of Materials
52. W.S. Coleman and M. Simpson, Residual Stresses in Carburized Steels, *Fatigue Durability of Carburized Steel*, American Society for Metals, 1957, p 47–67
53. C. Razim, Effect of Residual Austenite and Reticular Carbides on the Tendency to Pitting of Case-Hardened Steels, thesis, Technische Hochschule, Stuttgart, 1967

54. L.Y. Oshinya and V.A. Grishko, Effect of Heat Treatment on Changes in the Properties of the Surface Layer of Steel 12Kh N3A During Wear, *Met. Sci. Heat Treat. (USSR)*, (No. 4), April 1973
55. R.M. Matveevskii, V.M. Sinaisky, and I.A. Buyanovsky, Contributions to the Influence of Retained Austenite Content in Steels on the Temperature Stability of Boundary Lubrication Layers in Friction, *J. Lubr. Technol. (Trans. ASME)*, July 1975, p 521–525
56. W. Grew and A. Cameron, Role of Austenite and Mineral Oil on Lubricant Failure, Nature, Vol 217 (No. 5127), 1968, p 481–482
57. I.I. Kozlovskii, S.E. Manevskii, and I.I. Sokolov, Effect of Retained Austenite on the Resistance to Scoring of Steel, 20Kh2N4A, *Met. Sci. Heat Treat. (USSR)*, (No. 1), 1978, p 70–80
58. S.E. Manevskii and I.I. Sokolov, Resistance to Seizing of Carburised and Carbon Nitrided Steels, *Metalloved. Term. Obrab. Met.*, (No. 4), April 1977, p 66–68
59. J. Terauchi and J.I. Takehara, On the Effect of Metal Structure on Scoring Limit, *Bull. Jpn. Soc. Mech. Eng.*, Vol 21 (No. 152), Feb 1978, p 324–332
60. A. Naruse and S. Haizuka, Limiting Loads for Scoring and Coefficient of Friction on Disk Machines, *Bull. Jpn. Soc. Mech. Eng.*, Vol 21 (No. 158), Aug 1978, p 1311–1317

SELECTED REFERENCES

- A. Jordine, Increased Life of Carburised Race Car Gears by Cryogenic Treatment, *Mater. Aust.*, Vol 27 (No. 7), Aug 1995, p 8–10
- A. Jordine, Increased Life of Carburised Race Car Gears by Cryogenic Treatment, *IMMA Conference "The Heat Is On!"* 24–25 May 1995 (Melbourne, Victoria, Australia), Institute of Metals and Materials Australasia Ltd., 1995, p 107–111
- R.F. Spitzer, H.A. Chin, and D.A. Haluck, Cryogenic Turbopump Bearing Material Development Program, *Conf.: Creative Use of Bearing Steels*, 6–8 Nov 1991 (San Diego), STP 1195, ASTM, 1993, p 156–167
- B. Vinokur, The Isothermal Diagram of Transformation of the Residual Austenite by Tempering, *Metall. Trans. A*, Vol 24 (No. 12), Dec 1993, p 2806–2810

Chapter 8

Postcarburizing Mechanical Treatments

The mechanical treatments most commonly applied to case-hardened parts are shot peening and surface grinding, as well as roller burnishing to a very much lesser degree. Shot peening and roller burnishing plastically deform and texturally modify the surface and induce compressive residual stresses, thereby improving the fatigue resistance of the treated part. Parts are ground to obtain dimensional accuracy and to replace the heat-treated finish with one that is both clean and smooth. The process parameters for each of these mechanical treatments are controlled within fairly narrow, though not unreasonable, limits, and when correctly executed, these processes are beneficial to the parts treated. If, however, the processes are not carried out correctly they can adversely influence properties.

Grinding

When components are case hardened, some growth and distortion (size and shape changes) can be expected to occur as a result of the thermal processing. For example, with gears, the gear teeth themselves may thicken, diameters may increase or decrease, and worm threads may unwind. Squat cylinders may "barrel," and long cylindrical shapes may do the opposite. In their process schedules, manufacturers include measures that allow for such movements or contain them within tolerable levels, then correct the critical dimensions by grinding.

Besides restoring precision, grinding is also employed to remove the heat-treated surface. It also removes any potentially detrimental metallurgical features, such as carbide films, internal oxidation, and high-temperature transformation products (HTTP) that can otherwise adversely affect certain strength aspects of the component. Grinding for surface finish is important, because it too can influence the bending and contact fatigue lives of a component and the efficiency of a lubricant to separate mating surfaces.

Grinding Action

During gentle grinding (when the depth of cut and thrust forces are low), a particle of abrasive protruding from the working surface of the grinding wheel ploughs into the surface of the component, which produces a furrow. In deformable workpiece materials, the furrow is formed mainly by the displacement of metal from below and to either side, producing heaped edges (Ref 1). A second abrasive particle, slightly offset to the first, ploughs into one heaped furrow edge. This contact displaces metal and adds more deformation to an already plastically deformed material, and it possibly removes some metal particles by shear from the common edge between the two furrows. When ploughing is the main occurrence, the metal removal rate is low. With greater amounts of feed (depth of cut), metal is removed directly from the workpiece surface as chips without involving much ploughing. The metal removal rate under such conditions is high.

The ploughing or chipping action takes place simultaneously many times across a contact area between the workpiece and the grinding wheel and thousands of times during each wheel revolution. Acknowledging that the grinding wheel rotates at thousands of revolutions per minute, an appreciable amount of high-rate deformation takes place during grinding along with the generation of heat caused by that deformation. Not all the deformation and heat generated is produced by the furrowing and chipping action. The grit of the grinding wheel protrudes for only a limited distance from its bond and, therefore, rubbing and rolling actions by bond contact also produce deformation and heat. Further, as the grit of the wheel becomes duller or more loaded with attached metal debris and, therefore, more in need of a dressing operation, the amount of metal removed from the workpiece decreases, but the heat generation caused by friction increases.

Therefore, much of the energy used in the grinding action is converted into heat, and this heat is dissipated by conduction into the workpiece, the coolant (if used), and the abrasive. The heat is also carried away in the detached incandescent metallic particles that are continuously ejected from the wheel/workpiece contact zone. For successful grinding of materials having some measure of ductility, the objective should be to control the heat and ensure that the final surface is lightly plastically deformed.

Hard materials generally have little ability to be deformed, and the displacement of metal to form heaps at the sides of a furrow does not occur. Instead, for such materials, small cracks may radiate from the grinding furrow, even when grinding conditions are carefully controlled (Ref 2). This cracking is attributed primarily to mechanical causes. If some of the heat generated at the surface by the grinding action dissipates into the workpiece surface and thereby reduces the yield strength at the surface, then cracking at the furrow edges is eliminated. Mikhailov (Ref 2) suggested that the millions of variable loadings to which a workpiece surface is subjected during grinding cause the breakdown of the surface material at lower stresses. The mechanism for how this breakdown occurs is not clear. It could involve "crumbling" of the cracked furrow edges or a lowering of the surface yield strength.

Case-hardened surfaces are generally regarded as being hard and essentially brittle. However, such surfaces have the ability to be deformed (see the sections "Roller Burnishing" and "Shot Peening" in this chapter). Therefore, the martensite/austenite aggregate, typical of a case-hardened surface, should be regarded as being deformable with respect to grinding. Carbides within a case-hardened surface, on the other hand, are considered hard materials.

Grinding Burns and Cracks

By far the most common problem during grinding is a loss of control of the heat generated at the surface by the abrading action. This occurrence may be caused by not replacing or not renovating the grinding wheel at the correct time, by using the wrong grade of wheel, or perhaps by removing too much stock at each grinding pass. If the heat generated at a carburized surface is great and is not transferred away to any significant extent, then the temperature of the surface rises to a level sufficient to induce tempering of the predominantly martensitic surface. Such local tempering is termed "overtemper burning." The hardness of an overtemper burn area is generally several equivalent HRC points lower than that of an unburned area. The extent depends on the temperature and the duration of heating. In addition, because tempering involves the precipitation of carbides (accompanied by a volume contraction), the local tempered areas are in a state of reduced residual compression or possibly in a state of residual tension.

However, if during grinding the surface temperature exceeds the Ac_3, thereby producing a thin layer of austenite, then because of the large heat-sink effect of the component, the austenite rapidly cools to give a hard, light-etching martensite. Such an induced defect is known as a "rehardening burn." Because a thermal gradient must exist between the zone that is made austenitic and the underlying material, such a rehardening burn must be surrounded by an intermediate layer of overtempered material (Fig. 8.1) and of reduced hardness (Fig. 8.2). Reasoning suggests that the rehardened zone should be in residual compression due to the martensitic expansion, while the adjacent tempered areas should experience residual tension. However, at the instant of burning and just after, the surface is subjected to appreciable mechanical forces via the action of the grinding wheel (the surface may even be ruptured by the force). Therefore, it is possible that mechanically induced surface stresses significantly influence the eventual residual stress distribution associated with a burn. Consequently, it is difficult to predict the final residual stress distribution in a burn area.

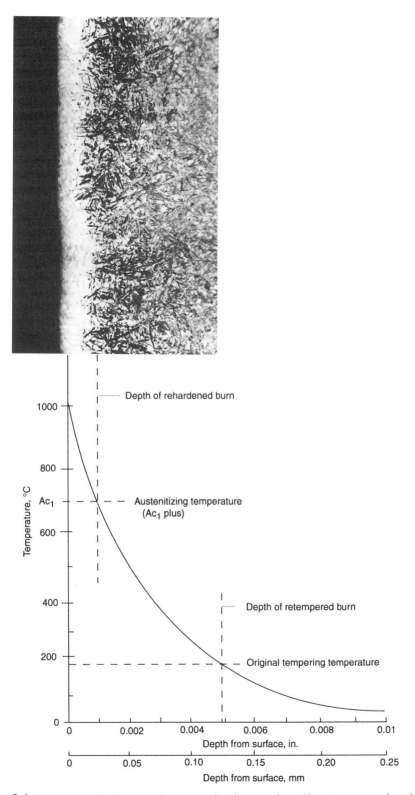

Fig. 8.1 Temperature distribution within a ground surface as indicated by microstructural modifications. The Ac temperature indicated is for slow heating; at high heating rates, e.g., in a grinding pass, the Ac temperatures will be elevated. Note: high-speed heating raises Ac temperatures. Micrograph, 500×

Another aspect to consider is the amount of plastic deformation induced in the immediate surface area by the action of the grinding wheel. Gentle grinding will produce a lightly deformed surface layer, which is desirable. Poor grinding techniques, on the other hand, will cause severe plastic deformation and smearing to a depth of about 10 µm. In extreme cases, smearing can produce laps or the heavily worked layer can rupture or even spall. In either case, there is damage that can develop into something more serious during service.

If the temperature rise accompanying deformation is sufficient to bring about a transformation to austenite, then metallographic examination will reveal layering within the rehardened part of the burn, thereby confirming that both plastic deformation and transformation have been involved in its formation. A heavily worked layer can be mistaken for a rehardening burn.

Grinding cracks are found in hard or hardened surfaces and are often associated with grinding burns. Narrow continuous or intermittent burn tracks are likely to produce essentially straight cracks (transverse to the grinding direction). Broad tracks or areas of burn, with appreciable surface overtempering, can produce surface network (mud) cracks. Grinding cracks form perpendicular to the grinding direction (Fig. 8.3a) and penetrate approximately at right angles to the surface. Also, the penetration of a crack is deeper, often many times deeper, than the depth of burning (Fig. 8.3b). This crack to burn depth relationship does not support the idea that cracking is caused solely by the stresses involved in overtempering and/or due to transformations taking place during the formation of a rehardening burn. If anything, relationship implies that tensile stresses above any residual stresses associated with the metallurgical events, or any thermal effects, are necessary to form such deep cracks. The tensile stresses that cause grinding cracks must therefore be caused by the forces exerted by the grinding wheel. In other words, the cracks form when the wheel is in contact with the workpiece. When a surface contains large amounts of retained austenite, a small crack may develop between a thin layer of martensite (created by the action of the grinding wheel) and the adjacent overtempered material during one grinding pass. This crack may become a much deeper grinding crack by the forces exerted during the next grinding wheel pass (Ref 5).

The small cracks referred to by Burnakov develop parallel to the surface at the interface between the rehardened and the overtempered layers. Similar cracks have been observed beneath heavily deformed surface layers produced during grinding. These cracks may result when the thin rehardened (or heavily worked) layer bows elastically to accommodate longitudinal tensile stresses, thereby increasing the radial tensile stresses. The cracks relieve those stresses.

In the absence of a crack from grinding, a burn is not necessarily harmful during service; it depends on the magnitude and direction of applied stresses. Corrosion is also a factor, because a corrosive environment might reduce the stress level required to produce a crack in a burned area.

Subsequent to the grinding operation, the presence of grinding burns is often not obvious, and therefore, some method is needed to determine the presence of burning (e.g., chemical etching according to MIL-STD-867A or equivalent). An example of grinding burns in the flanks of a gear, as detected by acid etching, is shown in Fig. 8.4. A burn formed in the first grinding pass may be completely removed during subsequent passes, whereas grinding cracks could still persist. The reason for this persistence is because grinding crack generally penetrate much deeper than burns.

Grinding cracks may or may not be obvious, and their detection is best achieved using a magnetic particle test. Grinding cracks can be removed by further grinding, assuming the tolerances on the workpiece permit it, although it is possible for the additional grinding to "chase" the crack to greater depths.

With abusive grinding, even an ideal surface can be ruined. With good grinding procedures, on the other hand, most case-hardened surfaces

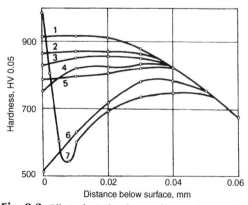

Fig. 8.2 Effect of grinding burns of increasing severity on microhardness. Curve 7 represents a rehardening burn. Source: Ref 3

can be ground without burning. However, microstructural features other than martensite present at the surface contribute to the degree of difficulty in achieving (or reestablishing) the correct grinding parameters. What is optimal for grinding martensite is not optimal for grinding austenite, bainite, and ferrite.

Effect of Grinding Variables

There is little reported work on the effect of grinding variables on residual stresses in case-hardened surfaces. However, a fair amount of information is available concerning through-hardened steels.

Influence on Depth of Cut (Feed). Increasing the depth of cut increases the depth to which the surface metal is deformed and in which heat is generated. In consequence, the depth of penetration of the ensuing residual stress distribution is also increased (Fig. 8.5). Excessive down feeds and cross feeds can cause burning and cracking.

Influence on Wheel Peripheral Speed. Schreiber (Ref 3) determined that wheel speed has no significance on residual stresses developed within a ground surface. Gormly (Ref 6), on the other hand, reported that a reduction of wheel speed causes a reduction of residual stresses, presumably because less heat is generated in the workpiece surface, which results in a reduced tempering effect. Conversely, a slower peripheral speed can lead to an inferior ground surface finish. It is considered more favorable to employ a softer wheel, because soft wheels tend to wear more readily, which exposes fresh abrasive cutting edges, and the residual stresses are confined more to the surface and to a lower magnitude.

Influence on Workpiece Speed. Within limits, an increase of workpiece speed can be beneficial, because it improves the feed of coolant, if used, into the wheel/workpiece contact zone. The contact time is reduced, and the quantity of heat removed is greater (Ref 3).

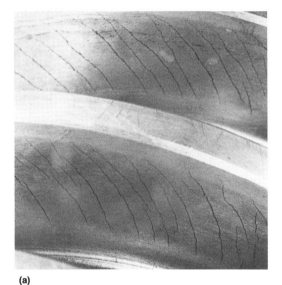
(a)

(b)

Fig. 8.3 Examples of grinding cracks. (a) Cracks on the flank of a worm thread. (b) Micrograph of grinding cracks in case-hardened 8620 steel showing several small cracks (arrows at right) that extended through the hardened case to the core, and the burned layer on surface (dark band indicated by arrow at left) that resulted from grinding burns. Note: nital and acidic ferric chloride are suitable etchants for grinding burns. Source: Ref 4

Influence on Wheel Grit and Hardness. Wheel selection for grinding is governed by the amount of stock to be removed from the workpiece surface and the quality of finish required. Two important wheel characteristics to consider are grit size and wheel hardness. The grit size (given as a number) quantifies the coarseness of the abrasive, whereas hardness (specified by a letter) refers to the hardness of the bond holding the grit.

Where relatively large amounts of stock are to be removed, a coarse grit wheel can be used initially, followed by a fine grit wheel to obtain the surface finish and size required. Finishing passes should remove any defects (assuming no cracks) induced by the roughing passes. Thus, the total thickness of material removed by the finishing passes should be roughly equivalent to the thickness of material removed by the last of the roughing passes.

The hardness of the wheel selected for a particular job is influenced by the hardness of the workpiece surface. According to Price (Ref 7), changes of wheel grade affect residual stresses induced by grinding less than the hardness of the workpiece surface; the latter controls the depth to which the grit penetrates. Hard wheels are employed for grinding softer steel surfaces, and soft wheels are better suited for grinding harder steel surfaces. Thus for case-hardened steels, wheel grades G, H, or even I may be used. The hardness of a wheel influences its self-sharpening characteristics. If the wheel surface loses abrasive particles during grinding, the bond must quickly wear away to expose fresh abrasive particles. Alternatively, if the abrasive particles become worn (dulled) by grinding, the bond must disintegrate (under the pressure of grinding) to expose fresh abrasive particles.

Fig. 8.4 Grinding cracks on the flanks of a small spur gear wheel.

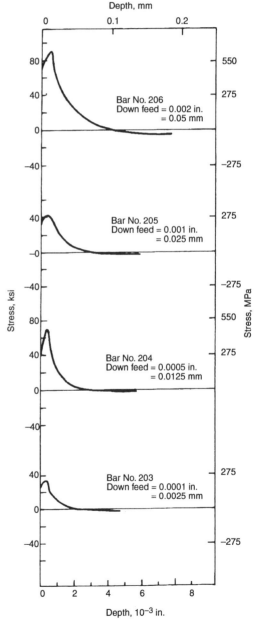

Fig. 8.5 Effect of depth of cut on residual stress distribution. Source: Ref 6

Conventional grinding refers to the process wherein the abrasive material (the grit) is aluminum oxide. During the many years that aluminum oxide grinding wheels have been utilized, their benefits and shortcomings have become fairly well understood. When the grinding technique is good, the product is good. On the other hand, deviations from good practice can lead to an inferior product or even expensive scrap. The importance of abrasive wheel dressing cannot be overemphasized. Continuing grinding after the wheel is worn beyond a certain level invites burning and cracking. The sound of the grind changes when the wheel is becoming (or has become) dulled or loaded, and often when that point is reached the workpiece may already have been burned. Note that it is important to ensure that the grinding wheel is true and balanced.

Cubic boron nitride (CBN) appears to have the potential to overcome many of the problems previously associated with aluminum oxide grinding. Cubic boron nitride is twice as hard as aluminum oxide; therefore, an exposed grain retains its cutting qualities better and longer. The wear rate of CBN is approximately fifty times less than that of aluminum oxide (Ref 8). This property alone makes for a more consistent ground product. The main benefit of CBN, however, stems from its high thermal conductivity relative to aluminum oxide. With aluminum oxide grinding, two thirds of the heat generated during the grinding action passes into the workpiece surface, whereas with CBN grinding, only about 4% of the heat produced by metal removal passes into the workpiece (Ref 9). This difference is to some extent reflected in Fig. 8.6, which compares the surface temperatures generated by grinding with aluminum oxide and CBN abrasives. In other words, CBN abrasives provide greater potential for controlling the thermal aspects of grinding; therefore, there is a better chance of avoiding temperature related problems, such as burning and cracking.

Even though CBN grinding wheels may be more tolerant to deviations from optimal grinding conditions, abusive grinding and damage are still possible. Therefore, for good quality grinding, the working procedures, equipment, and workmanship must each maintain a high standard, regardless of the type of abrasive used. Having selected a wheel specification for grinding case-hardened parts, it is wise to persevere with one supplier to avoid manufacturer to manufacturer variability.

Influence on Grinding Fluid. In wet grinding applications, the grinding fluid has three main functions. First, it is a lubricant that reduces the coefficient of friction between the abrasive wheel and the metallic surfaces; thus, frictional heat is controlled to some extent. Second, it acts as a coolant by extracting heat from the workpiece, although it does not much affect heat generation at the contact zone. Third, it helps keep the wheel clean and remove debris from the zone of action.

It is important to maintain the coolant in good condition, otherwise it might fail to adequately satisfy its functions and grind quality will suffer. Appropriate tests should be carried out periodically to ensure that the strength and the level of contaminants are maintained within set limits. Also, it is good practice to monitor the working temperature of the fluid. The careful choice of a fluid and a quality procedure to maintain good condition are meaningless unless the fluid is correctly directed and fed at a suitable rate to the working zone.

Degree of Difficulty. The term "abusive grinding" is often used when a ground surface is burned or cracked, and it may imply negligence on the part of the operator. This judgment is not altogether fair, because there are some design features that are difficult to grind and/or cool (e.g., the side walls of grooves or slots). Figure 8.7(a) shows the relative sizes of the grinding contacts at the side wall and at the bottom of a groove. With its small contact area, the bottom of the groove is easier to grind and cool than the side wall. At the side wall, because of the large

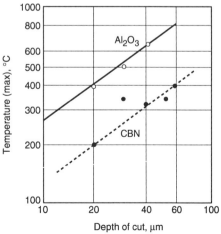

Fig. 8.6 Maximum surface temperature during dry grinding with aluminum oxide or cubic boron nitride (CBN) of a bearing steel. Source: Ref 10

contact area, much of the generated heat cannot escape except by passing into the workpiece surface, and the highest workpiece surface temperatures are reached at the exit side of the contact of the wheel. The high peripheral speed of the grinding wheel throws the coolant to the edge of the wheel, thereby favoring cooling at the bottom of the groove, but it works against effectively cooling the side wall.

In form grinding of gear teeth, although a similar process, all the surfaces of a tooth gash are ground at the same time, unlike the groove already described, which can be ground in three distinct operations (the base and the side walls individually). If the tooth profile is exactly the same as the grinding wheel profile, then the degree of difficulty is not too great. However, it is important not to over feed into the root (Fig. 8.7b). In practice, the as-case-hardened gear tooth does not have exactly the same profile as the grinding wheel. Both distortion and growth occur, and the degree of difficulty of grinding without damage is increased.

Influence of Workpiece Metallurgical Condition on Grinding. For standard grades of case-hardening steels, manufacturers generally aim to produce case-hardened surfaces of 0.75 to 0.95% C with tempered martensite along with small amounts of well-distributed retained austenite. Besides being a good all-round microstructure for most service conditions, martensite is probably the most suitable for successful grinding. The material should be tempered to at least 130 °C (265 °F) (preferably no more than 180 °C, or 355 °F) and to a maximum 60 HRC hardness (Ref 11). Low-temperature tempering

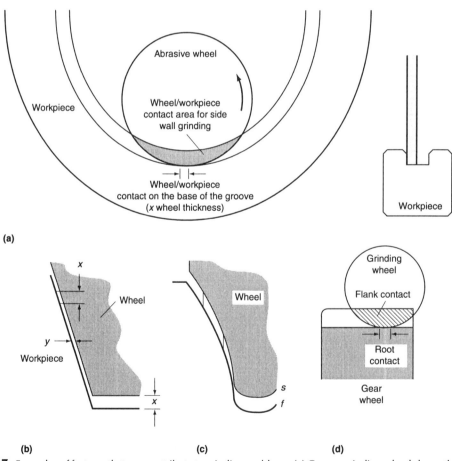

Fig. 8.7 Examples of features that can contribute to grinding problems. (a) Groove grinding wheel shows the differences of contact at groove base and at side wall. (b) For form grinding, a vertical feed of x at the root removes only a thickness y on the flank. The profile is simple, as in a rack. (c) For a gear tooth profile, s is the starting position of the grinding wheel and f is the final position. Therefore, the depth of cut at any point is represented by the gap between s and f, assuming that the gear tooth and wheel had the same original profiles. (d) Differences of contact area at tooth root and flanks present on a gear wheel.

in this range induces the precipitation of very fine coherent carbides, reduces the tetragonality of the martensite, and relieves to some degree both micro- and macroresidual stresses, all of which improve grindability.

The propensity for a case-hardened surface to burn during grinding increases as the surface carbon content in solution increases over about 0.5% C (Ref 5). It is as though the higher the carbon in solution in the martensite/austenite matrix is, the more ready carbon is to precipitate as carbides when excessively heated and strained during the grinding operation (i.e., increasing carbon in solution increases the instability of the surface).

Retained austenite in case-hardened surfaces contributes to the formation of grinding defects, such as burns, cracks, and poor finish. The conditions selected for commercial grinding are those most suitable for grinding martensitic surfaces and, therefore, are not optimal for grinding austenite. This departure from optimal only aggravates the problem. The main problem is that retained austenite is relatively soft and easily adheres to the abrasive particles (i.e., the grinding wheel quickly becomes "loaded"), which in turn favors excessive heat generation at the ground surface by deformation and friction. Burnakov et al. (Ref 5) found that with more than 20% austenite present in a surface, there is a greater tendency to crack than with only 5% austenite. They also determined that the crack faces have brittle features and are transcrystalline, and that fracture surfaces are not, in this instance, associated with carbide films or inclusions. For successful grinding, therefore, the retained austenite content of a case-hardened surface is best kept low.

Surface oxides and internal oxidation, or the HTTP associated with it, can cause the wheel to become "loaded" and make grinding difficult.

Fine, well-dispersed spheroidized carbides do not greatly influence the response of a surface to grinding. Such carbides might even be beneficial, because they tie up some of the carbon, which means that the matrix has a lower carbon content and, therefore, less retained austenite.

Coarse carbide particles and heavy network carbides in the surfaces of case-hardened lean-alloy steels would be expected to make grinding more difficult and perhaps contribute to cracking during grinding (Ref 12, 13). However, the overriding factor regarding burning and cracking must be the grinding conditions (wheel type, balance, feed, speed, periods between wheel dressing, etc.). Carbides might have an influence on grindability because of the several differences in properties and behaviors between carbides and the matrix material (tempered martensite). These microstructural features have different thermal conductivities so they will influence the conduction of heat during grinding; carbides, if anything, have a negative effect on the conduction of heat deeper into the surface.

The carbide is an intermetallic material which is harder and less ductile than the metallic matrix. During a grinding pass, the abrasive particle ploughs through the matrix, but tends to impact the carbide and either be dulled, shattered, or dislodged by the carbide.

Loss of abrasive particles and abrasive dulling each impair the grinding efficiency of the abrasive wheel, particularly if the wheel bond material is too hard. A reasonably soft wheel chosen for grinding martensite will likely be suitable for grinding martensite containing some carbide, but unsuitable for grinding predominantly carbide surfaces. Therefore, coarse and network carbides at the immediate surface of a part negatively influence the grinding efficiency and ground surface quality. How negative the influence is depends on how much carbide is present.

Summary. For case-hardened low-alloy steels, tempered martensite structures have the highest threshold against burning and cracking. Untempered martensites have a much lower threshold against burning and cracking. Surfaces with significant amounts of carbides have a lower threshold than an entirely martensitic surface but should grind better than a high-austenite surface. However, any case-hardened surface can be ground satisfactorily provided that the grinding parameters are correct for whatever the surface condition is and that the wheel dressing is carried out at the right time.

Residual Stresses Caused by Grinding

Gormly (Ref 6) categorized the measured residual stress distributions into three types (Fig. 8.8). Type I represents abusive grinding, when conditions become such that surface burning is likely. The residual stresses at the surface are decidedly tensile, although if cracking takes place, the residual stresses will likely be relieved to some extent. Type III, on the other hand, occurs when an extremely good grinding technique is employed. The surface residual stresses are compressive. The factors involved in surface heat generation are controlled sufficiently to curb microstructural changes; therefore, only the mechanical effect (surface deformation) occurs. Such a stress distribution does not impair the fatigue resistance of the surface and can, indeed, improve it. The type II curve suggests that heat has been generated to produce a tensile peak. Conversely, the effect of plastic deformation at the immediate surface has predominated and induced surface compressive stresses sufficient to more than counter the heat-induced tensile stresses. It is likely that the type II curve is more typical of production grinding in general, although drifting conditions (e.g., the wheel becoming duller) can lead to a type I situation.

Peak stresses are mainly found less than 0.025 mm (0.001 in.) from the surface, and in this layer, extreme changes are likely (Fig. 8.9). The magnitude of the stresses (of whatever sign) is determined by the direction of grinding. The total depth over which the stresses extend is about 0.2 mm, and in the affected layer, the tensile and compressive stresses should be in balance. The residual stresses parallel to the direction of grinding are more tensile than those at right angles to the direction (Fig. 8.10), which is a common trend with grinding. In Fig. 8.10, the tensile residual stresses are not balanced by compressive residual stresses, therefore, it is assumed that beyond about 0.2 mm, the stresses are compressive.

Case-hardened steels have a residual stress distribution extending through their surfaces into the core due solely to heat treatment processes; therefore, any residual stresses induced by grinding must somehow superimpose themselves on what was originally there. Good grinding should, therefore, increase the compression at the surface. It could also cancel some of the original compressive stresses just beneath the surface (Fig. 8.11). When grinding is abusive, the residual stress distribution beyond 0.2 mm can be affected (Fig. 8.12). Here, decreasing the number of grinding passes from 7 to 2 to remove 0.3 mm of stock increased appreciably the tensile residual stresses at the surface and appears to have modified the whole of the stress distribution.

Grinding with CBN wheels increases the compressive residual stresses at the surface as a result of surface deformation to a depth of about 10 μm without any significant heat generation. Differences of stress magnitudes, if not distributions, might be expected depending on the wheel properties (grit size and bonding, electroplated or resin bonded) and condition. For example, a freshly dressed wheel did not develop as much surface compression as a pre-used wheel (Fig. 8.13a and b).

The as-ground residual stress distribution can be modified by aging, tempering, and any postgrinding mechanical treatment, such as peening or rolling.

Effect of Grinding on Fatigue Strength

Influence on Bending Fatigue. Sagaradze and Malygina (Ref 16) determined the fatigue limit for case-hardened 8 mm diameter test pieces of a 20Kh2N4A steel (case depth, 1.35 mm) as 850 MPa (123 ksi). Burn-free grinding of comparable test pieces after carburizing and before hardening increased the fatigue limit after hardening to 990 MPa (143 ksi). This implies that the removal of surface defects introduced during carburizing was largely responsible for the increase of fatigue strength. When burn-free grinding was administered after carburizing and hardening, the fatigue limit was raised even further to 1090 MPa (158 ksi). The removal of adverse metallurgical features and surface roughening, which arise during the carburizing and hardening, is

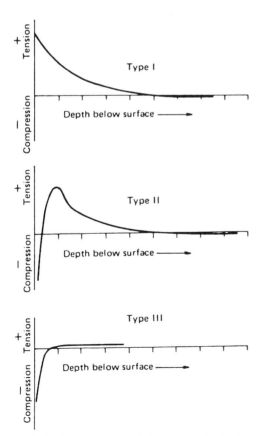

Fig. 8.8 Three types of grinding stress distribution. Source: Ref 6

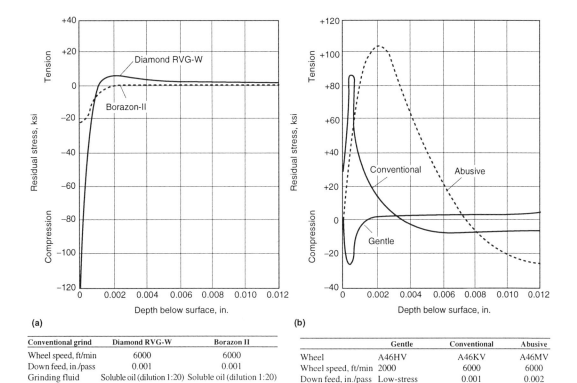

Fig. 8.9 Residual stresses in SAE 4340 steel (quenched and tempered, 50 HRC) after grinding (a) with CBN and diamond and (b) with alumina. Source: Ref 14

clearly involved in the improvement of the fatigue limit. It is also probable that a thin, worked surface layer due to a good grinding technique contributes to the improvement.

Kimmet and Dodd (Ref 18) demonstrated how the bending fatigue strength of a gear can be appreciably enhanced by fillet grinding with a CBN abrasive; Fig. 8.14 compares the unground with the ground condition. Drago (Ref 8), referring to bending fatigue tests on helicopter gears, concluded that there is little to choose between CBN tooth fillet grinding and conventional aluminum oxide grinding. This comparison, however, is based on a situation when high quality grinding conditions were utilized for both the CBN and the aluminum oxide grinding operations. Navarro (Ref 14) essentially confirmed the foregoing but illustrated how more severe grinding can significantly impair fatigue strength (Fig. 8.15).

In commercial gear-tooth fillet grinding where the emphasis is on output, the grinding conditions can be less than gentle; hence, there is a tendency to produce inferior quality components.

For example, trials on a production basis showed that gear-tooth fillet grinding reduces bending fatigue strength by 11 to 45% when compared to unground gears (Ref 15). The variability depends on how much of the tooth fillet was removed by grinding. This reduction of fatigue strength is confirmed by MIRA tests on 7 dp EN36 (BS 970, 665M13) spur gears. Those gears with

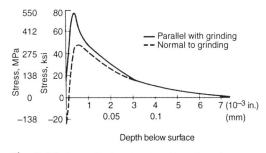

Fig. 8.10 Example showing that although peak stresses caused by grinding are located close to the surface, the balancing stresses can extend relatively deeply. Source: Ref 15

ground roots had fatigue strengths 25 to 50% below those with unground roots (Ref 19). When cracks are produced during grinding, the loss of fatigue strength can be as high as 66% (Ref 11).

The need to remove too much stock has an adverse effect on the fatigue life of the ground component, as Fig. 8.16 suggests for root ground gears. With flank-only ground gear teeth, the formation of a "step" in or along the edge of a tooth fillet can be detrimental to the bending fatigue strength of the gear.

With through-hardened steels, the coarseness of the ground finish and the direction of grinding subsequent to heat treatment affect the bending fatigue strength of the component. Coarse grinding where the grinding grooves are transverse to the direction of service loading is likely to be detrimental, whereas longitudinal grinding (i.e., the grooves are in the same direction as loading) would, if anything, have a much smaller effect.

With case-hardened surfaces, the situation is less clear. Tyrovskii and Shifrin (Ref 20) showed that transverse grinding before case hardening does not affect the bending fatigue strength of gears and plates. With longitudinal grinding, the rougher the surface (within the range of their tests) is, the higher the fatigue limit will be. However, the surface finish created by grinding after case hardening needs to be good.

If the conditions produce an acceptable work-hardened layer at the surface and are sufficient to cause any eventual failure to initiate beneath that layer, then the fatigue strength is enhanced. Table 8.1 conveys how modest changes in grinding conditions can affect both the residual stresses and the fatigue life. Although these data are not derived for case-hardened samples, the points are nevertheless still valid. Samples of test group AF had a type II verging on type I residual stress distribution (see Fig. 8.8). The only difference between groups AA and AB is that the grinding machine was overhauled between runs.

Influence on Contact Fatigue. Often the roots of case-hardened gears are not ground, whereas the tooth flanks are ground mainly to promote good line contact between meshing teeth (i.e., by removing general distortions and high spots on the contacting surfaces). For this reason, grinding is beneficial; it gives the teeth of a gear basically the same shape, pitch, and surface finish. Interestingly, though, Sheehan and Howes (Ref 21), using disk tests to determine the contact-fatigue strength

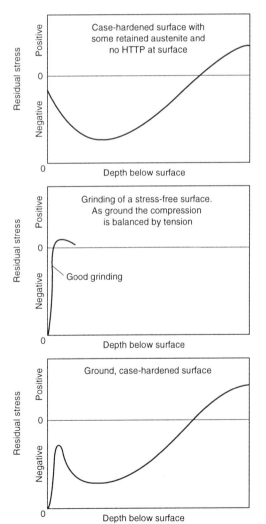

Fig. 8.11 Removal of a surface layer in compression modifies slightly the whole residual stress curve. Any stresses introduced by grinding further modify the curve

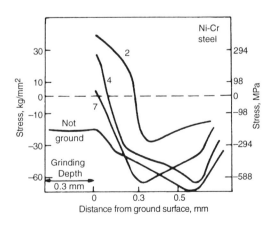

Fig. 8.12 Effect of the number of grinding passes on the residual stress distribution in case-hardened strips with a case depth of 1.6 to 1.7 mm. Source: Ref 16

of case-hardened surfaces, show that irrespective of surface carbon content and quenching technique, an as-heat-treated surface is superior to a heat-treated and ground surface. Taking all factors into consideration, the difference probably occurs because the unground surface supports a more stable oil film than the ground surface.

In a later work (Ref 22), Sheehan and Howes concluded that contact fatigue is influenced by the surface roughness of the two mating surfaces, the sign of the sliding action (positive or negative), and the type of lubricant. When the roughness of the loading member with positive slip approaches the oil film thickness (0.1 to 0.2 µm), the load carrying capacity increases appreciably. Thus, polished surfaces are superior to ground surfaces. With respect to lubricant type, the contact fatigue strength of ground surfaces is favored by base oil, whereas the best results for polished surfaces are obtained when an extreme pressure lubricant is used.

Nakanishi et al. (Ref 23), taking the onset of gray staining (micropitting) as the failure criterion, determined that initial surface roughness has an effect on surface durability (Fig. 8.17). The tangential load for surface-hardened gear teeth with a surface roughness (R_{max}) of ~1 µm is about ten times that of teeth where R_{max} is ~4 µm.

The presence of grinding burns might not have much effect in pure rolling contact situations, because the main stresses are subsurface. However, in slide/roll situations, the surface condition and the surface hardness are important, so that grinding burns and the changes of hardness associated with them have an adverse effect.

Influence on Wear. The influence of the as-ground roughness on wear of surfaces in relative motion depends on whether the combined roughness of the surfaces in contact exceed the oil film thickness. If the surfaces are kept apart by the lubricant, then essentially no wear can take place, assuming an absence of debris pass-

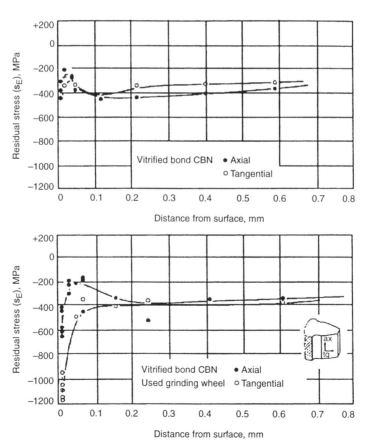

Fig. 8.13 Influence of wheel condition on the residual stress distribution. Infeeds (mm) and feed rates (mm/rev) for the used wheel were less than those for the newly dressed wheel. ax, axial; tg, tangential

ing through the gap. Where the combined roughness exceeds the oil film thickness so that the asperities of one surface make contact asperities with the other surface, then surface wear and deformation can take place. Thus, smoothing and surface conditioning (running-in) occur, normally without too much concern. If adhesive wear develops, which is dependent on a number of factors, then there is a need for concern. Therefore, the roughness of the ground finish can have a bearing on the wear processes, regardless of the method of grinding (aluminum oxide or CBN). If, however, the grinding has been on the abusive side, with heat generation sufficient to temper the surface, then both abrasive and adhesive wear processes are more likely to ensue during sliding contact.

Roller Burnishing

Rolling, or roller burnishing, is a metalworking technique employed to locally strengthen the surface of the component (particularly at fillets and in grooves) in much the same way that shot peening does. With peening, small projectiles are directed at the workpiece surface, whereas with burnishing a rolling force is applied to the surface using either rollers or spherical bearings. In both processes, the surface is worked, and the surface residual stress distribution is rendered more favorable.

Effect on Microstructure

When a case-hardened surface is cold worked, its microstructure is modified to a depth and degree dependent on the specific working conditions applied. In general, adequate cold working induces hardening in the worked layers. Excessive working with heavy pressures can induce microcracks, which might develop into microtears or even flaking at the surface. These defects are less likely to occur in hard materials; nevertheless, their formation is to be avoided, even though the fatigue limit could still be better than that of an unburnished counterpart (Ref 24).

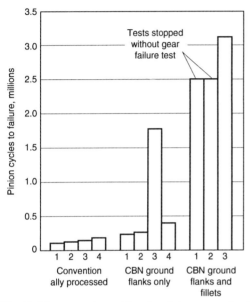

Fig. 8.14 Comparison of bending fatigue strength of conventionally processed (cut/harden/lap) versus CBN ground (cut/harden/lap) spiral bevel gears. Test gear design specifications: hypoid design, 4.286 dp, 11 by 45 ratio, 1.60 in. face. Gears were installed in axles using a 4-square loaded axle test machine. Torque applied was 70% of full axle torque rating. Source: Ref 18

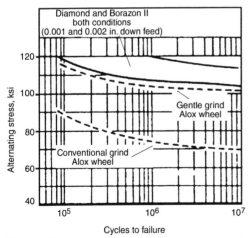

Fig. 8.15 S-N data for SAE 4340 steel ground with various abrasives. AISI 4340 conditions: quenched and tempered to 50 HRC, surface grinding, cantilever bending, zero mean stress, 75 °C. Source: Ref 14

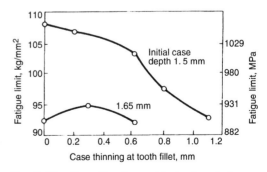

Fig. 8.16 Effect of local case thinning by grinding on the bending fatigue strength of Ni-Cr steel gear teeth. Source: Ref 16

Table 8.1 Effect on grinding conditions of residual stress and fatigue

Test group	Wheel grade(a)	Down feed mm	Down feed 10^{-3} in.	Grinding fluid	Residual stress Surface MPa	Residual stress Surface 10^3 lb/in.^{2}	Residual stress Peak MPa	Residual stress Peak 10^3 lb/in.^{2}	Fatigue limit MPa	Fatigue limit 10^3 lb/in.^{2}	Failure point
AA	H	0.0025	0.1	Soluble oil	90	13	159	23	503	73	Surface
AB	H	0.0025	0.1	Soluble oil	−172	−25	124	18	483	70	Surface
BC	I	0.025	1.0	Soluble oil	234	34	338	49	503	73	Surface
AC	M	0.05	2.0	Soluble oil	−200	−29	841	122	427	62	Surface
AF	I	0.025	1.0	Grinding oil 1	−662	−96	7	1	683	99	Subsurface
AG	I	0.025	1.0	Grinding oil 2	−600	−87	145	21	627	91	Subsurface
BA	I	0.025	1.0	Grinding oil 2	(b)	(b)	(b)	(b)	634	92	Subsurface
AD	M	0.025	1.0	Grinding oil 1	−462	−67	83	12	503	73	Surface
AE	M	0.05	2.0	Grinding oil 1	−738	−107	221	32	483	70	Surface

Steel: modified AISI 52100, 59 HRC. Wheels: white, vitrified bond, friable aluminum oxide. (a) I, wheel grade normally used; H, softer grade; M, harder grade. (b) Comparable with AG. Source: Ref 4

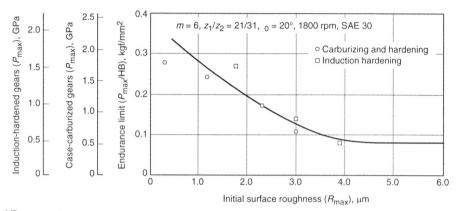

Fig. 8.17 Effect of surface roughness on the surface durability of surface-hardened gears (P_{max}/HB – R_{max} curves). Source: Ref 23

In deformed martensite-austenite structures, changes of orientation, slip line development in both constituents (Ref 25), and a decrease in the average martensite plate size (Ref 26) have been observed as evidence of the cold-working process.

At high rolling pressures, some of the austenite is transformed to martensite (Fig. 8.18), and this transformation, in part, could account for the reduction of the average martensite plate size. Figure 8.18 shows that the change in the amount of retained austenite close to the surface (0.1 mm) caused by high rolling pressures falls from around 20% down to about 5%. Papshev (Ref 26) observed austenite reductions in a case-hardened Ni-Cr-Mo steel from 30 to 13.5% and from 45 to 16% austenite. In tensile and bend tests, Krotine et al. (Ref 28) noted that straining reduces austenite and increases the martensite at the surfaces of carburized and hardened test pieces. There is an indication that the reduction of austenite content is greatest with higher nickel steels (Table 8.2).

Finely dispersed carbides have been observed within deformed austenite volumes, which should hinder slip along the slip planes. Razim (Ref 25), on the other hand, reasoned that similar precipitates observed in the deformed contact tracks of surface fatigue specimens are too large (probably incoherent) to hinder slip and, therefore, would not contribute to

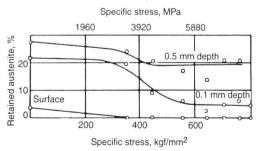

Fig. 8.18 Influence of roller burnishing on the retained austenite content at and beneath the surface of a case-hardened steel. Source: Ref 27

Table 8.2 Effect of straining on surface austenite content

Steel	Nickel, %	Pretest austenite, %	Posttest austenite, %
SAE 4080	0.1	19	14
SAE 4095	...	22	17
95MnCr5	...	33	24
105MnCr5	...	42	36
SAE 4675	1.65	16	4
SAE 4685	1.65	23	7
SAE 4875	3.42	23	6
SAE 4885	3.42	31	10

Source: Ref 28

the total hardening effect. The reason for this precipitation is not clear; it could be caused by straining only. On the other hand, much of the energy that causes plastic deformation is expended as heat. At very high rolling speeds and feeds, temperatures as high as 600 to 650 °C have been recorded (Ref 26), although normally the temperature is contained between 150 and 350 °C to obtain optimal resultant residual stresses. It is inferred that the precipitation could be caused by both thermal and mechanical means.

In terms of dislocations, plastic deformation increases the dislocation density and reduces the subgrain size. Balter (Ref 24) considered that work hardening of well-tempered martensite structures is solely the result of an additional hardening effect caused by the interaction of dislocations with the interstitial atoms, mainly carbon.

Effects on Material Properties

Influence on Hardness. Below a critical value of contact stress, roller burnishing does not affect the hardness of the material at the surface. Above the critical value, given as 3480 MPa (355 kg/mm²) for a case-hardened 20Kh2N4A steel (Ref 23), work hardening can be detected. The affected depth is only about 1 mm, whereas at high pressures (e.g., 7845 MPa, or 800 kg/mm²), it exceeds 1 mm. Figure 8.19 shows the influence of contact stress on hardness and depth. With high contact stress, the peak hardness is not at the surface but 0.05 to 0.1 mm beneath it. The greatest hardening effect is observed in martensite-austenite materials, and the higher the carbon content or the greater the initial hardness is, then the greater the hardening effect will be (Fig. 8.20).

Influence on Residual Stresses. Surface work hardening has a very marked influence on the residual stress distribution within a case-hardened surface (Fig. 8.21). In Fig. 8.21, a maximum specific stress increases compression at the surface. The danger (although not too obvious from the figure) is the magnitude of balancing tensile stress and its location beneath the case. Figure 8.22 shows how the number of passes affects the residual stress distribution, whereas Fig. 8.23 suggests the trend resulting from increasing the rolling speed (i.e., the higher the rolling speed, the lower the surface residual stress).

The value of the residual stress must relate to the changes incurred, which must, to some extent, depend on the initial structure. For example, Fig. 8.24 shows how the amount of retained austenite initially residing near the surface of a case-hardened 14Kh2N3MA steel influences the final residual stress distribution. In this instance, the surface containing the largest quantity of retained austenite developed the highest value of residual compressive stress due to roller burnishing. After burnishing, the amount

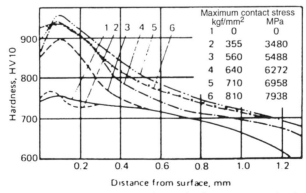

Fig. 8.19 Effect of roller burnishing of a 20Kh2N4A steel (case depth, 1.1 to 1.5 mm) on hardness and depth of hardening for various rolling pressures (maximum contact stress). Source: Ref 27

of austenite surviving in each was only about 15%.

Influence on Bending Fatigue. The bending fatigue strength reaches a maximum at an intermediate value of rolling pressure (Fig. 8.25), increasing about 22% for the material burnished with optimal process conditions compared with the unburnished one. Figure 8.25 represents results from unnotched test pieces, and it is conceivable that where stress concentrators exist (fillets, etc.), the improvement might even be better. This conjecture was correct with through-hardened test pieces (Ref 29), where the improvement was greater than 100%.

Influence on Contact Fatigue. In both laboratory and field tests (Ref 25), surface cold working improved the lives of case-hardened drill bits and bearing race grooves by retarding the formation and the development of contact fatigue cracks.

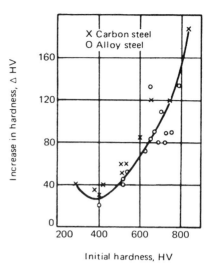

Fig. 8.20 Relationship between initial hardness and change of hardness due to roller burnishing. Source: Ref 26

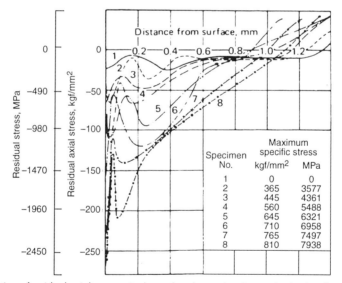

Fig. 8.21 Distribution of residual axial stresses in the surface layer of carburized cylindrical specimens subjected to roller burnishing at various maximum specific stress levels. Source: Ref 27

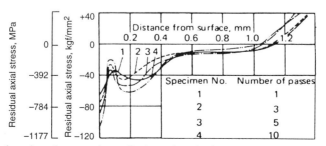

Fig. 8.22 Influence of number of passes when roller burnishing (with maximum specific stress of 4364 MPa, or 445 kgf/mm^2) on residual axial stresses in the surface layer of specimens. Source: Ref 27

Influence on Wear. Roller burnished surfaces are more resistant to wear than in the as-ground condition (Table 8.3).

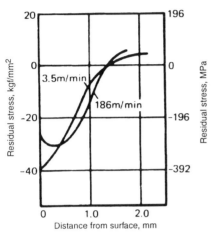

Fig. 8.23 Effect of rolling speed on the residual stress distribution at the surface of a 0.45% C steel. Source: Ref 26

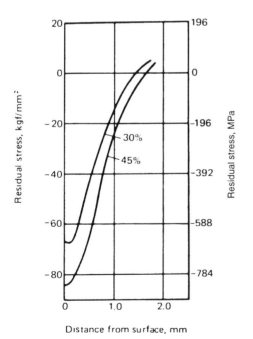

Fig. 8.24 An indication of how retained austenite content influences the residual stress distribution of a case-hardened and burnished Ni-Cr-Mo steel. Note: burnishing reduced the 30% austenite to 13.5% and the 45% austenite to 16%. Source: Ref 26

Shot Peening

Shot peening is a process in which the component surface is bombarded with a multitude of small spherical projectiles (shot) with sufficient velocity to produce a minute indentation with each impingement (a discrete zone of plastic deformation). The rate and duration of the bombardment need to be sufficient to saturate the entire surface of the target area with overlapping impingements. The treated surface is evenly cold worked to a uniform but shallow depth, and the hardness and strength of the material in that layer are increased. The primary purpose of shot peening, however, is to induce compressive residual stresses at the surface, which, among other things, improves the fatigue resistance. Shot peening is a strength improvement process in its own right, and it need only be applied to critical areas where peak stresses are anticipated during service, such as in the radius at a change of section of a multidiameter shaft or at the tooth fillets of gears. Shot peening is not to be confused with the overall cleaning process of shot blasting, which, although similar, does not have the same control or the same objectives.

Process Control

In order to obtain the maximum benefits from shot peening, it is necessary to maintain a strict

Table 8.3 Effect of burnishing on relative wear

Workpiece condition	Burnishing pressure MPa	kgf/mm²	Relative wear, %
Ground	...	...	100
Ground and burnished	1472	150	66
Ground and burnished	2256	230	67
Ground and burnished	2747	280	50
Ground and burnished	2943	300	55

Tests conducted on steel 14Kh2N3MA, 57 to 61 HRC before and 3 to 5 units harder after burnishing. Testing condition: rolling friction in clay solution. Source: Ref 26

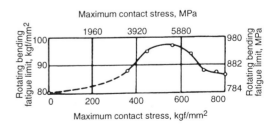

Fig. 8.25 Relationship between fatigue limit at 10^7 cycles and contact stress for case-hardened 20Kh2N4A test pieces (7.5 mm diam, 1.1 to 1.5 mm case depth). Source: Ref 27

policy of good "housekeeping" and tight process control. The type, hardness, size, shape, grading, and condition of the shot; its impact velocity and impingement angle; exposure time; nozzle size; and nozzle to workpiece distance must be carefully controlled at all times to ensure optimal results and reproducibility.

Machine settings for peening a particular workpiece are derived from peening standard test pieces (Almen strips) and also from past experiences of how shot-peened parts have behaved under test or service conditions. Details of the test strips and their use can be obtained from relevant specifications, such as SAE J442 and J443. Briefly, a set of test strips of appropriate thickness and hardness are attached to blocks and subjected to peening on one face for different exposure times. After peening and removal from the blocks, each strip deflects to a degree related to the induced residual stresses in the peened face. These deflections (arc heights) are then measured and plotted in terms of arc-height against exposure time to produce a curve (Fig. 8.26). From this information, the basis for the correct process settings can be derived.

A case-hardened surface is relatively resistant to indentation. When it is struck by a spherical shot, the indented area is smaller than a comparable collision area on the surface of an Almen strip (44 to 50 HRC). Therefore, for complete saturation of the surface with overlapping indentations, the exposure time for a case-hardened part is longer than for the Almen strip. It is important, nevertheless, not to overpeen to where small surface ruptures are induced. Such a condition can be detrimental to the service life of a component. The unexpected presence of a soft "skin" on a case-hardened surface (e.g., from HTTP that may accompany internal oxidation) could encourage overpeening. In such an instance one might consider grit blasting first to remove some, if not all, of the soft layer followed by a controlled shot peening.

The shot used for case-hardened parts should have equivalent, if not greater, hardness than that of the surface being peened. Thus, the intended increase of compressive residual stresses and the required consistency of peening will be achieved (Fig. 8.27). A softer shot might itself deform, assume an irregular shape, and thereby reduce the effectiveness of the process. The size of the shot must be considered in connection with the geometry of the surface to be peened. No improvement is obtained if, in the critical section of the component, there is a radius smaller than that of the shot being used; the shot must have good access.

Effect on Microstructure

The microstructural features resulting from shot peening mainly involve plastic deformation and are essentially the same as those described

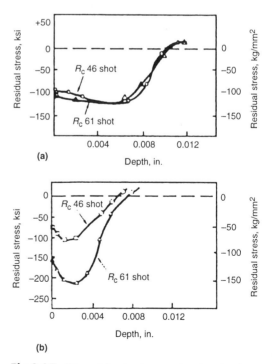

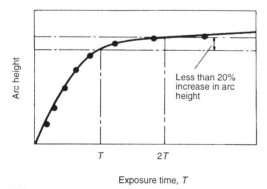

Fig. 8.26 Typical saturation curve from shot-peened Almen strips. The time to saturation is that which, when doubled, does not produce an increase of arc height greater than 20%.

Fig 8.27 Effect of shot hardness and surface hardness on the distribution of residual stresses. (a) 1045 steel hardened to R_c 48. (b) 1045 steel hardened to R_c 62 peened with 330 shot. Source: Ref 30

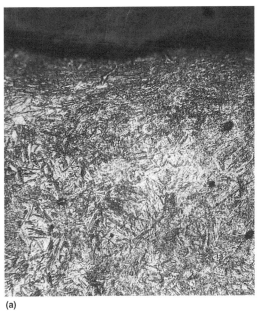
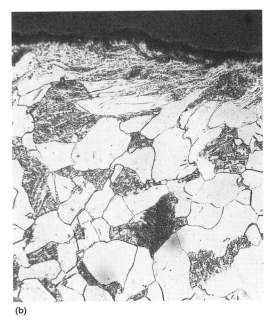

Fig. 8.28 Plastic deformation produced at the (a) case-hardened surface and (b) non-case-hardened surface of shot-peened steels. Both 270×

for roller burnishing without any specific directionality. Further, just as over burnishing can produce surface defects, so too can over-peening. The modifications due to shot peening generally are in the first 0.25 mm depth of surface. Micrographs of peened surfaces are presented in Fig. 8.28.

Effect on Material Properties

Influence on Hardness. The surface hardness is influenced by peening (Fig. 8.29). In this instance, the initial material has a high retained austenite content of about 80%; it is soft and responds readily to cold working by both peening and fatigue stressing. Case-hardened surfaces containing typical amounts of retained austenite (less than 30%) and having hardnesses in excess of 59 HRC increase in hardness by 1 to 2 HRC when peened. However, such an increase can be accompanied by a slight softening beneath the peened layer, which might reflect on a modification to the residual stresses near to the surface.

Influence on Residual Stresses. Surface deformation by peening increases the surface area. However, because the worked layer must remain coherent with the underlying, undeformed material, a new stress distribution develops in which the surface is in residual compression and the subsurface is in tension. Such is the case when a material is initially in a stress-free condition. Carburized and hardened parts already have compressive-residual stresses in their surfaces, and peening results in increased compression at the peened surface and reduced compression beneath the peened layer (Fig. 8.30).

The condition of the initial material influences the eventual (as peened) residual stress distribution. Figure 8.31 shows this for a case-hardened steel in the tempered and untempered conditions. Even so, in this example, the immediate surface compression for each condition is essentially the same.

The maximum compressive stress achieved by peening is related to the tensile strength of the

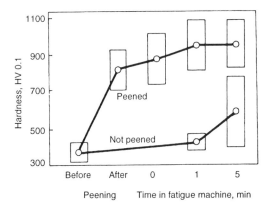

Fig. 8.29 Effect of shot peening and fatigue stressing on surface hardness. Source: Ref 31

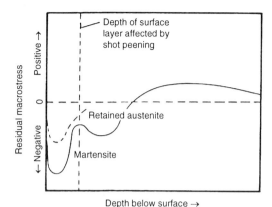

Fig. 8.30 Effect of shot peening on residual macrostress distributions in a carburized surface (initially with tensile residual stress at the surface). Source: Ref 32

Table 8.4 Effect of peening intensity on fatigue resistance

Peening intensity	Fatigue life at a given load, cycles	
	Mean	B-10
Unpeened	9,850	56
12A	16,571	7,708
18A	23,800	11,512
24A	33,167	11,953
8C	24,750	3,568
24A (conventional)	16,375	2,663

Test material: SAE 4023, carburized and quenched, 200 °C temper, 58 HRC. Peening: 110% saturation for 2 × 100% Almen saturation. Source: Ref 34

Table 8.5 Fatigue limits determined for carburized steel specimens

Condition	Fatigue limit, MPa
EN353	
Carburized/quenched	621
Carburized/quenched + subzero treatment	542
Carburized/quenched + 0.008A shot peen	686
Carburized/quenched + 0.014A shot peen	718
Carburized/quenched + 0.025A shot peen	686
EN16	
Carburized/quenched	542
Carburized/quenched + 0.008A shot peen	605
Carburized/quenched + 0.014A shot peen	671
Carburized/quenched + 0.025A shot peen	608

EN353 composition: 0.6 to 0.9 Mn, 0.8 to 1.2 Cr, 0.1 to 0.2 Mo, 1.2 to 1.7 Ni. EN16 composition: 0.32 to 0.4 C, 1.3 to 1.7 Mn, 0.22 to 0.32 Mo. Source: Ref 32

material, and it is typically 0.5 to 0.6 times the ultimate tensile stress (Ref 30). The increase of surface and near-surface compressive residual stresses in a given steel is determined by the process parameters. For instance, Table 8.4 shows the effect of peening intensity, including an example of slight overpeening, on both the residual stresses and the mean fatigue life (low cycle). Table 8.5 provides data relating to peening intensity and fatigue limit (high cycle). Figure 8.27 shows the need to use hard shot for case-

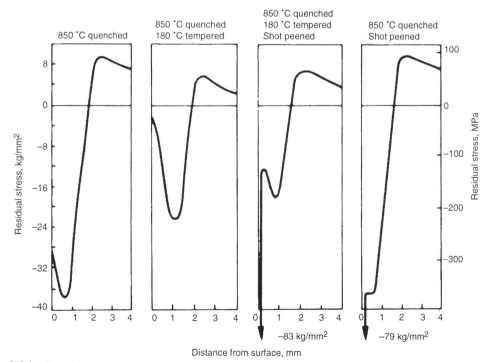

Fig. 8.31 Effect of shot peening on the residual stress distribution in 20KhNM steel rings (case depth, 1 mm). Source: Ref 33

hardened surfaces if the desired residual stress distributions are to be achieved.

Influence on Fatigue. Whereas the axial fatigue strength of a part is unlikely to be improved by controlled shot peening, the bending fatigue limit is improved (Fig. 8.32) about 20% (Ref 33, 36, 37). However, the current understanding of, equipment for, and refinement of the shot-peening process indicates that much higher quality products are possible, consistent with extremely good fatigue lives in the high-cycle regime. Low-cycle fatigue strength is not greatly affected by peening (Fig. 8.33).

The influence of peening intensity on fatigue resistance is presented in Table 8.4. The effect of peening duration is shown in Fig. 8.34, where the fatigue limit increases with peening time, although at greater times it levels off before diminishing when the point of overpeening is reached.

Influence on Contact Fatigue. There are few data regarding the effect of peening on the contact fatigue resistance of case-hardened parts. Further, it is difficult to predict what effect peening might have. For contact loading situations, the surface roughness plays an important role, and for surfaces roughened by peening, the asperities might penetrate the lubricating oil film, thereby encouraging contact damage. By removing or reducing the roughness by polishing, for example, the real benefit of peening can be realized. Nevertheless, the presence of compression at the peened and polished surface should inhibit failures that normally initiate at the immediate surface from sliding.

Gerasimova and Ryzhov (Ref 40), however, found no correlation between contact endurance and residual stresses or hardness; improvement is more influenced by subgrain size and dislocation stability. Tempering further improves the contact fatigue strength of peened parts, which is improved yet again by electropolishing to remove 12 μm (to remove surface roughness). Altogether, the combined processes of peening, tempering, and electropolishing raise the contact endurance by several hundred percent. A NASA study (Ref 41) concluded that the contact fatigue life is improved 60% by shot peening, presumably by strengthening the surface against sliding damage. Overpeening, on the other hand, is expected to have a detrimental effect on contact-fatigue strength because it weakens the surface and encourages sliding damage. Shot peening has no beneficial influence on deep spalling fatigue processes.

Shot Peening for Reclamation. To some degree, shot peening might be employed as a corrective treatment. The presence of HTTP associated with internal oxidation has already been mentioned. Other features, such as grinding damage or decarburization, which can render a part unfit for service, are still potential candidates for peening. If the extent of the defect is known, then with the approval of the design engineer, it may be possible to salvage the defective item by shot peening. With respect to damage caused by abusive

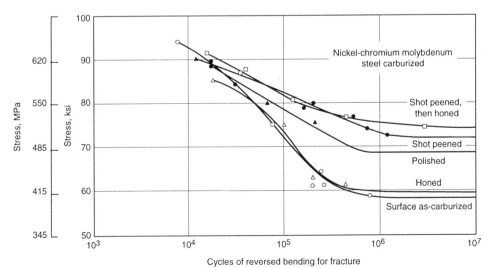

Fig. 8.32 Effect of peening, honing, and polishing on the reversed bending fatigue strength of a carburized alloy steel. Source: Ref 35

grinding, Fig. 8.35 show that reclamation by peening is possible. Figure 8.36, which considers decarburization, also suggests that, to a point, reclamation might be possible.

Apart from reclamation work, peening can be applied between the carburizing and the hardening stages (while the case is relatively soft) to remove internal oxidation, to smooth out coarse machining marks, and to effect some degree of grain refinement at the immediate surface.

The compressive stresses developed by good grinding or peening are of a similar magnitude,

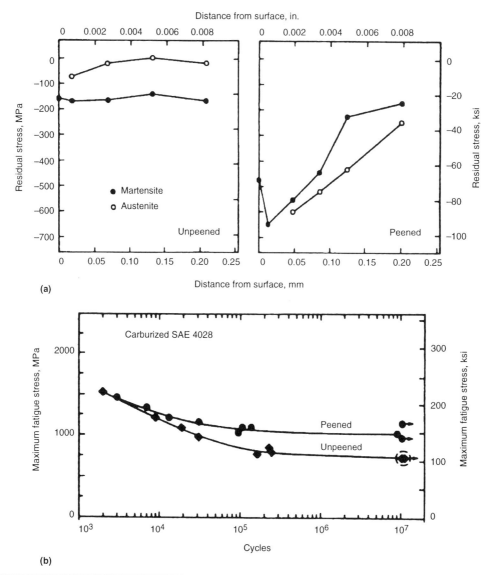

Steel	Condition	Surface hardness(a), HRC	Core hardness, HRC	Effective case depth(b)		Impact fracture stress		Fatigue endurance limit	
				mm	in.	MPa	ksi	MPa	ksi
Carburized SAE 4028	Unpeened	60.5	35	1.1	0.044	2240	325	730	105
	Shot peened	62.2	36	1.1	0.044	2265	329	1035	150

(a) Converted from HRA. (b) Distance to 510 HV

Fig. 8.33 Influence of shot peening on (a) residual stresses within austenite and martensite of a case-hardened surface and (b) fatigue strength. Table shows influence of shot peening on impact fracture stress. Source: Ref 38

but the depth of compression is greater after peening. Fig. 8.37 shows that with good grinding techniques, there is little to choose between conventional (aluminum oxide) grinding and CBN grinding, at lease not as far as the residual stresses go.

Summary

Grinding

Grinding removes adverse surface features present after heat treatment (e.g., inaccuracies due to distortion and growth, surface roughness, internal oxidation). It is done primarily to provide accuracy and finish. The advantages of grinding can be lost if the grinding operation is not well executed.

- *Preprocess considerations:* A properly designed case-hardening process should provide surface microstructures containing essentially fine, tempered martensite with a fair level of part-to-part and batch-to-batch consistency. Distortion and growth control are important. For grinding, the wheel grade, wheel speed, and depth of cut are important for a successful operation, and each must be selected with care.
- *In-process considerations:* Find the high spots, and proceed with light cuts. If wet grinding is done, ensure that coolant feed is adequate and correctly directed. Dress the abrasive wheel early; if operators wait until the sound of grinding changes, it may be too late for the workpiece.
- *Postprocess considerations:* Inspect for grinding burns, grinding cracks, and grinding steps in gear tooth fillets. Investigate reasons for any grinding damage. Corrective grinding is possible in some instances, in which case the use of a freshly dressed wheel should be considered. Tempering and peening are sometimes applied to ground surfaces to further improve their quality.
- *Effect on properties:* Accuracy of contact with static or moving surfaces is essential for good load distribution. Ground surfaces vary from good quality (with compressive-residual stresses) to poor quality (with burns or even cracks). These quality levels determine whether bending-fatigue strength is improved or seri-

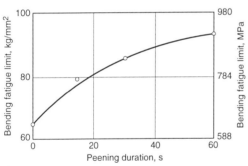

Fig. 8.34 Effect of saturation time on the bending fatigue strength of pinion teeth (pack carburized to 1.1 mm case depth). Source: Ref 39

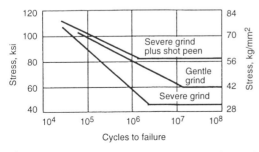

Fig. 8.35 Shot peening improves endurance limits of ground parts. Reversed bending fatigue of flat bars of 45 HRC. Source: Ref 42

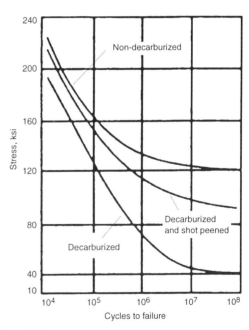

Fig. 8.36 Effect of shot peening on decarburization of SAE 4340 steel (ultimate tensile strength, 280 ksi), $k_t = 1$, $R = -1$, decarburization 0.003 in. to 0.03in., shot peening with 0.28 in. diam shot, 0.12A density. Source: Ref 30

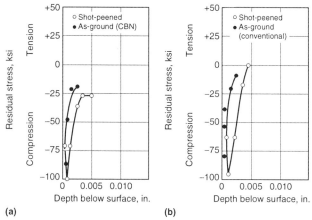

Fig. 8.37 Residual stresses in (a) CBN ground and (b) ground and shot peened surfaces for 9310 steel, 10 in. dp. Source: Ref 8

ously impaired. For contact fatigue, grinding improves the accuracy of contact and provides surfaces smooth enough to be separated by standard lubricating oils.
- *Standards:* Refer to magnetic-particle crack detection standards for grinding cracks, for example, ASTM E 1444-94a. Refer to temper-etch standards for grinding burns, for example, MIL-STD-867A.

Shot Peening

Shot peening is a surface treatment that increases skin hardness and induces compressive stresses into the immediate surface of the workpiece. It can also remove directional machining marks, thereby providing a more random finish. The process can be confined by masking to improve only the most critically stressed areas.

- *Preprocess considerations:* The benefits of peening are more assured if the initial surface is relatively smooth. Rough machined and ragged surfaces may lead to folded-in defects. Prior to treating, the type of shot and the precise process parameters must be determined, for example, by Almen strip, to suit the material and the hardness of the workpiece. Check shot condition; shot should be round.
- *In-process considerations:* Strict control of process settings is essential.
- *Postprocess considerations:* Check for complete coverage of the peened area.
- *Effect on properties:* Under bending conditions, fatigue cracks generally initiate at the surface. Metalworking by shot peening induces compressive-residual stresses within the surfaces of treated parts, thereby reducing the possibility of fatigue crack initiation at the surface. A 20% improvement in fatigue limit or high-cycle fatigue life has been quoted. Improvements in contact-fatigue resistance are also reported.
- *Standards:* ANSI/AGMA recommends that grade 2 gears be shot peened if the tooth roots are ground. For grade 3 gears, it requires that tooth roots and fillets are shot peened. Guidelines for shot peening gears are provided in AGMA 2004-BXX 1/88.

REFERENCES

Grinding

1. H. Grisbrook, H. Moran, and D. Shepherd, Metal Removal by a Single Abrasive Grit, *Machinability,* Special report 94, The Iron and Steel Institute, 1967, p 25–29
2. A.A. Mikhailov, The Origin of Grinding Cracks, *Russ. Eng. J.,* Vol XLVIII (No. 9), 1968, p 73–76
3. E. Schreiber, Residual Stress Formation during Grinding Hardened Steels, *Härt.-Tech. Mitt.* (BISI 13783), Vol 28 (No. 3), 1973, p 186–200
4. L.P. Tarasov, W.S. Hyler, and H.R. Letner, Effect of Grinding Conditions and Resultant Residual Stresses on the Fatigue Strength of Hardened Steel, *Proc. ASTM,* Vol 57, 1957, p 601–622
5. K.K. Burnakov et al., Reasons for Crack Formation in Grinding, *Russ. Eng. J.,* Vol LV (No. 9), 1975, p 52–54

6. M.W. Gormly, Residual Grinding Stresses, *Grinding Stresses—Cause, Effect and Control,* Grinding Wheel Institute, p 7–21
7. J.E. Price, Improving Reliability of Ground Parts by Avoiding Residual Stress, *Grinding Stresses—Cause, Effect and Control,* Grinding Wheel Institute, p 27–30
8. J.R. Drago, Comparative Load Capacity Evaluation of CBN-Finished Gears, *Gear Technol.,* May/June 1990, p 8–16, 48
9. D.V. Kumar, "Technological Fundamentals of CBN Bevel Gear Finish Grinding," SAE paper MR85-273, SAE Conf. on Superabrasives (Chicago), Society of Automotive Engineers, 1985
10. K. Neailey, Surface Integrity of Machined Components—Microstructural Aspects, *Met. Mater.,* Feb 1988, p 93–96
11. H. Staudinger, Bending Fatigue Strength of Carburised and Nitrided Steels Containing Grinding Cracks, *Z. VDI,* Vol 88, 1944, p 681–686
12. A.V. Podzei and A.V. Yakimov, Grinding Defects and Their Remedy, *Russ. Eng. J.,* Vol LII (No. 3), 1972, p 66–68
13. J.E. Varnai, Effect of Tempering on Grinding Cracks in Case-Hardening Parts, *Heat Treating '76,* Proc. 16th Int. Heat Treatment Conf. (Stratford, UK), Metals Society, 1976, p 95–98
14. N.P. Navarro, "The Technical and Economic Aspects of Grinding Steel with Borazon Type II and Diamond," SME paper MR70-198, Society of Manufacturing Engineers, April 1970
15. A.L. Ball, Controlling Grinding Stresses, *Grinding Stresses—Cause, Effect and Control,* Grinding Wheel Institute, p 63–67
16. V.S. Sagaradze and L.V. Malygina, The Fatigue Strength of Carburized Steel after Grinding, *Met. Sci. Heat Treat.,* Vol 12/13, Dec 1971, p 1050–1052
17. W. König, G. Mauer, and G. Röber, CBN Gear Grinding—A Way to Higher Load Capacity?, *Gear Technol.,* Nov/Dec 1993, p 10–16
18. G.J. Kimmet and H.D. Dodd, "CBN Finish Grinding of Hardened Spiral, Bevel, and Hypoid Gears," AGMA Fall Technical Meeting, American Gear Manufacturers Association, Oct 1984
19. R.J. Love, "Bending Fatigue Strength of Carburised Gears," MIRA report No. 1953/4, Motor Industry Research Association, 1953
20. M.L. Tyrovskii and I.M. Shifrin, Stress Concentration in the Surface Layer of Carburized Steel, *Russ. Eng. J.,* Vol L (No. 111), 1970, p 47–50
21. J.P. Sheehan and M.A.H. Howes, "The Effect of Case Carbon Content and Heat Treatment on the Pitting Fatigue of 8620 Steel," SAE paper 720268, Society of Automotive Engineers, 1972
22. J.P. Sheehan and M.A.H. Howes, "The Role of Surface Finish in Pitting Fatigue of Carburized Steel," SAE paper 730580, Society of Automotive Engineers, 1973
23. T. Nakanishi, Y. Ariura, and T. Ueno, Load Carrying Capacity of Surface-Hardened Gears (Influence of Surface Roughness on Surface Durability), *JSME Int. J.,* Vol 30 (No. 259), 1987, p 161–167

Roller Burnishing

24. M.A. Balter, Effect of the Structure on the Fatigue Limit of Steel after Surface Hardening, *Met. Sci. Heat Treat.,* Vol 13 (No. 3), March 1971, p 225–227
25. C. Razim, "Effects of Residual Austenite and Recticular Carbides on the Tendency to Pitting of Case-Hardened Steels," thesis, Techn. Hoschscule Stuttgart, 1967
26. D.D. Papshev, Increasing the Fatigue Strength of High-Tensile Steels by Work-Hardening, *Russ. Eng. J.,* Vol L (No. 1), 1970, p 38–42
27. M.A. Balter et al., Roller Burnishing of Carburized Steels, *Russ. Eng. J.,* Vol XLIX (No. 9), 1969, p 69–72
28. F.T. Krotine, M.F. McGuire, L.J. Ebert, and A.R. Troiano, The Influence of Case Properties and Retained Austenite on the Behaviour of Carburized Components, *Trans. ASM,* Vol 62, 1969, p 829–838
29. I.V. Kudryavtsev and N.M. Savvina, The Effect of Ten Years Holding on the Fatigue Strength of Parts with Residual Stresses, *Met. Sci. Heat Treat.,* (No. 4), April 1964, p 225–226

Shot Peening

30. M. Lawerenz and I. Ekis, Optimum Shot Peening Specification I, *Gear Technol.,* Nov/Dec 1991, p 15–22
31. C. Razim, Influence of Residual Austenite on the Strength Properties of Case-Hardened Test Pieces during Fatiguing, *Härt.-Tech. Mitt.* (BISI 6448), Vol 23, April 1968, p 1–8
32. D. Kirk, The Role of Residual Stress Measurement in Improving Heat Treatment, *Heat Treat. Met.,* (No. 3), 1985, p 77–80

33. D.A. Sveshnikov, I.V. Kudryavtsev, N.A. Gulyaeva, and L.D. Golubovskaya, Chemicothermal Treatment of Gears, *Met. Sci. Heat Treat.,* (No. 7), July 1966, p 527–532
34. R.P. Garibay and N.S. Chang, Improved Fatigue Life of a Carburised Gear by Shot Peening Parameter Optimization, *Proc. Processing and Performance* (Lakewood, CO), ASM International, 1989, p 283–289
35. H.F. Moore, Strengthening Metal Parts by Shot Peening, *Iron Age,* 28 Nov 1946, p 70–76
36. J.B. Seabrook and D.W. Dudley, Results of a Fifteen Year Program of Flexural Fatigue Testing of Gear Teeth, *Trans. ASME, Series B, J. Eng. Industry,* Vol 86, Aug 1964, p 221–239
37. G.I. Solod et al., Improving Bending Fatigue of Large-Module Gear Teeth, *Russ. Eng. J.,* Vol LII (No. 1), 1972, p 19–22
38. C. Kim, D.E. Diesburg, and R.M. Buck, Influence of Sub-Zero and Shot Peening Treatments on Impact and Fracture Properties of Case-Hardened Steels, *J. Heat Treat.,* Vol 12 (No. 1), June 1981, p 43–53
39. M.L. Turovskii, Residual Stresses in the Clearance Curve of Case-Hardened Pinions, *Russ. Eng. J.,* Vol LI (No. 9), 1971, p 46–49
40. N.G. Gerasimova and N.M Ryzhov, Effect of Shot Peening on the Contact Endurance of Case-Hardened Steel, *Russ. Eng. J.,* Vol 58 (No. 6), 1978, p 26–30
41. D.P. Townsend and E.V. Zaretsky, "Effect of Shot Peening on Surface Fatigue Life of Carburised and Hardened AISI 9310 Spur Gears," NASA paper 2047, Aug 1982
42. M. Lawerenz and I. Ekis, Optimum Shot Peening Specification II, *Gear Technol.,* Jan/Feb 1992, p 30–33

Index

A

Abrasive wear
   decarburization influence 47
   grinding influence......................... 212
   and internal oxidation .... 29–30(F)
   refrigeration influence 193–194
   retained austenite influence 89
   tempering influence 183
Ac temperature .... 201(F)
Ac$_{cm}$ temperature .... 53, 54(F), 56, 66, 67(F)
   austenite stability 77
   quenching and microcracking 112, 113
Ac$_1$ temperature .... 53, 54(F), 69
Ac$_3$ temperature .... 54(F), 69
   austenite stability 77
   heat treatments above, effect on
      coarsening temperature................ 101
Adhesive wear
   decarburization influence 47
   grinding influence......................... 212
   and internal oxidation .... 29–30(F)
   refrigeration influence 193–194
   retained austenite influence 89–90, 91, 92, 93, 94
   tempering influence 183
Aerospace gear steels 5
Aging 174, 208
   room temperature 78
   tempering influence .... 184–185(F,T)
AGMA yield number 142
Air tempering........................... 184(T)
Almen strips...................... 217(F), 223
Alumina .... 123, 124(F), 125
   effect on machinability of steels 128
Alumina carbonitrides, as nonmetallic
   inclusions....................... 125, 126(F)
Aluminum
   addition for grain refinement by
      alloying..................... 100(F), 101
   content effect with internal oxidation 16
   oxidation potential .... 11–12(F)
   with titanium, grain coarsening
      temperature affected 100
Aluminum nitride, as grain refining agent..... 100
Aluminum oxide
   as grit material .... 205(F), 209(F),
                              212, 222, 223(F)

   hardness, as nonmetallic inclusion .... 121(T)
   as nonmetallic inclusion .... 120(T), 128(T)
Aluminum treated steels, grain size...... 100–101
Ammonia, to restore hardenability and
   strength, in carburizing chamber........... 32
Anisotropy effect........... 120, 121(T), 122, 125
Apparent eutectoid 9
Apparent eutectoid carbon.................. 56
Applications, high-temperature service 135
Applied cyclic stresses 86
Applied stresses 1
   and case depth 163
   and cycles in fatigue life stages 161
   and effective case depth .... 158(F), 159
Arsenic, segregation susceptibility..... 114, 116(T)
Atmospheres
   effect of decarburization .... 39, 40(T)
   endothermic...................... 11–12(F)
   exothermic-based 32
   nitrogen-base 32
   oxygen-free gas-carburizing................. 11
Austenite
   case hardenability 149–150
   case transformation 159
   grinding defects.......................... 207
   grinding of.......................... 200, 202
   in case-hardened surface with contact
      damage.............................. 155
   internal oxidation and wear resistance..... 29–30
   surface carbon content................ 54–55(F)
   tempering effect............. 173–174(T), 177,
                     179–180, 181, 183
   thermally stabilized 77
   transformed by refrigeration........ 186, 187(F),
                      188(T), 190–191,
                      192, 193–194
   untransformed, volume of................... 79
Austenite formers....................... 51–52
Austenite grain size 81, 102
Austenitic stainless steels, heat-treatment
   deformations after quenching........... 166(F)
Austenitizing................. 100–101(F), 165
Austenitizing temperature...... 66, 67(F), 201(F)
   effect on microcrack sensitivity .... 109(F)
   equilibrium conditions..................... 53
Autotempering .... 172, 173(T), 175(F)

B

Bainite 55–56
    after decarburization .... 43(F)
    case hardenability 149–150
    case transformation 157
    of core 145
    and decarburization 57
    and decarburization after internal oxidation ... 24
    in case-hardened surface with contact
        damage 155
    in cores during tempering 172
    in cores of case-hardened parts .. 136, 137(F), 138
    internal oxidation and wear resistance 30
    lower .... 173(T), 174
    tempering effect .... 173(T), 174,
                   175, 178–179, 180
    upper .... 173(T), 174
Balanced composition 51
Ball bearings, contact fatigue and
        nonmetallic inclusions .... 127(F)
Bar end, carburized, microsegregation 116, 117(F)
Bearing fatigue tests 128
Bearing steels. *See also* Steels, SAE,
    specific types, 52100.
    nonmetallic inclusions
        effect on fatigue fractures .... 125(F)
        stress-raising properties .... 124(F)
Bending, internal oxidation influence on 28–29(F)
Bending fatigue limit 84, 154
    surface microhardness effect .... 25, 27(F)
Bending fatigue strength 1, 2
    carbide influence ...... 63–66(F,T), 67(F), 71–72
    case carbon effect 155
    case depth influence 167
    decarburization influence .... 45–46(F), 47(T)
    effective case depth influence ..... 159–163(F,T),
                          164(F)
    grain size effect .... 106(F), 107, 108(T)
    grinding influence ..... 208–210, 212(F), 213(T)
    microsegregation influence .... 118(F)
    nonmetallic inclusions influence .... 120, 121(T)
    refrigeration influence 193
    retained austenite influence .. 89(F), 90(F), 91(F)
    roller burnishing influence .... 215, 216(F)
    rotating, overheating effect 129
    shot peening influence .... 219(T), 220(F),
                       221(F), 222(F)
    tempering influence .... 178–181(F,T)
    vacuum carburizing effect .... 29(F)
Bending stress limits, for gears .... 4(T)
Bend testing .... 89, 90(F), 213
    internal oxidation effect .... 29(F)
Boost-diffuse method of carburizing ...... 151(F)
Boron, as grain refining agent 101
Bowden-Leben machine .... 91, 92(F)
"Butterfly" inclusions .... 127(F), 128(T)

C

Calcium aluminates .... 120, 123, 124(F)
    as nonmetallic inclusions .... 125, 126–127(F)
Calcium oxide .... 124(F)
Calcium sulfide .... 124(F)
Calcium treatments 128
    for decreasing number of nonmetallic
        inclusions 121
Carbide films, removed by grinding 199
Carbide formers 51–52
    as alloying elements 109–110
    defined 51
Carbide network 4, 57
Carbide precipitation 104, 183
    and case hardenability 150
Carbides .... 51–73(F,T)
    bending fatigue strength influence .. 63(T), 71–72
    bending fatigue strength
        influenced by .... 63–66(F), 67(F)
    of case-hardened surface 200
    chemical composition .... 51–53(F,T)
    contact fatigue influenced by .... 66–67(F), 72
    corner buildup 55
    corner, heat treatment effect on
        bending fatigue strength .... 64, 65(F)
    crack propagation .... 65, 66(F)
    critical crack size influenced by .... 68(T)
    cyclic tensile stressing 65
    discontinuous 73
    dispersed .... 53–60(F), 73
        effect on properties .... 62–68(F,T)
    equilibrium .... 52(T)
    equilibrium conditions .... 53(F)
    fatigue life affected by .... 63(T)
    film .... 70(F), 71(T)
        effect on properties 70–71
        formation of .... 56(F), 57(F), 61, 62(F)
        HTTP formation and 72
        internal oxidation 72
    flake .... 70(F), 71(T)
    formation of .... 55(F), 56(F),
                    57(F), 60–61, 62(F)
    forming elements 69–70
    forms 51
    fracture strength affected by .... 63(T)
    fracture toughness influenced by .... 68(F,T)
    free 1, 54, 56, 73
        case depth increases and 158
        defined 51
        white etching constituent 81
    geometric models formed during
        case-hardening .... 62(F)
    globular 54, 62
        carbide influence on contact fatigue 72
        deposits .... 69–70(F)
        effect on properties 70–71
        heavy dispersions and .... 69–70(F)
    grain-boundary 54, 62, 65, 67
    grinding defects and 207

hardness influenced by.....................71
intragranular ....61, 62(F)
load carrying capability influenced by .....68(T)
massive ....53–60(F), 65, 73
massive, formation of ....55(F), 61, 62(F)
massive, heat treatment effect.............65(F)
near-equilibrium 52
network...........53–60(F), 65, 67–68, 73, 207
  effect on properties...............62–68(F,T)
  heat treatment effect....................65(F)
  heat treatment effect on bending
    fatigue ....64, 65(F)
  impact damage and......................68
  influence on contact fatigue.........66(F), 72
  metallographic examination ......102, 103(F)
  network/spheroidal heat treatment effect ...65(F)
  nonequilibrium cooling..............54–60(F)
  residual stresses influenced by ....63(T), 65,
                                  66(F), 71(F), 72(F)
  spheroidal.....................52, 62, 63, 65
    influence on contact fatigue.........66(F), 72
  spheroidized 63, 67
  standards................................73
  toughness influenced by...........63(T), 72–73
  wear influenced by 73
  with decarburization......................41
Carbide segregation 177–178
Carbon
  as austenite former 52
  content effect on fracture toughness ....89, 91(F)
  content effect with internal oxidation......30–31
  effect on hardness after internal oxidation ..24(F)
  released by gas-metal reactions 13
Carbon case hardening 2
  drawbacks..................................2
Carbon-chromium steels, fatigue resistance 86(F)
Carbon clustering 183
Carbon dioxide, providing oxygen for
  internal oxidation 11
Carbon gradient ....151(F), 156
  negative, promoted by decarburization 37
  positive, promoted by carburization 37
  with decarburization.................41, 42(F)
Carbon-manganese-boron alloys
  internal oxidation 16–17
Carbon monoxide, content effect on internal
  oxidation of manganese chromium steels ...15
Carbon-nickel-chromium steels
  microstructures after cooling........136, 137(F)
Carbon potential........................32, 156
  decarburization and ....38, 39(F),40–41(F)
  during carburizing.....................38(F)
  of endothermic gas as function of
    temperature ....69(F)
  in film carbides ....70(F)
  in internal oxidation 18
  residual stress influenced by ....158, 159(F)
Carbon segregation.............104, 172, 173(T)
Carbon steels
  grain size and normalizing effect 104–105, 106(F)

grain size influence on impact
  strength ....106(F), 107
impact strength influenced by
  grain size.....................107, 108(T)
nitriding....................................2
Carburizing
  cycle...................................38(F)
  duration, and case depth 156
  microstructural features......................1
  variables 1
Carburizing steels
  chemical compositions..................136(F)
  heat treatment deformations after
    quenching.........................166(F)
  strength vs. section diameter.............136(F)
Case carbon..............................168
  standards................................168
Case crushing.....................148(F), 159
  contact damage influenced by case depth 162(T)
Case-crushing resistance 162
Case depth 167–168
  bending fatigue influenced by.....159–163(F,T),
                                    164(F)
  carbon content 168
  contact damage influenced by ......162–163(F),
                                    164(F)
  and core properties...............135–168(F,T)
  dependence on shape and size ....156–158(F)
  effect on internal oxidation 14
  and mechanical properties................158(F)
  and residual stresses ....158–159(F), 160(F)
  specifications...................6(F), 7–8(F)
  standards..............................167–168
**Case depth-to-section thickness
  ratio (*CD/t*)** ....159–160, 161(F)
**Case depth-to-tooth diametrical
  pitch relationship** ....8(F)
Case factors ....148–163(F,T), 164(F)
  carbon content ....150–155(F,T), 156(F)
  carbon effect on
    bending fatigue 155
    case toughness ....152–153(F,T)
    contact damage.................155, 156(F)
    impact fatigue..................153–154(F)
    residual stresses.................154–155(T)
    surface hardness.................151–152(F)
  effective case depth ....155–161(F,T)
  hardenability................138(F), 149–150(F)
Case hardenability...........138(F), 149–150(F)
  level categories 149
Case-hardened steels
  retained austenite effect on contact fatigue 87
  rotating beam fatigue strength ....85(F)
Case hardening 164
  compressive-residual stresses 1–2
CBS-600 5
CBS-1000M VIM-VAR 5
Cementite................................173(T)
  coarsening..................................52
  microstructure 52

Cerium, as grain refining agent 101
Charpy impact tests 144
Charpy toughness (shelf energy)
    nonmetallic inclusions influence 128
Chi(χ)-carbides . . . . . 173(T), 174(F), 175, 178, 185
"Chilled" surface layer 113
Chip breaking 128
Chromium
    carbide formation affected by 52
    content effect on microstructure 109, 110
    content effect with internal oxidation 15–16
    depletion
        and surface decarburization 21
        with internal oxidation . . . . 22, 23(F)
    effect on case fracture toughness 152–153
    as grain refining agent 101
    hardenability effect with internal oxidation 31
    impact resistance effect with internal
        oxidation presence 28–29
    in equilibrium carbides . . . . 52(T)
    microsegregtion behavior . . . . 114, 116(F)
    oxidation 13
    oxidation potential . . . . 11, 12(F)
    segregation susceptibility . . . . 114, 115(F), 116(T)
    with manganese sulfide as
        nonmetallic inclusion . . . . 120, 121(T)
Chromium-manganese steels
    carbide influence on contact fatigue 66, 67(F), 72
    internal oxidation . . . . 17(F)
    vacuum carburizing 102
Chromium-manganese-titanium steels
    core material effect on residual stresses . . . 145(F)
    decarburization . . . . 45, 46(F)
    fatigue limit effect on surface residual
        stress . . . . 145, 146(F)
    internal oxidation . . . . 16, 17, 24(F)
    oxygen penetration after carburizing
        in endothermic atmosphere . . . . 13(F)
Chromium-manganese-vanadium steels
    internal oxidation 17
Chromium-molybdenum steels
    carbide influence . . . . 70(F)
    impact strength, as-carburized 145
    internal oxidation 31
    microsegregation effect on hardness . . . . . . 117(F)
Chromium-molybdenum-vanadium steels
    fatigue curves . . . . 118(F)
    mechanical properties of bars after
        heat treating . . . . 123, 124(T)
    microsegregation effect on properties . . . . . 118(F)
    microsegregation influence on fatigue . . . . 118(F)
    nitriding 2
Chromium-nickel-molybdenum steels
    carbide assistance of crack propagation 65
    carbide effect on fatigue-crack initiation 66
    composition variations effect on
        transformation behavior . . . . 22(F)
    hardenability effect . . . . 19–21, 22(F)
    internal oxidation . . . . 19–21, 22(F)
    refrigeration effect on hardness and
        fatigue . . . . 188, 189(T)
    sulfide inclusions influence on
        mechanical properties . . . . 121(T)
Chromium-nickel steels
    core properties and fatigue strength . . . . . . 146(F)
    fatigue limits 26
    fatigue resistance 86
    internal oxidation 17
    retained austenite influence on bend and
        impact fracture strength . . . . 89(T)
    retained austenite influence on
        fatigue resistance 84
Chromium oxide 24
    chromium content by electron
        probe analysis . . . . 17(T)
Chromium steels
    carbide formation 52
    cementite coarsening 52
    fatigue resistance 86
    refrigeration effect on hardness
        and fatigue . . . . 188, 189(T)
Cleanliness 1
Clutch hub, sliding, microsegregation 116, 117(F)
Cobalt, with manganese sulfide, as
    nonmetallic inclusion . . . . 121(T)
Cold working 1
Composition gradients
    with internal oxidation . . . . 17–18(F,T), 19(F), 31
Compressed oxides 24
Compressive-residual stresses
    effect on rolling contact 8
    grinding as cause of . . . . 208, 209(F), 210(F)
    internal oxidation effect . . . . 23(F,T), 24–25(F)
    tempering influence . . . . 173(T), 177–178(F),
        181(F), 185
Constitutional diagrams . . . . 114(F)
Contact damage
    case carbon effect . . . . 155, 156(F)
    case depth influence . . . . 162–163(F), 164(F)
    pitting fatigue resistance . . . . 155, 156(F)
Contact fatigue
    carbide influence . . . . 66–67(F), 72
    decarburization influence 46
    grinding influence . . . . 210–211, 213(F)
    internal oxidation effect . . . . 27–28(F)
    nonmetallic inclusions influence . . . 127–128(F,T)
    refrigeration influence . . . . 192–193(F,T)
    retained austenite influence . . . . 87–89(F)
    roller burnishing influence 215
    shot peening influence 220
    tempering influence . . . . 181(F)
Contact fatigue resistance 1
    achievement of . . . . 88(F)
Contact-fatigue strength 2
Contact loading 90
Contact stress limits, gears . . . . 4(T)
Continuous-cooling diagrams . . . . 149(F)
    retained austenite presence . . . . 77, 78(F)

Continuous-cooling transformation
    (CCT) curve . . . . 56, 58(F)
Continuous-cooling transformation
    (CCT) diagrams 54
    hardenability of carburizing steels 137(F), 138(F)
    and high-temperature transformation
        product formation . . . . 21, 22(F), 32
Coolants, for grinding 203, 205–206
Cooling, with internal oxidation of gears . . . . . 32(F)
Copper, oxidation potential . . . . 11, 12(F)
Core, microstructures . . . . 136–137(F)
Core factors . . . . 135–148(F,T)
    ductility . . . . 143(F)
    elongation . . . . 143(F)
    fatigue limit vs. surface residual
        stress . . . . 145, 146(F)
    fatigue strength . . . . 146(F)
    hardenability . . . . 135–140(F,T)
    impact resistance . . . . 143(F)
    material effect on
        contact-damage resistance . . . . 148(F)
        impact-fatigue resistance . . . . 147–148(F)
        residual stresses . . . . 145(F)
    microstructure and hardness . . . . . . 140(F), 141(F)
    quenching temperature effect on
        fatigue strength . . . . 147(T)
    reduction of area . . . . 143(F)
    strength effect on bending fatigue
        resistance . . . . 145–147(F,T)
    tensile strength . . . . 140, 141(F), 142(F)
    toughness . . . . 143–145(F)
    ultimate tensile strength . . . . 140, 141(F),
        142(F), 143(F)
    upper limit of desirable core strength 145
    yield strength . . . . 140–143(F), 147
Core properties . . . . 1, 135–168(F,T)
    standards 166
Corrosion, as factor in grinding 202
Corrosion products, with internal oxidation 31
Corundum 120
Crack growth rate 86
    overheating effect 129
Cracking, grain size influence 105
Crack initiation 81, 161
    grain size effect 106
    overheating effect 129
    refrigeration influence . . . . . 190, 191(F), 192, 193
    retained austenite influence . . . . 89(F)
    subcase, with low-carbon cores 145
    tempering influence . . . . 179(F), 181–182,
        184(T), 186
Crack propagation 161, 179
    grain size effect 106
    hydrogen content effect . . . . 184–185(F)
    nonmetallic inclusions influence . . . . 125(F)
    quenching temperature effect . . . . 147, 148(F)
    rate of 89
    refrigeration influence 192
    and retained austenite . . . . 87(F)
    tempering influence 181–182
    under load 81
Cracks, from grinding . . . . 200–203(F)
Cratering 128
Cristobalite 120
Critical crack size 89
Critical defect size 65
Crumbling 200
Cubic boron nitride (CBN)
    as grit material . . . . 205(F), 209(F),
        212(F), 222(F), 223(F)
    wear rate 205
Curvature, relative radius of . . . . 7(F)
Cyclic loading
    grain size effect on fatigue strength . . 105–106(F)
Cyclic straining 143

D

Decarburization . . . . 1, 37–48(F,T)
    agents 37, 38
    atmosphere effect . . . . 39, 40(T)
    austenitization temperature effect on
        contact fatigue . . . . 61(T)
    bending fatigue strength
        influenced by . . . . 45–46(F), 47(T)
    boost/diffuse method 47
    carbon potential effect on microstructure,
        hardness and residual stresses . . . . 40(T)
    chemical reactions, decarburizing 37, 38
    conditions for . . . . 37–40(F,T), 47
    contact fatigue influenced by 46
    control of 47
    decarburized layer depth 40
    defined 37
    detected by
        macrohardness testing 43
        microhardness testing 43
    dew point effect . . . . 38–39(F)
    effect on austenite layering 80
    examples . . . . 38–40(F)
    fatigue strength influenced by . . . . 47(T)
    hardness influenced by . . . . 24(F), 43–44(F)
    high-strength steels . . . . 39, 40(T)
    holding time variable . . . . 39(F)
    in fluidized bed in air . . . . 39(F)
    and internal oxidation 18, 21–22, 23, 24
    material properties influenced by . . . . 40(T),
        43–47(F,T)
    metallography of . . . . 42–43(F)
    partial 39, 43
    physical metallurgy . . . . 40–41(F), 42(F)
    processes . . . . 37–41(F,T)
    residual stresses influenced by . . . . 40(T),
        44–45(F), 46(F)
    shot peening influence . . . . 220, 221, 222(F)
    standards 48
    temperatures 37
    testing . . . . 41–43(F)
    wear influenced by . . . . 30(F), 47

Decarburization (continued)
   with internal oxidation 31
Dedendum-pitch line area 7, 156
Deformation
   from grinding 200
   refrigeration influence on contact fatigue 192
Deformation rolling
   deforming austenite and martensite 159
Dendrites . . . . 113–114(F), 115(F)
Design
   and internal oxidation limitations 32–33
Dew point . . . . 38–39(F)
Diamond, as grit material . . . . 209(F)
Diffusion processes, retained austenite 77
Direct hardening steels, heat-treatment
   deformations after quenching . . . . 166(F)
Direct quenching 130
Disk testing 194, 210–211
Dislocations 104
Distortion . . . . 164–165, 166(F), 167(F)
   from tempering . . . . 177(F)
   hardenability influence on trends 165
   related to grain size 105
   types 164
Double knee effect . . . . 26, 28(F)
Double quenching 89, 107
   and microcracking 111
Double reheat quenching, and carbides . . 59, 62(F)
Dry ice 195
Ductility
   as core factor . . . . 143(F)
   defined 143
Durability 2

E

Effective case depth . . . . 8(F), 148, 156–158
   after carbon case hardening 2
   bending fatigue influenced by . . . . 159–161(F)
   defined . . . . 8(F), 167
Elastic limit, true 143
Electric furnace (conventional) melting
   and nonmetallic inclusions . . . . 123(T)
   nonmetallic inclusions influence on
      steel bars . . . . 124(T)
Electrochemical machining
   with internal oxidation 32
Electron beam remelting
   and nonmetallic inclusions . . . . 121, 123(T)
Electron probe analysis
   of internal oxides . . . . 17(T), 18(F)
Electropolishing
   with internal oxidation 32
Electroslag remelted (ESR) steels
   microsegregation influence . . . . 118(F)
Electroslag remelting (ESR)
   and nonmetallic inclusions . . 121, 123(T), 124(T)
Elongation
   of core . . . . 143(F)

nonmetallic inclusion influence . . . . . 121, 124(T)
   refrigeration influence . . . . 188(T)
Endo-gas . . . . 184(T)
Equilibrium carbides, chromium content . . . 52(T)
Equilibrium diagrams, iron-carbon . . . . . 53(F), 54
Eta (η)-carbides . . . . 172, 173(T), 174(F),
                175, 177–178, 185, 193
Eutectoid carbon content 9
   defined 9
   in pure iron-carbon alloys 53
Eutectoid point . . . . 54(F)
Eutectoid temperature
   equilibrium conditions 53

F

Failure analysis, gear design . . . . 4–5(F,T)
Fatigue
   internal oxidation and surface
      microhardness effect . . . . 25, 27(F)
   microsegregation influence . . . . 118(F), 119(F)
Fatigue life
   carbide state effect . . . . 63(T)
   and high-temperature transformation
      products presence 25–26
   and impact resistance 154
   stages, of a part 161
Fatigue limits 26
   and impact resistance 154
   refrigeration influence . . . . 189, 190(F)
   tempering influence . . . . 178–181(F,T)
Fatigue resistance
   nonmetallic inclusions influence 123
   overheating effect 128–129
   refrigeration influence . . . . 188–190(F,T), 191(F)
   retained austenite influence . . . . 84–87(F)
   shot peening influence . . . . 219(T), 220(F),
                  221(F), 222(F)
Fatigue strength
   decarburization influence . . . . 47(T)
   grain size effect . . . . 105–107(F)
   quenching temperature effect . . . . 147(T)
Ferrite
   after decarburization . . . . 43(F)
   of core 145, 146–147
   and decarburization after internal
      oxidation 24
   developed with decarburization . . . . 41(F)
   free, production with internal oxidation 21
   in cores during tempering 172
   in cores of case-hardened parts 136,
                 137–138(F), 143
   recrystallization . . . . 173(T)
   tempering effect 183
Ferrite formers 51–52
Ferritic steels, hardness and grain size 104
File hardness test 24
Four-ball test . . . . 66(F), 72, 194
Fractures, intergranular overload 106–107

Fracture strength, carbide state effect....... 63(T)
Fracture toughness
   carbide influence .... 68(F,T)
   case carbon effect .... 152(T)
   microsegregation influence .... 117–118(F)
   overheating effect......................... 129
   refrigeration influence .... 189, 190(F), 193
   tempering influence .... 180(T)
Fracture toughness testing
   grain size influence...................... 107
   retained austenite influence .... 89, 91(F)
   tempering influence 182
Free-machining steels 125
Fretting..................................... 29
Friction, coefficient of
   for ground or electropolished
      surface vs. unground surface........ 27–28
   internal oxidation and lubrication 29
   and refrigeration influence 194
Friction testing............................ 194
Furnaces, fluidized bed, and decarburization 39(F)
FZG spur gear test 194

G

Gas carburizing .... 153(F)
   distortion due to 165
   endothermic carrier gas 11
   tempering influence on unnotched
      Charpy bars .... 182(T)
Gas flow rate, and case depth 156
Gears
   aerospace.................................. 4
   aerospace, retained austenite permitted 194
   automotive 4
   basic allowable stress numbers
      AGMA 2001-C95 .... 5(T)
      ISO 6336-5 1996...................... 4(T)
   carbide effect on bending
      fatigue strengths................. 63, 64(F)
   carburized, surface carbon requirement 9
   case hardenability.......................... 150
   contact damage 162
   cooling times .... 32(F)
   core properties and fatigue strength
      after case hardening .... 146(F), 147(T)
   deep spalling failures 163
   effective case depth measurement 156,
                                     157(F), 158
   fatigue cracking and nonmetallic inclusions .. 125
   form grinding of teeth.................. 206(F)
   grinding...................... 199, 209–210,
                                211, 212(F), 213(F)
   grinding cracks .... 202, 204(F)
   hardness tests............................ 140
   impact resistance.......................... 154
   lubricants for 6
   marine, retained austenite permitted......... 194
   microsegregation .... 115, 117(F)

   nonmetallic inclusions..................... 128
   oil-hardened, residual stresses .... 85(F)
   refrigeration............................. 193
      and adhesive wear 194
   residual stresses on
      case-hardened teeth........ 23(T), 24–25(F)
   retained austenite
      control procedures.................... 93–94
      effect on pitting resistance.............. 88(F)
      standards 194
   rotating beam fatigue strength .... 25, 26(F)
   "safe" design .... 4–5(F)
   shot peening standards..................... 223
   surface microhardness and internal
      oxidation effect on fatigue .... 25, 27(F)
   tempering......................... 180–181(F)
Gear sets, spur and helical
   case hardening .... 2(F)
   through hardening .... 2(F)
Gear standards.............................. 3
Gear steels, rolling-contact fatigue limit,
   core strength and case depth effects....... 6(F)
Gear tests............................. 4, 194
   full-scale .... 4(F)
Gibbs energy, vs. interaction energy for
   alloying elements in steels .... 51(F)
Grain coarsening........................... 105
Grain size .... 1, 99–107(F,T), 108(T)
   ASTM grain numbers and
      their dimensions.................. 103(T)
   bending strength influenced by .... 106(F),
                                      107, 108(T)
   coarsening 100–101, 102
   control of .... 100–102(F)
   effect on martensite strength............. 105(F)
   evaluation .... 100–102(F)
   fatigue strength influenced by .... 105–107(F)
   grain refinement by alloying................. 99
   growth mechanisms 100
   hardenability influenced by 103
   hardness influenced by .... 104, 105(F)
   heat treatments .... 99–100(F,T)
   impact fracture strength influenced by ... 106(F),
                                      107, 108(T)
   and internal oxidation...................... 27
   internal oxidation influenced by........ 102–103
   metallographic examination......... 102, 103(F)
   as microstructural feature 129
   properties influenced by..... 104–107(F), 108(T)
   refinement by alloying..................... 100
   residual stresses influenced by.............. 105
   tensile strength
      influenced by.......... 104–105(F), 106(F)
   versus hardness................... 104, 105(F)
   with retained austenite, effect on
      fatigue strength .... 106, 107(F)
   yield strength influenced by 104–105(F), 106(F)
Graphitizers............................ 51–52
   defined................................... 51
Gray staining 211

Grinding ...... 199–212(F), 213(F,T), 222–223(F)
  abusive.............. 202, 205, 207, 208, 210(F)
  action 199–200
  bending fatigue strength influenced by 208–210,
      212(F), 213(T),
  burns .... 200–203(F), 207, 211
  contact fatigue influenced by ... 210–211, 213(F)
  corrective 164
  cracks 200–203, 207, 210
  degree of difficulty................. 205–206(F)
  depth of cut (feed) influence .... 203, 204(F)
  and double-stage carburizing .... 151(F)
  effect on surface carbon................... 168
  fluid influence 205
  followed by tempering..................... 195
  form grinding............................. 206
  for surface finish......................... 199
  microsegregation and precautions followed . . 119
  problems caused by retained austenite 94
  purposes 199
  residual stresses influenced by ...... 203, 204(F),
      207–208(F), 209(F), 210(F)
  standards................................. 223
  stress distribution types............. 207, 208(F)
  surface................................... 199
  tempering influence................. 185–186(T)
  variables, influence of.............. 203–207(F)
  wear influenced by.................... 211–212
  wheel grit influence................. 204–205(F)
  wheel hardness influence .... 204–205(F)
  wheel peripheral speed influence 203
  with electrochemical machining 32
  workpiece metallurgical condition
      influence....................... 206–207
  workpiece speed influence 203
Grit blasting
  before shot peening 217
  with electrochemical machining 32
Grit size, for grinding wheels.................. 204
Growth, conditions influencing............... 165

H

Habit planes........................... 109, 110
Hagg carbides, monoclinic 193
Hardenability
  grain size influence....................... 103
  with internal oxidation..................... 31
Hardenability effect
  chromium-nickel-molybdenum steel 19–21, 22(F)
Hardening, with decarburization 46
Hardness................................... 1
  carbide influence.................... 62–63, 71
  of core...................... 140(F), 141(F)
  decarburization influence....... 24(F), 43–44(F)
  grain size effect.................... 104, 105(F)
  internal oxidation influenced by .... 23(T), 24(F)
  microsegregation influence.............. 117(F)
  refrigeration influence............ 186–188(F,T)
  related to carbon for untempered
      martensite in case-hardened steels .... 83(F)
  retained austenite influence ..... 81–82(F), 83(F)
  roller burnishing influence .... 214(F), 215(F)
  shot peening influence .... 218(F)
  tempering influence................. 175–176(F)
  versus ferrite grain size .... 104, 105(F)
  versus grain size .... 105(F)
Hardness profile............................. 8
Hardness traverse.......................... 156
Heat affected zone (HAZ) liquation cracking 129
Heat-sink effect 200
Heat treating
  alternative cycles for hardening
      carburized components .... 99(F)
  case and core characteristics resulting .... 100(T)
  and retained austenite....................... 89
  to control retained austenite 93
H grade steels......................... 165, 166
High-cycle fatigue 26
  CD/t ratio 161
  damage from, and case depth 158
  internal oxidation and HTTP effect..... 26, 28(F)
  microsegregation influence................. 118
  nonmetallic inclusions effect 123
  refrigeration influence............ 188, 189, 190
  shot peening influence..................... 220
  tempering effect 179
High-speed steels, carbide influence .. 70(F), 71(T)
High-strength nitrided steels, retained
      austenite effect on contact fatigue.......... 87
High-strength steels, decarburization.... 39, 40(T)
**High-temperature carburize-and-diffuse
  treatment**............................... 55
**High-temperature transformation
  products (HTTP)**........................ 11
  associated with carbides.............. 62–63, 65
  avoidance techniques with internal oxidation . . 33
  carbides, globular influence 69
  case carbon effect on contact damage........ 155
  case depth increases and 158
  compared to decarburization 37
  contact fatigue..................... 27–28(F)
  double knee effect................. 26, 28(F)
  fatigue life 25–26
  formation of 77
  from internal oxidation .... 19, 20–26(F,T)
  grain size effect 27
  grain size influence....................... 103
  and grinding defects..................... 207
  and internal oxidation............... 19–30(F,T)
  molybdenum reducing amount of ...... 22, 23(T)
  quench severity effect on cooling rate 22
  removed by grinding 199
  section size effect on cooling rate 22
  softening measured by microhardness
      testing .... 24(F)
  soft skin effect with decarburization.......... 42
  with decarburization and internal oxidation 43–44
  with nonmetallic inclusions 129

Index / 235

Holding time, effect on decarburization...... 39(F)
Honing, with electrochemical machining....... 32
Hot-oil quenching
    to improve toughness............ 150(F), 153(F)
Hot shortness 120
Hot working 101
    microsegregation influenced by........ 114–115, 116(F,T)
Hydrogen
    content effect on tempering....... 183–184(F,T)
    released by gas-metal reactions.............. 13
Hydrogen cracking.......................... 110

I

Impact fracture strength
    grain size effect............ 106(F), 107, 108(T)
    internal oxidation effect.............. 28–29(F)
    nonmetallic inclusions influence............ 128
    retained austenite influence 89(F,T), 90(F), 91(F)
    tempering influence.............. 181–183(F,T)
Impact resistance
    of core................................ 143(F)
    tempering influence.................... 182(F)
Impact testing
    grain size influence....................... 107
    internal oxidation effect................. 29(F)
Impact toughness, refrigeration influence..... 193
Impingements..................... 108, 109, 110
    defined................................ 216
Impurities, surface with internal oxidation...... 31
Inclusions, nonmetallic................. 130–131
    bending fatigue strength
        influenced by................. 120, 121(T)
    Charpy toughness (shelf energy)
        influenced by........................ 128
    contact fatigue influenced by...... 127–128(F,T)
    elongation influenced by........... 121, 124(T)
    exogenous...................... 119–120, 124
    fatigue limit vs. inclusion limit........... 125(F)
    fatigue resistance influenced by............. 123
    impact fracture strength influenced by....... 128
    inclusion chemistry effect.......... 123–125(F)
    indigenous..................... 119–120, 124
    machinability influenced by................ 128
    melting processes influence on
        mechanical properties......... 122, 123(T)
    microsegregation influenced by..... 118, 119(F)
    as microstructural features........ 119–129(F,T)
    number of............................. 127(F)
    origin of.......................... 119–121(T)
    overheating influenced by............. 128–129
    quantity effect........................ 125(F)
    reduction of area influenced by..... 120, 121(T), 122, 123(F), 124(T)
    residual stresses from.............. 126–127(F)
    shape control and anisotropy........ 120–121(T)
    size and location effects................ 125(F)
    stability............................... 121

    standards.............................. 130
    steelmaking parameters and their effects..... 121
    steel strength influence...... 121(T), 125, 126(F)
    stress-raising properties................. 124(F)
    ultimate tensile strength
        influenced by......... 120, 121(T), 122(F)
    yield strength influenced by................ 121
Induction hardening.......................... 2
Interaction energy, vs. Gibbs energy for
    alloying elements in steels............. 51(F)
Intergranular fracture ratio................. 107
Internal oxidation................. 1, 11–33(F,T)
    alloy depletion and eutectoid
        carbon content.................. 22–23(F)
    alloy depletion within the matrix....... 19, 21(T)
    ammonia introduction into carburizing
        chamber............................ 32
    ANSI/AGMA standards..................... 33
    bending fatigue strength
        effect................... 23(T), 25–28(F,T)
    bending influenced by................. 28–29(F)
    case depth increases and............ 158, 159(F)
    characteristics imparted.................... 12
    chromium depletion.................. 22, 23(F)
    composition gradients..................... 31
    contact fatigue influenced by.......... 27–28(F)
    cooling times.......................... 32(F)
    and decarburization....................... 23
    and decarburization of surface............ 21–22
    degrees of................................ 4
    design limitations...................... 32–33
    effect on
        austenite layering........................ 80
        fatigue strength................... 25, 26(F)
        local microstructure............. 18–23(F,T)
        microstructure..................... 23–24(F)
    elimination measures................. 30–33(F)
    factors promoting.................... 11–12(F)
    fatigue strength, loss of................ 25, 27(T)
    grain size effect........................... 27
    and grinding defects...................... 207
    hardenability effect............... 19–20, 22(F)
    hardness influence................. 23(T), 24(F)
    and high-temperature transformation
        products..................... 19–30(F,T)
    impact fracture strength influenced by.. 28–29(F)
    influence on
        material properties............... 23–30(F,T)
        residual stresses....... 23(T), 24–25(F), 26(F)
    in-process considerations................... 33
    low-carbon surfaces.................. 18, 19(F)
    manganese depletion................. 22, 23(F)
    postprocess considerations.................. 33
    preprocess considerations................... 33
    process........................... 13–18(F,T)
    reduction controls.................... 30–33(F)
    removed by grinding...................... 199
    removed by shot peening.............. 220, 221
    semiquantitative analysis of elements in
        material adjacent to oxides....... 19, 21(T)

Internal oxidation (continued)
  and surface microhardness . . . . 25, 27(F)
  wear resistance influenced by . . . . 29–30(F)
  with decarburization 31
  without HTTP 27
Internal oxidation process . . . . 13–18(F,T)
  aluminum content effect 16
  case depth effect 14
  chromium content effect 15–16
  commercial case-hardening alloys 16–17
  composition gradients . . . . 17–18(F)
  grain size effect 14
  grain size effect on penetration
    depth . . . . 14, 15(F)
  manganese content effect 14, 15–16
  multicomponent alloys 15–16
  oxide composition . . . . 17(F)
  oxide morphology . . . . 13–15(F)
  oxygen penetration . . . . 13(F)
  silicon content effect 15–16
  steel composition effect 14
  temperature effect on penetration depth . . . 13(F),
    14, 15(F)
  titanium content effect 16
  two-component alloys, atomic number
    and size effect . . . . 15, 16(F)
  vanadium content effect 17
Internal twinning 104
Iron
  oxidation potential . . . . 11, 12(F)
  with manganese sulfide, as nonmetallic
    inclusion . . . . 120, 121(T)
Iron carbide
  carbide influence on residual stresses . . . 63, 64(T)
Iron-carbon alloys, microcracking . . . . 108–109(F)
Iron-carbon diagram
  of reheat condition . . . . 57, 59(F)
Iron-carbon equilibrium diagram . . . . . 40, 41(F),
    53, 54
Iron-carbon phase diagram . . . . 55(F)
Iron-chromium-carbides
  equilibrium states . . . . 52(T)
Iron-manganese silicates 120
Iron oxides . . . . 124(F)
Isothermal heat treatment 101
Izod impact tests 144

J

Jobbing 164
Jominy diagrams . . . . 167(F)
Jominy hardenability curve 54

K

Killed steels, inclusions, nonmetallic 120
Koistinen/Marburger relationship 79

L

Laboratory testing 3–4
Laboratory test pieces, design aspects . . . 4–7(F,T)
Lath martensite 161
  tempering effect . . . . 171, 172(F), 173(T)
Lath width 142
Latrobe CFSS-42L 5
Lead, in nonmetallic inclusions 124–125
Lean-alloy steels 73
  carbide content effects 66
  case carbon effect on bending fatigue 155
  decarburization 57, 58
  fatigue strength/case depth
    relationship . . . . 159–160(F)
  hardenability and grain size 103
  internal oxidation 31
  retained austenite contents 80
Lean-alloy steel parts 56–57
Leaner alloys 33
Limit of proportionality (LoP) 140–142
  tempering effect . . . . 176(F), 177(F)
Linear expansion coefficients
  for carbides and matrices . . . . 63(T)
Liquation, at grain boundaries 129
Liquefied natural gas (LNG)
  for subzero treating . . . . 186, 192(T), 195
Liquid nitrogen
  refrigeration treatment . . . . 191–192(F)
Load amplitudes
  effect on crack propagation behavior 87
Load/extension curves
  nonmetallic inclusions effect . . . . 121, 122(F)
Load-time curves . . . . 89, 90(F)
Low-cycle fatigue
  CD/t ratio 161
  core material effect . . . . 147, 148(F)
  internal oxidation and HTTP effect . . . . . 26, 28(F)
  microsegregation influence 118
  nonmetallic inclusions effect 123
  refrigeration influence 188, 189, 190
  shot peening influence . . . . 220, 221(F)
  tempering effect 179
Low-cycle impact-fatigue tests 154
Low-speed roller test 92
Lubricants. *See also* Lubrication.
  extreme pressure 211
  for grinding 205, 211
  high-temperature limitations for gearing 6
  refrigeration influence of adhesive wear 194
  with internal oxidation 31
Lubrication. *See also* Lubricants.
  adhesive wear with internal oxidation 29
  effect on contact damage 162
  effect on wear resistance of carbides 73
  retained austenite influence 91
  and rolling contact fatigue 89
  wear resistance control with tempering 183

M

Machinability
  nonmetallic inclusions influence 128
Machining
  microsegregation influence on tool wear 119
McQuaid-Ehn test 102
Macrochemical analysis surveys 113
Macrohardness .... 187, 188(T)
  carbide effect on properties 62–63
Macrohardness testing
  to detect decarburization 43
Macroresidual stresses .... 188(T), 191
Macrostraining, and retained austenite 81
Macrostresses 105, 111
  carbide influence 71
  and carbides 63
  microsegregation influence .... 118, 119(F)
Macroyielding 143
Magnesium oxide .... 124(F)
Magnetic fields, cyclic 190
Manganese
  carbide formation affected by 52
  content effect on
    internal oxidation 14
    microstructure 109, 110
  content effect with internal oxidation 15–16
  depletion
    and surface decarburization 21
    with internal oxidation .... 22, 23(F)
  effect on case fracture toughness 152–153
  hardenability effect with internal oxidation 31
  microsegregation behavior .... 115, 116(F)
  oxidation 13
  oxidation potential .... 11, 12(F)
  segregation susceptibility .... 114, 116(T)
Manganese-chromium-boron steel
  internal oxidation .... 17, 18(F)
Manganese-chromium-nickel-
  molybdenum steels
  internal oxidation .... 26, 28(F)
Manganese-chromium-nickel steels
  composition gradients, internal
    oxidation .... 17, 18(F)
Manganese-chromium steels
  fatigue limits 26
  impact strength, as-carburized 145
  internal oxidation 15
  nickel addition effect on case toughness 153
Manganese-chromium-titanium steels
  crack propagation rate 87
Manganese-nickel-chromium-molybdenum
  steels, carbide content 56
Manganese oxide .... 24, 124(F)
  hardness, as nonmetallic inclusion .... 121(T)
  manganese content by electron
    probe analysis .... 17(T)
Manganese silicon dioxide
  hardness, as nonmetallic inclusion .... 121(T)
Manganese steels, internal oxidation 31

Manganese sulfide .... 124(F), 125
  effect on machinability of steels 128
  as nonmetallic inclusions .... 120(T), 121(T),
                                           125, 128(T)
Manganese sulfide + aluminum oxide
  as nonmetallic inclusions .... 128(T)
Martensite 55–56
  after decarburization .... 43(F)
  case hardenability 149–150
  case transformation 156–157, 159
  of core 145, 146–147
  created by grinding wheel action 202
  and decarburization 56–57
  grinding of 200
  in case and cores during tempering 172
  in case-hardened surface with
    contact damage 155
  in cores of case-hardened parts .. 136, 137(F), 138
  internal oxidation and wear resistance 29–30
  on carburized surfaces 82
  refrigeration influence on formation 190–191, 192
  tempering effect .... 171, 172(F), 173(T),
                                    177–178, 183, 185(T)
  transformation product done by
    refrigeration .... 186, 187, 188(T)
  type effect on fracture toughness 68
  untempered 207
Martensite finish (M_f) temperature 56, 77, 78(F)
  carbon content influence 108
  and retained austenite 79
Martensite plate size 102
Martensite start (M_s) temperature 56, 77
  carbide influence on residual stresses ...... 63(T)
  carbon content influence 108
  carbon effect .... 79(F)
  determination factors for retained
    austenite content 93
  determining degree of autotempering 172
  difference with quenchant temperature 79, 80(F)
  related to retained austenite ..... 78–79(F), 80(F)
Martensite transformation
  range (M_s-M_f) 77, 154
Martensitic stainless steels, heat-treatment
  deformations after quenching .... 166(F)
Matrix recovery 172
Maximum bending strength 162
Maximum shear stress-to-shear yield
  strength ratio 162
Maximum stress, tempering effect .... 176(F)
Mechanical working 78
Medium-carbon steels, toughness .... 144(F)
Medium-speed roller test 92
Methane, as addition to endothermic gas 11
MF, grain size .... 99–107(F,T), 108(T)
Microcracking .... 107–113(F), 129–130
  alloying elements effect 109–110
  carbide formation 112–113
  carbon content of the steel .... 108–109(F)
  carbon in the martensite .... 109(F)
  control 112

Microcracking (continued)
   detection 107
   factors influencing .... 108–113(F)
   fatigue influenced by .... 111–112(F)
   formation conditions 107–108
   from refrigeration 195
   grain size influence 105
   mechanisms in case depth 161
   plate size and grain size influence .... 110(F)
   properties affected 111
   quench severity 110
   roller burnishing influence 212
   standards for prevention of 130
   tempering influence .... 110, 111(F)

Microcracks
   and carbide formation 59
   in the martensite 1

Microdistortions 190

Microflaking 91

Microhardness, surface
   effect on fatigue of gears .... 25, 27(F)
   and internal oxidation .... 25, 27(F)

Microhardness testing .... 24(F), 187
   to detect decarburization 43
   to measure softening with internal
      oxidation .... 24(F)

Microhardness traverse
   after decarburization 43

Micropitting 211

Microplastic yielding 143

Microsegregation .... 1, 113–119(F), 130
   alloying element tendencies .... 114, 116(T)
   bending fatigue strength influenced by .... 118(F)
   cooling after carburizing, and
      residual stresses 119
   defined 113
   development of .... 114(F), 115(F), 116(F)
   fatigue influenced by .... 118(F), 119(F)
   formation .... 113–117(F,T)
   fracture toughness influenced by ..... 117–118(F)
   grinding precautions 119
   hardness influenced by .... 117(F)
   homogenization 115
   hot working influence .... 114–115, 116(F,T)
   ingot solidification .... 113–114(F)
   and internal oxidation 17
   macrostructures .... 115–117(F)
   mechanical and thermal
      treatment effects .... 114–115, 116(F,T)
   microstructures .... 115–117(F)
   properties influenced by .... 117–119(F)
   reduction of area influenced by 117–118
   tensile strength influenced by .... 117–118(F)
   yield strength influenced by .... 117–118(F)

Microstresses 105, 190–191, 192, 193
   carbide influence 63, 71
   microsegregation influence .... 118, 119(F)
   structural 111

Microstructural features .... 99–131(F,T)
   grain size 129

   inclusions, nonmetallic .... 119–129(F,T)
   microsegregation .... 113–119(F)

Microstructure
   internal oxidation effect on hardness .... 24(F)
   roller burnishing influence .... 212–214(F,T)
   shot peening influence .... 217–218(F)

MIRA tests .... 146(F), 209

Molybdenum
   carbide formation affected by 52
   content effect on grain coarsening in steels ... 101
   content effect on toughness 145
   content effect with internal oxidation 30–31
   effect on case fracture toughness 152–153
   and formation of high-temperature
      transformation products 19
   microsegregation behavior .... 118(F)
   oxidation potential .... 11, 12(F)
   reducing amount of high-temperature
      transformation products .... 22, 23(T)
   segregation susceptibility .... 114, 116(T)

Molybdenum-chromium steels
   vacuum carburizing 102

N

Necking 121
Necklace effect 61
Needle martensite .... 172, 173(T)
Nickel
   as alloying element determining
      hardenability 135
   carbide formation affected by 52
   content and case depth effect on
      bending fatigue strength .... 154(F)
   content effect on
      critical crack size 89
      fracture toughness .... 89, 91(F)
      grain coarsening in steels 101
      microstructure 109, 110
      toughness 145, 147
   content effect with internal oxidation 30–31
   effect on case fracture toughness 152–153
   effect on fatigue-crack initiation life 65
   and formation of high-temperature
      transformation products 19
   impact resistance effect with internal
      oxidation presence 28–29
   lowering the overheating temperature 128
   oxidation potential .... 11, 12(F)
   segregation susceptibility .... 114, 115(F), 116(T)
   with manganese sulfide, as
      nonmetallic inclusion .... 120, 121(T)

Nickel-chromium-molybdenum steels
   bending-fatigue strength 159
   carbides produced 69
   carburizing temperature effect on oxide
      penetration depth .... 14, 15(F)
   core microstructure and hardness
      relationship .... 140, 141(F)
   impact strength, as-carburized 145

nonmetallic inclusions influence
    after melting .... 122, 123(T)
roller burnishing .... 216(F)
ultimate tensile strength vs.
    proof stress .... 140–142(F)
wear resistance influenced by
    retained austenite .... 91, 92(F)
Nickel-chromium steels
carbide influence 71
case hardenability .... 138(F), 149–150(F)
composition effect on Ac_1, Ac_3, and
    Ac_{cm} phase boundaries .... 54(F)
continuous-cooling transformation
    diagrams .... 137, 138(F)
core material effect on residual stresses ... 145(F)
grinding and bending fatigue strength 210, 212(F)
hydrogen content effect on toughness
    and tempering .... 184(T)
internal oxidation......................... 14(F)
martensite transformation from austenite 81
microcracking .... 108(F)
refrigeration influence on wear resistance.... 194
retained austenite and
    contact fatigue .... 88(F)
    martensite in carburized surfaces .... 81, 82(F)
retained austenite influence on
    surface hardness................. 81–82(F)
retained austenite in relation to
    carbon content..................... 80(F)
tempering .... 175(F), 176(F)
tensile strength .... 140, 141(F)
Nickel-molybdenum-chromium steels
continuous cooling diagram .... 77, 78(F)
Nickel-molybdenum steels
grain size .... 102, 103(F)
microsegregation influence .... 118, 119(F)
Nickel oxide, nickel content by electron
    probe analysis .... 17(T)
Nickel steels
refrigeration and quenching......... 186, 187(T)
roller burnishing .... 213(F), 214(T)
toughness .... 144(F)
Niobium
as grain refining agent 101
segregation susceptibility........... 114, 116(T)
Nitrides................................ 124(F)
Nitriding, carbon steels 2
Nitrogen
as austenite former 52
liquid, refrigeration treatment .... 191–192(F)
Nonmetallic inclusions. *See* Inclusions, nonmetallic.
Normalizing................... 101, 102(F), 129
decarburization and 39, 46
prior to grain size testing.................. 103
Notched impact bend tests 144
Notched impact testing .... 72–73, 89(F),
                                    90(F), 152(F), 180(F)
Notched rotating beam tests................. 179
Notched slow bend tests.................... 144
Nucleation sites 102

O

**Orthogonal shear stress to Vickers
    hardness ratio**.................. 163, 164(F)
Overaging 78
Overheating 130
nonmetallic inclusions influence 128–129
Overpeening......................219(T), 220
Overtemper burning 200
Overtempering............................202
Oxidation, penetration depth of .... 13(F)
Oxidation potentials, of alloying elements
    and iron in steel, endothermic
    atmosphere...................... 11–12(F)
Oxide removal, with internal oxidation 31, 32
Oxides
and inclusions, stress-raising effect 127
simple .... 124(F)
surface, and grinding defects 207
Oxygen
inclusion size influenced by
    content of .... 119, 120(T)
penetration depth....................... 13(F)
released by gas-metal reactions 13
Oxygen potential, of atmosphere 13

P

Pack carburizing..................... 3, 153(F)
distortion due to 165
Packet size................................ 142
Pase diagram .... 55, 56(F)
Pearlite................................55–56
and decarburization 57
and decarburization after internal oxidation ... 24
in cores of case-hardened parts 136
Peening. *See also* Shot peening................... 1
with electrochemical machining 32
Penetration depth 203
Phase diagrams, iron-carbon .... 55(F)
Phosphorus
carbide formation affected by................ 52
effect on overheating and burning........... 129
segregation susceptibility........... 114, 116(T)
Pitting.................................. 162
fatigue resistance 159
resistance 155
Plasma carburizing......................... 32
Plastic deformation
from grinding........................... 202
from roller burnishing 214
from shot peening .... 218(F)
grinding as cause of................. 207–208
retained austenite influence 80–81, 87,
                                        89, 90, 92
tempering-induced .... 177(F)
Plate martensite........................... 161
tempering effect............ 171, 172(F), 173(T)
Plates, grinding and bending fatigue strength... 210

P

Ploughing 199–200
Poisson's ratio, for carbides and matrices .... 63(T)
Postcarburizing mechanical
    treatments .... 199–223(F,T)
Postcarburizing thermal
    treatments .... 171–195(F,T)
Post-case-hardening refrigeration
    to control retained austenite 93–94
Post-case-hardening shaft-straightening
    operation 177
Power-to-weight ratios 2, 5
Precipitaiton clustering .... 172, 173(T), 177–178
Precracked fracture tests 144
Precracked impact tests 152
Proeutectoid ferrite
    internal oxidation and wear resistance 29–30
Proof stress, tempering effect .... 176(F)
Proof stress (offset yield)
    of core .... 140–142(F), 143(F)
Propane, as addition to endothermic gas 11
Pyrowear Alloy 53 5

Q

Quartz 120
Quenchant temperature (T_q) 79
    difference with martensite-start
        temperature .... 79, 80(F)
Quenching
    direct 164
    double reheat 164
    methods 164, 168
    reheat 164
    reheat, reasons for use 164
    severity, effect on cooling rate 22
    temperature 86

R

Rare earth (RE) treatments 128
    for decreasing number of nonmetallic
        inclusions 121, 123
Reaustenitizing
    microsegregation influenced by 116
Reclamation
    shot peening for .... 220–222(F), 223(F)
Reclamation heat treatment
    to recover decarburized parts 47
Recrystallization .... 172, 173(T), 178
Reduction of area
    of core .... 143(F)
    microsegregation influence 117–118
    nonmetallic inclusions influence .... 120, 121(T),
        122, 123(F), 124(T)
    refrigeration influence .... 188(T)
    tempering influence .... 184(T)
Refrigerants 77
Refrigeration 77, 78, 130,
    186–194(F,T), 195

bending fatigue strength influenced by 193
contact fatigue influenced by ...... 192–193(F,T)
double subzero cooling .... 192(F)
effect after quenching .... 187, 188(T)
elongation influenced by .... 188(T)
fatigue limit influenced by .... 189, 190(F)
fatigue resistance
    influenced by .... 188–190(F,T), 191(F)
fracture toughness
    influenced by .... 189, 190(F), 193
hardness influenced by .... 186–188(F,T)
impact toughness influenced by 193
"in-line" treatment 186
properties influenced by 195
reduction of area influenced by .... 188(T)
residual stresses
    influenced by .... 188(T), 190–192(F)
standards 194, 195
to alter austenite/martensite proportions 1
ultimate tensile strength influenced by .... 188(T)
wear resistance influenced by .... 193–194(T)
with liquid nitrogen .... 191–192(F)
with tempering .... 186–187(F)
yield strength influenced by .... 188(T)
Rehardening burn .... 200, 202(F)
Reheat quenching 102, 129
reasons for use 164
with refrigeration .... 186, 187(T)
Reheat temperature 128
Residual stresses. *See also* Compressive-
    residual stresses 1
carbide influence .... 63(T), 71(F), 72(F)
case carbon effect .... 154–155(T)
case depth influence 167
case hardening 1–2
core material effect .... 145(F)
decarburization influence 40(T), 44–45(F), 46(F)
and effective case depth ..... 158–159(F), 160(F)
from grinding burns 200
from nonmetallic inclusions .... 126–127(F)
grain size influence 105
grinding influence .... 203, 204(F),
    207–208(F), 209(F), 210(F)
involvement in contact fatigue process 163
microsegregation as cause 119
on case-hardened gear teeth ..... 23(T), 24–25(F)
profiles .... 25, 26(F)
refrigeration influence ...... 188(T), 190–192(F)
retained austenite influence .. 82–84(F), 85(F), 89
roller burnishing influence .. 214–215(F), 216(F)
shot peening influence .... 217(F),
    218–219(F), 221(F)
tempering influence .... 173(T),
    177–178(F), 181–182
through hardening 1–2
versus rolling contact fatigue 8
Retained austenite .... 1, 77–94(F,T)
associated with carbides 63
bending strength influenced by .... 89(F,T),
    90(F), 91(F)

carbide influence.............................65
carbon content for zero level in steels 154, 155(T)
case depth increases and...............158, 159
contact fatigue influenced by..........87–89(F)
control procedures......................93–94
and decarburization.......................56
direct quenching effect on hardness....82, 83(F)
and double reheat quenching................59
effect on bending-fatigue strength..............3
fatigue resistance influenced by........84–87(F)
formation of austenite.................77–81(F)
fracture toughness influenced by............107
and grinding cracks.......................202
grinding defects, in case-hardened
  surfaces.........................206, 207
hardness influenced by..........81–82(F), 83(F)
impact fracture strength influenced by...89(F,T),
  90(F), 91(F)
in as-quenched microstructure...............171
in carburized steel........................44(F)
in microstructure...........................81
layering of austenite............78(F), 80–81(F)
lowered by decarburization..................43
microsegregation influence...............119(F)
properties influenced by....................94
related to martensite-start
  temperature...............78–79(F), 80(F)
residual stresses influenced by........82–84(F),
  85(F), 89
roller burnishing influence...............216(F)
roller burnishing (straining)
  influence..................213(F), 214(T)
stabilization of austenite...............77–78(F)
standards............................94, 194
tempering effect..........171, 172, 173(T), 174,
  175, 177–181(F), 183
tensile strength influenced by...82, 83(T), 84(F)
tolerated by refrigeration..186, 187(F,T), 188(T),
  189, 190(T), 192–193, 194
wear resistance influenced by........89–93(F,T)
with
  decarburization..........................41
  grain size, effect on fatigue strength 106, 107(F)
  internal oxidation..................31, 32–33
  yield strength influenced by...........82, 83(T)
Rockwell macrohardness tests.................24
Roller burnishing.............199, 212–216(F,T)
  bending fatigue strength
    influenced by................215, 216(F)
  contact fatigue influenced by...............215
  hardness influenced by..........214(F), 215(F)
  microstructure influenced by......212–214(F,T)
  residual stresses
    influenced by..........214–215(F), 216(F)
  wear influenced by......................216(T)
Roller testing...............................192
Rolling............1, 127, 192, 208, 212–216(F,T)
Rolling and sliding..............127, 192, 194
Rolling contact disc tests.....................87
Rolling-contact fatigue tests...............8(F)

Rolling-contact tests.......................7(F)
Roll-slide contact fatigue....................155
Roll/slide tests..............................92
Rotating beam fatigue strength.........25, 26(F)
Rotating beam test pieces..............26, 28(F)
Rotating beam tests........................190
Rotating-bending fatigue strength......25, 26(F)
"Running-in" process......................194
  retained austenite influence..............92–93
Rupturing.................................202

S

Safety factors, incorporated into a design...4–5(F)
Scale, oxide, with internal oxidation........31, 32
Scaling, and decarburization............39, 40(F)
Scoring................................155, 183
  damage............................92, 93(F)
  limiting load for..........................194
  resistance......................92, 93(F), 194
Scuffing........................91, 92–93, 155
Section size, effect on cooling rate and HTTP...22
Segregation and dislocation pinning..........77
Seizure.....................92, 93(F), 183, 194
  hardness with internal oxidation.......29–30(F)
Seizure tests................................194
Service life..............................88–89
Shear deformation..........................161
Shear-fatigue endurance.....................163
Shear-fatigue strength, vs. shear stresses.....7(F)
Shearing..................................127
Shear stresses.......................162, 163(F)
  and contact damage.......................135
  core material effect on contact-damage
    resistance...........................148
  45 shear stress.............162, 163(F), 164(F)
  maximum hertzian..........................8
  orthogonal.........7–8(F), 162, 163(F), 164(F)
  versus shear-fatigue strength...............7(F)
Shear yield values..........................162
Shot blasting..........................31, 216
  with electrochemical machining.............32
Shot peening.................129, 130, 158, 199,
  208, 216–222(F,T), 223
  bending fatigue strength influenced by...219(T),
    220(F),
    221(F), 222(F)
  contact fatigue influenced by...............220
  decarburization influenced by..220, 221, 222(F)
  defined.................................216
  deforming austenite and martensite.........159
  for reclamation............220–222(F), 223(F)
  hardness influenced by..................218(F)
  internal oxidation removal............220, 221
  microstructure influenced by........217–218(F)
  NASA study of fatigue life.................220
  overpeening......................219(T), 220
  process control.....................216–217(F)
  purpose.................................216

Shot peening (continued)
   residual stresses influenced by .... 217(F), 218–219(F), 221(F)
   saturation curve .... 217(F)
   standards 223
   standards (specifications) 217
   to control retained austenite 94
   versus roller burnishing 212
Shrinkage, conditions influencing 165
Silica 120
Silicates .... 123, 124(F)
   effect on machinability of steels 128
Silicon
   carbide formation affected by 52
   content effect with internal
      oxidation 15–16, 30–31
   as deoxidizer for grain size control .... 100(F)
   oxidation potential .... 11, 12(F)
   segregation susceptibility 114
Silicon dioxide 120
   hardness, as nonmetallic inclusion .... 121(T)
Silicon oxide 24
   oxide composition 17
Slide-roll testing .... 67(F), 72
Slide/roll wear tests .... 193(T)
Sliding 194
   and contact damage 135
Smearing 202
S-N curve .... 26, 28(F)
   bending fatigue strength and
      tempering effect .... 179, 180(F)
   carbide effect on bending fatigue ...... 63–64(F)
   case-hardened steel .... 86(F)
   failures initiated at nonmetallic inclusions ... 127
Soft surfaces, due to decarburization 42
Solubility limit 53
Spalling 46, 88, 202
   contact damage influenced by
      case depth .... 162(T)
   deep 167
     contact 159
     failures 162, 163
     fatigue .... 148(F)
     resistance 162
   resistance 155
   and shot peening 220
Spheroidization .... 173(T)
   of carbides 60
Spinels .... 123, 124(F)
Stainless steels, heat-treatment
   deformation after quenching .... 166(F)
Stainless steels, specific types
   X40Cr13, heat-treatment deformations
      after quenching .... 166(F)
   X5NiCrMo$_{10}^{18}$, heat-treatment
      deformations after quenching .... 166(F)
Standards
   core properties 166
   for
      carbides 73

   case carbon 168
   case depth 167–168
   case hardening 9
   decarburization 48
   gears, ANSI/AGMA 2001 or
      ISO 6336 .... 4(T), 5(T)
   grinding 223
   microcracking prevention 130
   nonmetallic inclusions 130
   refrigeration 195
   refrigeration, and retained austenite 194
   retained austenite 194
   retained austenite content 94
   shot peening 223
   tempering 181, 195
Static bending tests 72–73
Static bend strength
   versus bending fatigue limit .... 153(F)
   versus case depth .... 153(F)
Steels
   cleanness, consequences of production .. 128–129
   cleanness effect on load/extension
      curves .... 121, 122(F)
   H grade, to control distortion 165, 166
Steels, British designations, specific types
   BS 970 832M13, continuous-cooling
      transformation diagrams
      compared .... 136, 137(F)
   En 16, shot peening and
      fatigue limits .... 219(T), 220
   En19 (705M40), toughness .... 144(F)
   En29 (722M24), toughness .... 144(F)
   En30 (835M30), toughness .... 144(F)
   En34 (665M17)
     carbon content for zero retained
       austenite .... 155(T)
     case hardenability .... 150(F)
     decarburization .... 44, 45(F)
     grain size control .... 100(F)
   En36 (832M13)
     decarburization .... 45–46(F)
     grinding 209
     tensile strength .... 140, 141(F)
     toughness .... 144(F)
   En36A (655M13)
     case hardenability .... 150(F)
     decarburization .... 44(F)
     tensile strength .... 140, 141(F)
   En36C, case hardenability .... 150(F)
   En39, microsegregation .... 115, 116(F)
   En39B, case hardenability .... 150(F)
   En 352
     case hardenability .... 150(F)
     tempering .... 179(T)
   En 353
     case hardenability .... 150(F)
     distortion .... 167(F)
     shot peening and
       fatigue limits .... 219(T), 220
   En 354, case hardenability .... 150(F)

Steels, German designations, specific types
Ck 15
   heat-treatment deformations after
      quenching.........................166(F)
   tempering ....178–179(F), 180(F)
Ck 45, heat-treatment
   deformations after quenching ....166(F)
10 CND 6, case hardenability............150(F)
20Cr, tensile properties and tempering....184(T)
25CrMo4, microsegregation effect on
   hardness ....117(F)
42CrMo4, heat-treatment
   deformations after quenching ....166(F)
18CrNi8
   case hardenability...................150(F)
   tempering ....180(F)
17CrNiMo6, carbon content for
   zero retained austenite ....155(T)
42CrNiMo6, heat-treatment
   deformation after quenching ....166(F)
18CrNiW, tensile properties and
   tempering..........................184(T)
DVM/DIN 50115, load-time curves ....89, 90(F)
16MnCr5
   case hardenability...................150(F)
   time to bainite nose temperature ....150(F)
95MnCr5, roller burnishing effect on
   retained austenite ....213, 214(T)
105MnCr5, roller burnishing
   effect on retained austenite .....213, 214(T)
20Mo5, time to bainite nose temperature ..150(F)
20MoCr4
   carbon content for zero retained
      austenite........................155(T)
   case hardenability...................150(F)
   tempering ....185(T)
   time to bainite nose temperature ....150(F)
14NiCr14
   heat-treatment deformations after
      quenching.........................166(F)
   Jominy diagram.....................167(F)
   refrigeration and fatigue resistance........189
   tempering ....180(F)
   time to bainite nose temperature ....150(F)
10NiCrMo7, heat-treatment
   deformations after quenching ....166(F)
15NiCrMo 16 5, nonmetallic inclusions
   influence, after melting ....122, 123(T)
20NiCrMo6
   time to bainite nose temperature ....150(F)
   case carbon effect on impact fatigue....153(F)
   case hardenability...................150(F)
Steels, Japanese designations, specific types
   SCM415, internal oxidation 25
Steels, miscellaneous, specific types
815A16, case hardenability..............149(F)
16CD4, refrigeration influence on
   contact fatigue ....192(T)
EX36, case hardenability................150(F)
EX55, refrigeration influence on
   residual stresses ....191(F)
18GHT, crack propagation rate............87(F)
20HNMh, crack propagation rate..........87(F)
300M, decarburization ....39, 40(T)
15NCD2, case hardenability.............150(F)
20N3MA, impact resistance vs. temperature
   of second hardening ....153, 154(F)
PS55
   carbon content for zero retained
      austenite........................155(T)
   impact strength, as-carburized............145
SCR22, case hardenability ....150(F)
ShKh15
   refrigeration effect on tensile
      properties ....188(T),
                        190(F), 191(F)
   tempering ....177(F)
SMC 21, case hardenability ....150(F)
SNCM 22, case hardenability............150(F)
X38 CrMoV 51, microsegregation effect
   on properties ....118(F)
637M17, fatigue strength ....147(T)
822M17, yield strength ....142(F)
835M 15, carbon content for zero
   retained austenite ....155(T)
Steels, Russian designations, specific types
40Kh
   refrigeration influence on
      fatigue resistance..............188, 189(T)
   tempering ....178(F)
50Kh, refrigeration effect on
   tensile properties ....188(T)
25Kh2GHTA, contact-fatigue strength   66, 67(F)
90KhGNMFL, retained austenite and
   contact fatigue resistance ....88(F)
25KhGT, internal oxidation ....18, 25, 26(F)
30KhGT, tempering ....175–176(F), 182(F)
12KhN3, carbide effect on
   bending fatigue strengths ....63, 64(F)
50KhN, refrigeration effect on
   tensile properties ....188(T), 190(F)
12KhN3A, refrigeration
   influence on wear..................193(T)
20KhN3A, impact resistance vs.
   temperature of second hardening 153, 154(F)
20Kh2N4A
   decarburization..............45, 46(F), 47(T)
   grinding and bending fatigue strength 208
   roller burnishing ....214(F)
   roller burnishing and
      contact fatigue..............215, 216(F)
20KhNM
   refrigeration influence on
      fatigue resistance ....188, 189(T)
   shot peening and residual
      stresses......................218, 219(F)
14Kh2N3MA, roller burnishing and
   residual stresses ....214–215, 216(F)
18KhNVA, microstructure 101

Steels, Russian designations, specific types (continued)
  18Kh2N4VA
    internal oxidation 16
    refrigeration and quenching .... 186, 187(T), 189(F)
  20KhNV4MF, residual stresses and carbide influence .... 71(F)
Steels, SAE, specific types
  1015
    grain size .... 102, 103(F)
    oxide morphology .... 13–14(F)
  1017, carbon content for zero retained austenite.......................... 155(T)
  1018
    carbide influence on surface tensile stresses .... 71, 72(F)
    decarburization .... 42(F)
  1040, hardness and coarsening.............. 104
  1045, shot peening and residual stresses .. 217(F)
  1060, austenitizing treatment .... 101(F)
  1526
    refrigeration............................ 187
    residual stresses....................... 85(F)
  4023, shot peening and fatigue resistance............... 219(T), 220
  4028, shot peening .... 220, 221(F)
  4080, roller burnishing effect on retained austenite .... 213, 214(T)
  4095, roller burnishing effect on retained austenite .... 213, 214(T)
  4130, refrigeration 187
  4140
    nitriding 2
    residual stresses....................... 85(F)
  4320
    tempering........................... 182(T)
    tensile testing........................ 83(T)
    vacuum carburizing.................... 102
    volumetric transformation strain due to tempering................... 174(F)
  4330, decarburization .... 39, 40(T)
  4340
    decarburization 39
    grinding and fatigue strength .... 212(F)
    grinding and residual stresses ..... 208, 209(F)
    shot peening and decarburization......... 220, 221, 222(F)
  4615
    grain size .... 102, 103(F)
    internal oxidation.............. 17, 19, 21(T)
    residual stress profiles.............. 25, 26(F)
  4675, roller burnishing effect on retained austenite .... 213, 214(T)
  4685, roller burnishing effect on retained austenite .... 213, 214(T)
  4815, tempering .... 182(T)
  4820, tempering .... 182(T)
  4875, roller burnishing effect on retained austenite .... 213, 214(T)
  4885, roller burnishing effect on retained austenite .... 213, 214(T)
  5115, vacuum carburizing.................. 102
  6120, bending fatigue strength to carbides .... 63, 64(F)
  8600, core material effect on residual stresses .... 145(F)
  8615, tempering .... 182(T), 183(F)
  8617
    carbon content for zero retained austenite.................... 155(T)
    tempering........................... 182(T)
  8620
    alloy depletion and carbon content effect, internal oxidation............. 22(F)
    bending fatigue endurance after case hardening .... 64, 65(F)
    carbide influence on contact fatigue........ 72
    case hardenability.................... 150(F)
    composition gradients, internal oxidation .... 17–18(T)
    continuous-cooling transformation diagrams compared............ 136, 137(F)
    crack propagation rate 87
    decarburization................. 39, 40(F), 46
    grinding .... 202, 203(F)
    internal oxidation ..... 17, 19, 21(T), 27, 28(F)
    microcracking .... 110, 111, 112(F)
    refrigeration and fatigue lives ..... 190, 191(F)
    refrigeration effect on bending-fatigue strength .... 188, 189(F)
    residual stress profiles.............. 25, 26(F)
    surface curvature effect on case depth .... 156, 157(F)
    tempering........................... 182(T)
    vacuum carburizing..................... 102
  8620H, hardenability range.............. 139(F)
  8822, case hardenability .... 150(F)
  9310
    aerospace applications................... 5–6
    carbon content for zero retained austenite.......................... 155(T)
    case-hardened, S-N curves .... 86(F)
    case-hardening and fitness for service 9
    refrigeration .... 186, 187(F)
    refrigeration and fatigue resistance 189, 190(T)
    refrigeration, double subzero cooling... 192(F)
    shot peening .... 222, 223(F)
  9310H, hardenability range.............. 139(F)
  94B17
    chromium content in carbides 52
    internal oxidation .... 18, 19(F)
  52100
    carbide influence on contact fatigue........ 67
    decarburization .... 40(T), 44–45
    fracture toughness..................... 68(F)
    microcracking..................... 110, 113
    nonmetallic inclusions influence .. 125, 126(F)
Steven and Haynes formula .... 78, 79(F)
Strain aging process..................... 77–78

Strain-induced martensite 89
Stress concentrators 180
Stresses, applied cyclic. *See also* Applied stresses; Compressive-residual stresses; Residual stresses. 86
Stress raisers .... 126–127(F), 144, 155
Stress-strain curves
    nonmetallic inclusions influence 121
Stringers
    effect on anisotropy effect................... 120
    and nonmetallic inclusions 122
Subcase cracking 158, 167
Subcritical annealing..................... 69, 138
    and carbides 60
    prior to grain size testing................... 103
    with tempering 172
Subsurface fracture initiation points.......... 26
Subzero temperature treatment. *See* Refrigeration.
Sulfides .... 124(F), 128
    effect on machinability of steels 128
    nonmetallic inclusions..................... 120
Sulfur
    effect on overheating and burning........... 129
    inclusion size influenced by content of .... 119, 120(T)
    segregation susceptibility 114
"Super Carb" process 71
Super carburizing 69
    carbide influence........................... 71
Surface asperities, removal of 90–91
Surface flaking............................. 162
Surface grinding 1
Surface network (mud) cracks 202
Surface overtempering 202
Surface pitting
    contact damage influenced by case depth .... 162(T)
Surface rolling, to control retained austenite 94
Surface working, to control retained austenite .. 94

T

Temperature
    carburizing, effect on oxide penetration depth 13
    tempering influenced by .... 172–173(F,T), 174(F)
Tempered martensite 206, 207
    microsegregation detected by etching 117
Tempered steels 2
Tempering....................... 78, 81, 130, 171–186(F,T), 194–195
    advantages 171
    after grinding............................. 208
    aging influenced by .... 184–185(F,T)
    autotempering .... 172, 173(T), 175(F)
    bending and impact-fracture strength influenced by .... 181–183(F,T)
    bending fatigue strength influenced by .... 178–181(F,T)
    causing microcracking............. 110, 111(F)
    compressive-residual stresses influenced by .... 173(T), 177–178(F), 181(F), 185(T)
    contact fatigue influenced by .... 181(F)
    fatigue limit influenced by .... 181(F)
    grinding influenced by .... 185–186(T)
    hardness influenced by .... 175–176(F)
    high-temperature.......................... 171
    hydrogen content effect........... 183–184(F,T)
    impact resistance influenced by .... 182(F)
    low-temperature...................... 171, 174
    and machinability 194–195
    properties influenced by 195
    reactions .... 171–175(F,T)
    reduction of area influenced by .... 184(T)
    residual stresses influenced by .... 173(T), 177–178(F), 181–182
    retained austenite influence on bending strength........................ 89
    retained austenite influence on wear resistance 91
    secondary hardening 172
    short-duration 173
    stages .... 172–173(F,T), 174(F)
    standards 181, 195
    structural changes in martensite ... 172–173(F,T)
    structural features............... 171–172(F)
    temperature influence ..... 172–173(F,T), 174(F)
    temperature range 171
    tensile strength influenced by.......... 176–177(F), 184(T)
    tensile yield strength influenced by .. 176–177(F)
    time influence .... 174(F), 194
    to improve toughness...................... 153
    to modify the martensite 1
    ultimate tensile strength influenced by 176–177(F)
    volume changes during................. 174–175
    wear influenced by .... 183(F)
    with refrigeration .... 186–187(F)
    yield strength influenced by .... 176–177(F), 186
Tensile residual stresses
    and retained austenite...................... 81
Tensile strength
    of core .... 140, 141(F), 142(F)
    grain size effect .... 104–105(F), 106(F)
    microsegregation influence .... 117–118(F)
    overheating effect......................... 129
    retained austenite influence ..... 82, 83(T), 84(F)
    tempering influence .... 176–177(F), 184(T)
Tensile testing............................. 213
    carburized, hardened and tempered steel ... 83(T)
Tensile yield strength
    tempering influence................ 176–177(F)
Tetragonality of the martensite............. 207
    loss of .... 173(T), 174, 177–178
    tempering effect 185
Thermal energy, application during tempering 173
Thermal expansion coefficients
    nonmetallic inclusions influence .... 123, 124(F)

Thermal gradient 113, 114, 165, 200
Thermally stabilized austenite 77
Thermal stabilization 77
    effect on austenite layering 81
Thermal stresses 165
Theta (θ)-carbide . . . . 173(T), 174(F),
                                                                             175, 178, 185
Through-hardened steels 2
    decarburization sensitivity 45
    rotating beam fatigue strength . . . . 85(F)
Through hardening, residual stresses 1–2
Time, tempering influenced by . . . . 174(F)
"Time to bainite nose" curves . . . . 150(F)
Tin, segregation susceptibility 114
Titanium
    carbide formation not affected by 53
    content effect with internal oxidation 16
    as grain refining agent 101
    oxidation 13
    oxidation potential . . . . 11, 12(F)
    with manganese sulfide as
        nonmetallic inclusion . . . . 121(T)
Titanium carbonitrides
    as nonmetallic inclusions . . . . 125, 126(F)
Titanium nitride . . . . 123, 124(F)
    as nonmetallic inclusions . . . . 128(T)
Titanium treatments, for decreasing
    number of nonmetallic inclusions 121
Tolerances, working 137
Tool steels, rotating beam fatigue strength . . . . 85(F)
Torque-speed plots . . . . 2(F)
Torsional fatigue limit 86
Total case depth 14, 148, 156, 163, 167
    defined . . . . 8(F)
Total depth of carbon
    penetration (TPD) . . . . 156, 157(F)
Toughness
    carbide influence . . . . 63(T), 67–68(F), 72–73
    carbide state effect . . . . 63(T)
    of core . . . . 143–145(F)
    defined 143
    lack in through-hardened steels 2
    tempering influence . . . . 180(T)
    tests for . . . . 144(F)
Tridymite 120
Tungsten
    content effect with internal oxidation 30–31
    oxidation potential . . . . 11, 12(F)
Tungsten carbide
    carbide influence on residual stresses . . . 63, 64(T)
Twinned martensite 104
Two-stage (boost-diffuse) carburizing . . . . 151(F)

U

Ultimate tensile strength
    of core . . . . 140, 141(F), 142(F), 143(F)
    nonmetallic inclusions influence 120,
                                                     121(T), 122(F)

    refrigeration influence . . . . 188(T)
    tempering influence . . . . 176–177(F)
Ultrasonic surveys
    to detect nonmetallic inclusions 130
Unnotched impact bend tests 144
Unnotched impact energy
    vacuum-carburizing effect . . . . 129(F)
Unnotched slow bend tests 144

V

Vacuum arc remelting (VAR)
    and nonmetallic inclusions 121
Vacuum carburizing 32
    bending fatigue strength . . . . 63–64(F)
    carbide influence on bending fatigue . . . 63–64(F)
    effect on impact resistance . . . . 29(F)
    grain size influenced by 102
    processes 11
Vacuum induction melting/vacuum arc
        remelting (VIM/VAR) 3
Vacuum tempering . . . . 184(T)
Vanadium
    addition for grain refinement
        by alloying 100, 101
    carbide formation not affected by 53
    content effect with internal oxidation 17
    oxidation potential . . . . 11–12(F)
    with manganese sulfide, as
        nonmetallic inclusion . . . . 121(T)
Vasco X2-M 5
Volume, changes during tempering 174–175

W

Warpage, conditions affecting 165
Water vapor, providing oxygen for
    internal oxidation 11
Wear
    and contact damage 135
    decarburization influence . . . . 30(F), 47
Wear resistance 1
    carbide influence . . . . 66(F), 69, 73
    contact damage influenced by case depth 162
    grinding influence 211–212
    internal oxidation influence . . . . 29–30(F)
    refrigeration influence . . . . 193–194(T)
    retained austenite influence . . . . 89–93(F,T)
    roller burnishing influence . . . . 216(T)
    tempering influence . . . . 183(F)
    tools, and microsegregation 119
Weldments
    microcracking 110
    overheating, and nonmetallic inclusions 129
White etching . . . . 81, 82(F)
White irons
    carbide influence on toughness 72–73
    wear resistance and carbide influence 70–71
    wear resistance and case-hardening process . . . 51

Work hardening 143
    refrigeration and adhesive wear 194

X

X_w **factor** 91

Y

Yield ratio (YR) .... 142–143(F), 147

Yield strength
    of core .... 140–143(F), 147
    grain size effect .... 104–105(F), 106(F)
    microsegregation influence .... 117–118(F)
    nonmetallic inclusions
        influence 121
    refrigeration influence .... 188(T)
    retained austenite influence ..... 82, 83(T), 84(F)
    tempering influence .... 176–177(F), 186

Young's modulus, for carbides
    and matrices .... 63(T)